I0056384

Plant Stem Anatomy: An Illustrated Atlas

Plant Stem Anatomy: An Illustrated Atlas

Chrissy Bradford

R CALLISTO REFERENCE

www.callistoreference.com

Callisto Reference,
118-35 Queens Blvd., Suite 400,
Forest Hills, NY 11375, USA

Visit us on the World Wide Web at:
www.callistoreference.com

© Callisto Reference, 2023

This book contains information obtained from authentic and highly regarded sources. All chapters are published with permission under the Creative Commons Attribution Share Alike License or equivalent. A wide variety of references are listed. Permissions and sources are indicated; for detailed attributions, please refer to the permissions page. Reasonable efforts have been made to publish reliable data and information, but the authors, editors and publisher cannot assume any responsibility for the validity of all materials or the consequences of their use.

ISBN: 978-1-64116-754-3 (Hardback)

Trademark Notice: Registered trademark of products or corporate names are used only for explanation and identification without intent to infringe.

Cataloging-in-Publication Data

Plant stem anatomy : an illustrated atlas / Chrissy Bradford.
   p. cm.
Includes bibliographical references and index.
ISBN 978-1-64116-754-3
1. Plant anatomy--Pictorial works. 2. Stems (Botany)--Pictorial works.
3. Plants--Development. I. Bradford, Chrissy.
QK641 .P53 2023
581.4--dc23

Table of Contents

Permissions

Index

A stem refers to one of the primary structural axes of a vascular plant, while the other part is root. It provides support to fruits, leaves and flowers. It is also responsible for generating new living tissue, storing nutrients, and transporting water and dissolved substances among the shoots and roots in the phloem and xylem. There are two parts of the stem which include nodes and internodes. The majority of plants have stems that are above ground whereas certain plants contain underground stems. There are three types of tissues found in a stem, namely, vascular tissue, dermal tissue and ground tissue. The dermal tissue normally covers the exterior surface of the stem and is responsible for protecting, waterproofing and regulating gas exchange. The ground tissue is primarily made up of parenchyma cells and surrounds the vascular tissue. It occasionally plays a role in the process of photosynthesis. Vascular tissue is responsible for long-distance transport as well as structural support. This book provides comprehensive insights into the anatomy of the plant stem. It is an essential guide for both academicians and those who wish to pursue this discipline further.

This book is a comprehensive compilation of works of different researchers from varied parts of the world. It includes valuable experiences of the researchers with the sole objective of providing the readers (learners) with a proper knowledge of the concerned field. This book will be beneficial in evoking inspiration and enhancing the knowledge of the interested readers.

In the end, I would like to extend my heartiest thanks to the authors who worked with great determination on their chapters. I also appreciate the publisher's support in the course of the book. I would also like to deeply acknowledge my family who stood by me as a source of inspiration during the project.

Chrissy Bradford

Understanding Plant Stems

Terrestrial life forms made their move on land about 400 million years ago. Plants crossed the barrier between life in water to life in the atmosphere. With the invention of stable stems, plants overcame hydrological and mechanical problems. The construction of plant stems is the focus of this book. It demonstrates that nature created a framework in which plant stems evolved—annual herbs as well as century-old, 100 m-tall trees, from tropical to arctic environments.

The book offers a very wide view of stem anatomy. Chapter 2 explains simple anatomical preparation techniques. The six following chapters present basic, cell-based anatomical traits. Two chapters deal with taxonomically related anatomical stem characteristics in living and fossil plants. Anatomical structures which are related to short- and long-term external influences all over the globe are intensively discussed. The general part of the book ends with a section about wood decay and wood conservation.

One major objective of this book is to show that nature principally does not distinguish between plant stems of different growth forms, e.g. between small herbs and very tall trees. The following two microscopic cross-sections demonstrate that the basic stem construction of vascular plants, such as ferns, monocotyledons and dicotyledons, consists of the pith and the cortex, the xylem and phloem, and often a periderm.

Why a new book about plant anatomy? Isn't it repeating knowledge already previously demonstrated by great botanists? That is partially correct. However, this book has different goals.

Firstly, this book connects a variety of fields of research. Stem anatomy today is no longer a science per se. Dendrochronologists, ecologists, taxonomists, plant pathologists, foresters, archaeologists, paleobotanists, historians, criminologists, and technically oriented wood scientists use stem anatomical structures to solve specific problems. However, not all of them have

Principal stem construction of most dicotyledonous vascular plants

Sections stained with Astrablue/Safranin. Red-stained cell walls indicate an intensive lignification, blue-stained cell walls a purely cellulosic composition.

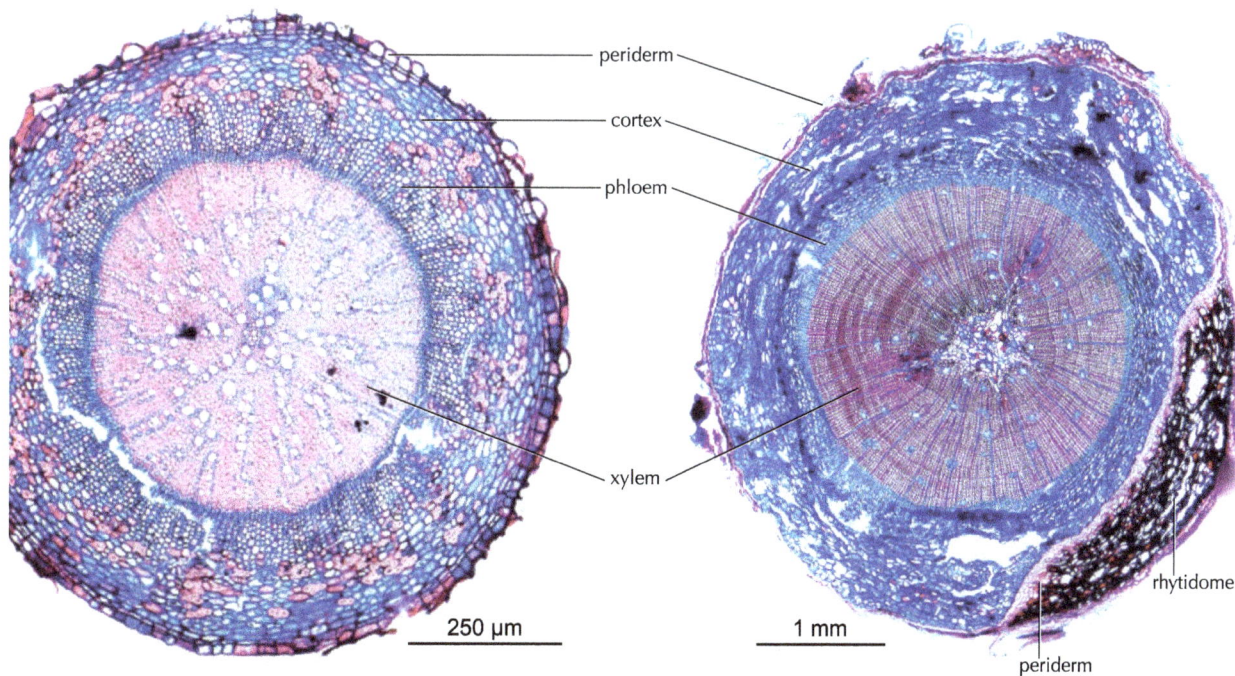

1.1 Main root of the annual, 5 cm-tall, dicotyledonous alpine herb *Polygonum plebeium*.

1.2 Fifteen-year-old twig of the 10 m-tall subalpine coniferous tree *Pinus mugo*.

an anatomical training. Each concept in this book is introduced by presenting well-known objects in macroscopic images before explaining their microscopic structures, e.g. an orange, followed by microscopic details of oil ducts, or an *Arabidopsis* plant followed by its anatomical stem structure. Basic anatomical knowledge is presented so it can be understood by readers with different academic training.

Secondly, this book addresses a worldwide multilingual auditorium—even when the knowledge of the English language might be limited. International dendroanatomical training classes have shown that pictures overcome many language problems. This book is therefore extensively illustrated, and introductory texts are kept short. Photographs are presented where possible instead of abstract drawings, and the images are captioned and labeled in an easily understandable manner.

The book builds a bridge between basic and detailed anatomical and physiological studies. Most concepts are of common knowledge, and can already be found in many botanical textbooks, e.g. Beck 2010, Bresinsky *et al.* (Strasburger) 2008, Carlquist 2001, Crivellaro & Schweingruber 2015, Cutler *et al.* 2008, Eschrich 1995, Evert 2006, Fahn 1990, Fink 1999, Herendeen *et al.* 1999, Mauseth 1988, Nabors 2004, Schweingruber 2007 and Taylor *et al.* 2009. However, the here presented color photographs of stained microscopic slides enhance previous knowledge about lignification, and in consequence the relations between anatomy, physiology and plant stability. All microscopic slides have been made recently with modern sledge microtomes, and have been analyzed with light microscopes.

Since most of the scientific content is of common knowledge, citations of sources occur only sparsely in the text. The reader can find a summary of sources and recommended reading at the end of the book.

Anatomical Preparation Techniques

As the title of this book suggests, the main objective of the book is to explain anatomical structures on the basis of microscopic slides. Since comparability is one of its major goals, most micro-photographs are based on recently prepared microscopic slides. Gärtner & Schweingruber 2013 explain sample design and preparation techniques in detail.

Taxonomically and ecologically labeled fresh material is collected in the field and stored in 40% ethanol before sectioning. The heart of the laboratory work is the newly developed sledge microtome (Gärtner *et al*. 2015), a modified copy of the Reichert sledge microtome. New are its low weight, two very stable knife guides and the blade holder, which is modified for disposable blades. These innovations made it possible to section the majority of stems without prior embedding.

Almost all sections are stained for a few minutes with a one to one mixture of Astrablue/Safranin. Staining and dehydration with 96% ethanol, absolute ethanol and xylene occurs directly on the glass slide. Permanent slides, lasting for more than 100 years, are embedded in Canada balsam. Photographs were made under transmitting, normal and polarized light with an Olympus BX51 microscope. Slides are conventionally stored in preparation boxes. A digital databank allows queries by any taxonomical, morphological, geographical or anatomical characteristic of each slide.

2.1 Plastic bag containing 40% ethanol with a sample of a herb stem. The label contains information about the taxon and its morphology, ecology and the sampling date.

2.2 A new laboratory microtome designed by Gärtner *et al*. 2015; distributed by the Research Institute WSL, CH-8903 Birmensdorf, Switzerland.

2.3 Holder with chemicals for preparation of permanent slides. It contains dyes, ethanol and xylene and corresponding pipettes.

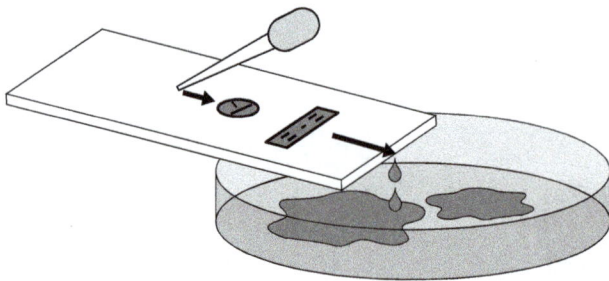

2.4 Principle of staining and dehydrating the cut sample on the glass slide.

2.5 Flattening of microscopic sections with magnets on an iron plate.

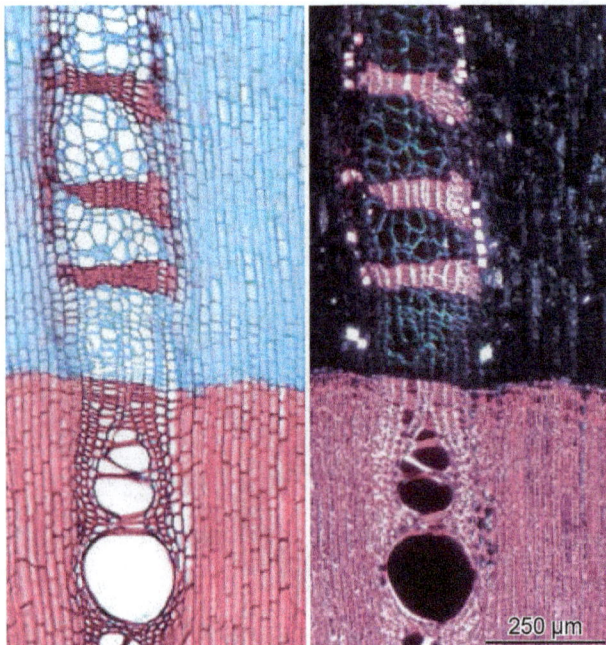

2.6 Section of *Vitis vinifera*, stained with Astrablue/Safranin, photographed under normal and polarized light.

2.7 Old-fashioned, but safe and handy storage of labeled microscopic slides in prefabricated cardboard boxes.

3

The Plant Body: Forms and Structure

The first plants on land appeared some 400 million years ago in the geological periods of the late Silurian and early Devonian. The morphological structure of these spore plants is similar: The tiny leaf-less plants consist of roots, stems and branches. These plants represent the initial point for the evolution of a vast taxonomic and morphological diversity over the next 250 million years. At the beginning of the Early Cretaceous, approximately 140 million years ago, seed plants developed a variety of growth forms, and occupied most terrestrial habitats on earth. The image below demonstrates that many different growth forms exist in ecologically similar ecotones.

Various growth forms in an ecotone

3.1 Riparian zone of the Danube in Slovakia.

3.1 Growth forms and life forms

Life forms and growth forms are principally synonymous terms. The image on the following page shows that the currently existing plant forms are based on a principle that was invented 400 million years ago. It also demonstrates that plant age does not correspond with the morphological classification; perennial and annual plants can have the same size.

A few annual (blue) and perennial (red) growth forms in Fig. 3.2 on the following page relate to branching and plant height:

- Straight, self-supporting, and poorly branched plants (monopodial at least at the base) with heights of 5 cm to >100 m.
- Intensively branched plants (sympodial) with heights between 3 cm and 5 m.

- Intensively branched, cushion-like plants with heights between 1 cm and 50 cm.
- Liana-like, not self-supporting annual and perennial plants with lengths between 1 m and 50 m.
- Soft plants in wet environments (hydrophytes and helophytes) with various morphological adaptations.

For further growth and life form classifications refer primarily to the Raunkiaer growth forms, which have been defined by Ellenberg & Mueller-Dombois 1967.

Upright growth forms with one stem at the base

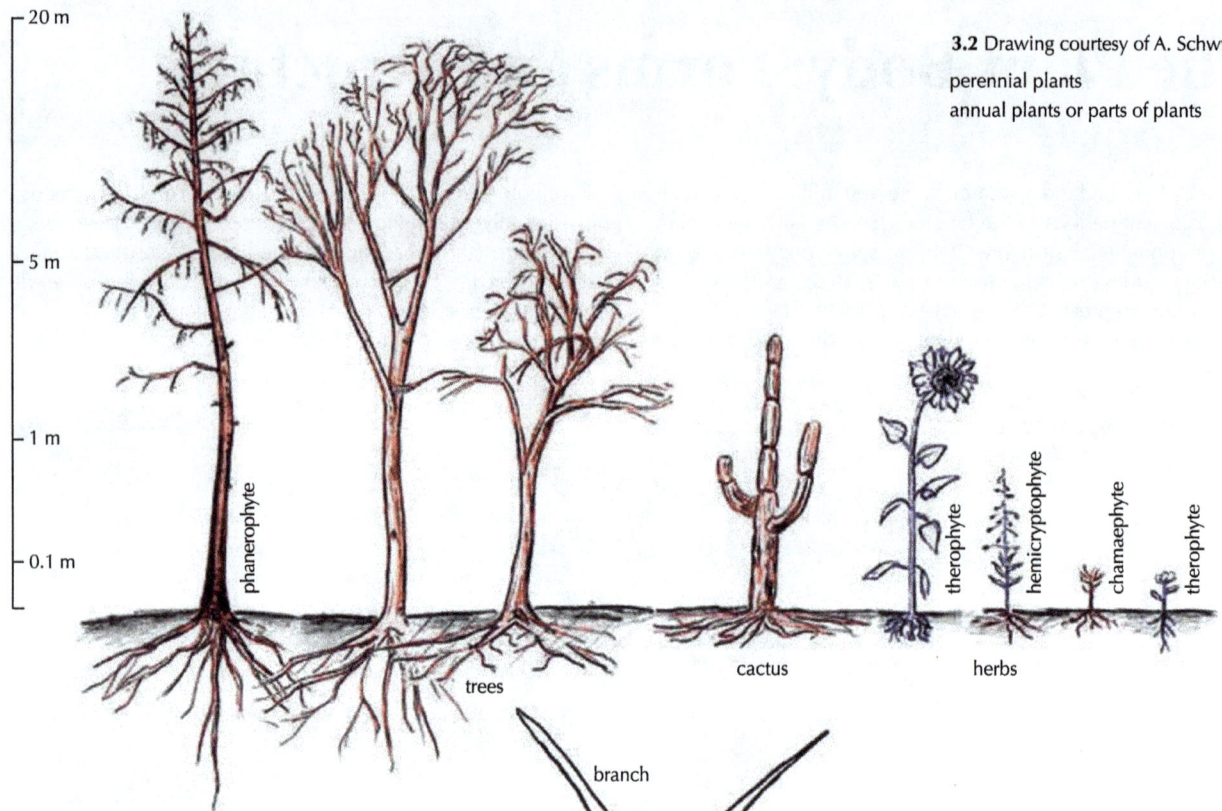

3.2 Drawing courtesy of A. Schwyzer
perennial plants
annual plants or parts of plants

Upright growth forms with several stems at base

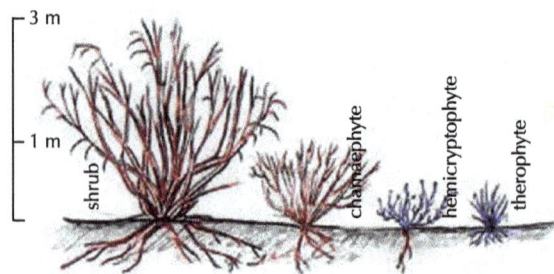

Cushion plants

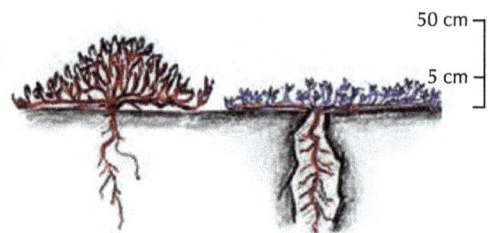

The ideal plant

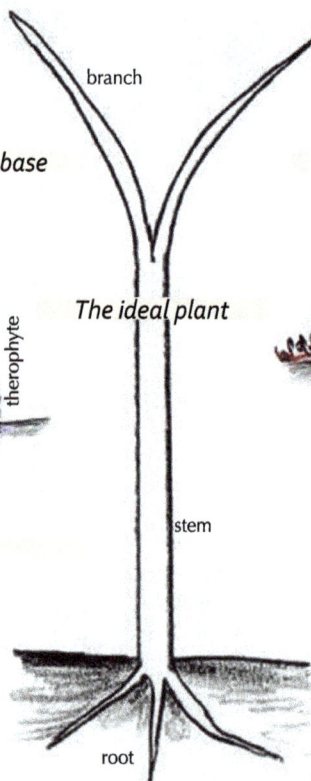

Lianas

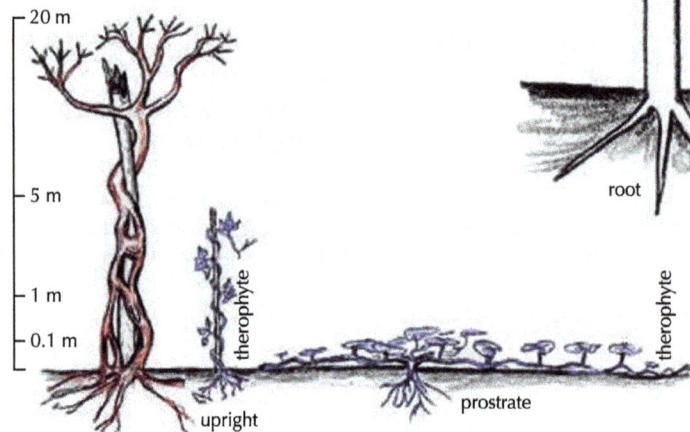

Water plants

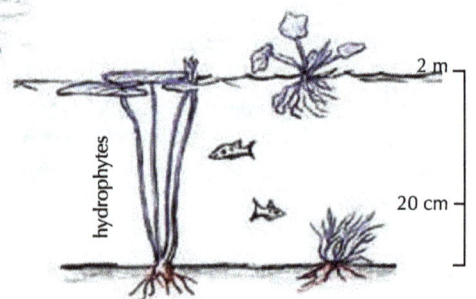

3.2 Parts of the stem and definition of bark terms

The principal structure of stems is defined in this section, generally following Junikka 1994 and Crivellaro & Schweingruber 2015.

Bark is a general term, which includes all tissues outside of the wood (*Borke* in German).

  Inner bark (colloquial) or phloem (scientific term) includes the living part of the bark.

  Outer bark (colloquial) or rhytidome (scientific term) includes all dead parts outside the phellogen.

Bast is a colloquial term for phloem. Annual increments in the phloem consist of an early and a late bark.

Cambium is the growth zone (meristem) between the xylem and phloem.

Cambial zone contains the initial meristem, the xylem and phloem mother cells.

Cork is the colloquial term for phellem.

Cortex is a product of the primary meristem. It is located between the epidermis and the phloem. Rays are absent in the cortex.

Heartwood is the dead, non-conducting part of the xylem.

Periderm consists of phellogen, phelloderm and phellem.

Pith is the central parenchymatic part of shoots.

Phellem is the product of the phellogen, often called cork. It consists of suberized dead cells.

Phelloderm is the product of the phellogen. It is normally a very small zone which consists of living parenchyma cells.

Phellogen is a growth zone (tertiary cambium). Its origin are parenchyma cells within the cortex or the phloem.

Phloem is the (peripheral) product of the cambium. Rays are characteristic for the phloem.

Rhytidome is the outer bark. It consists of all dead parts outside of the periderm.

Sapwood is the living and water-conducting part of the xylem.

Wood is a colloquial and technical term for all components inside of the cambium of which most are lignified.

Xylem is the (centripetal) product of the cambium.

Macroscopic aspect of old bark

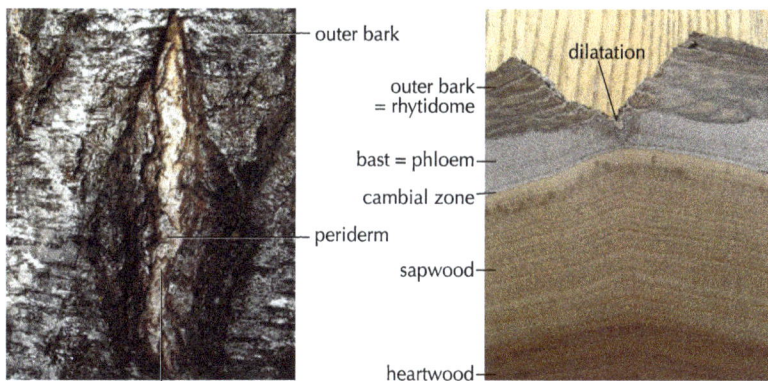

3.3 Old bark with dilatation in *Betula alba*.

3.4 Old bark with dilatation in *Juglans regia*.

Microscopic aspect of young bark

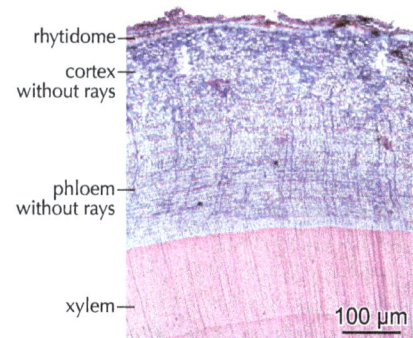

3.5 Young bark of *Juniperus communis*.

Microscopic aspect of old bark

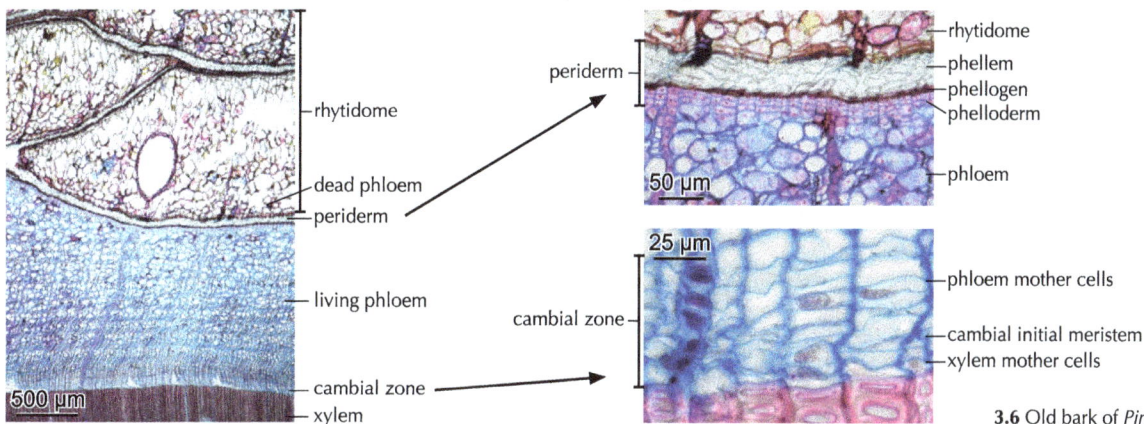

3.6 Old bark of *Pinus sylvestris*.

Plant Cells and Tissues

4.1 The individual cell

Single cells can only be observed through microscopes. The cell is anatomically and physiologically complex. It principally consists of protoplasts, which contain various organelles, the vacuole and the cell wall. The following section introduces elements that are visible under normal and polarized light without special microscopic equipment.

Shown here are different cell types, cell walls, nuclei and plastids and ergastic substances (non-protoplasm material such as crystals, resins, tannins etc.).

The following figure schematically shows all components. Nuclei, plastids, vacuoles and cell walls are recognizable by light microscopy.

Nuclei

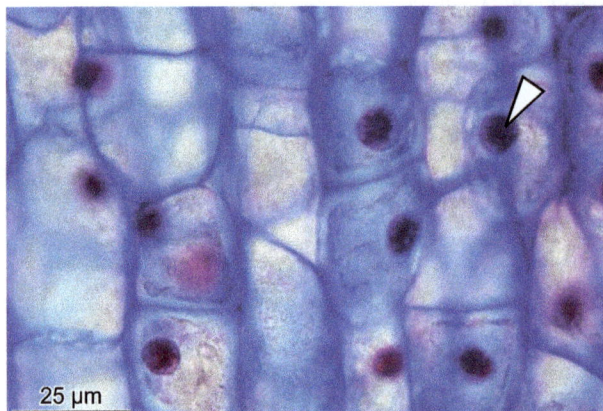

4.1 Nuclei in meristematic cells with unlignified cell walls in *Viscum album*.

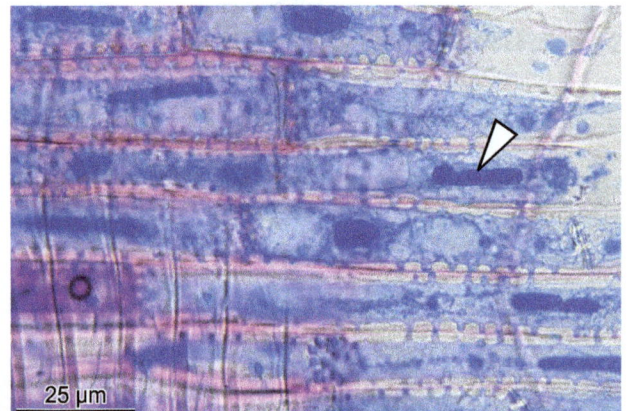

4.2 Nuclei in adult cells in a xylem ray with lignified, thick walls in *Picea abies*.

Plastids

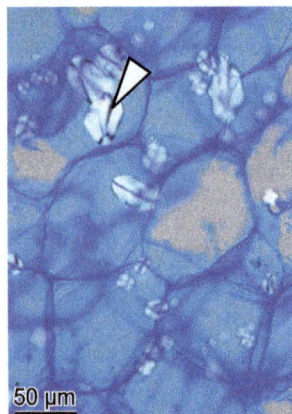

4.3 Starch grains in large unlignified cells of *Solanum tuberosum*, polarized light.

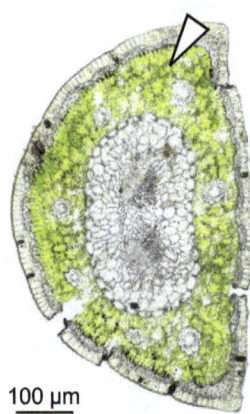

4.4 Chloroplasts in peripheral cells of a needle of *Pinus nigra*.

Ergastic substances

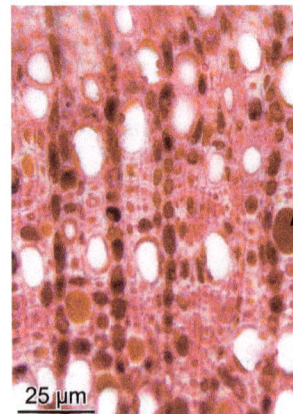

4.5 Tannins in heartwood cells of the dwarf shrub *Globularia cordifolia*.

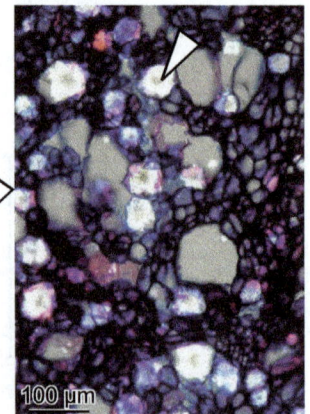

4.6 Calcium oxalate crystals in enlarged parenchyma cells of the herb *Gypsophila repens*, polarized light.

4.2 Meristematic initials – The source of new cells

Meristematic cells are anatomically undifferentiated and capable of dividing. Meristematic initials are living cells and have exclusively unlignified, thin primary walls. Primary meristematic initials are arranged in the tip of longitudinal shoots and roots (vegetation point). Secondary meristematic initials are arranged around the shoot and are the initial part of the cambium. Tertiary meristematic initials are a product of parenchyma cells in the bark. They represent the cork cambium (phellogen).

Primary meristems produce the cortex and the pith in stems. Secondary meristems (cambium) are a product of primary meristems and occur around stems where they produce the xylem and phloem. Tertiary meristems are a product of living parenchyma cells in the cortex, the phloem, and rarely in the xylem, where they produce mainly cork cells. Meristematic cells have no pits.

Primary meristems in shoots & roots

4.7 Primary meristems in a sapling of *Carpinus betulus*.

Embryonic cells in shoot tips, primary meristem

4.8 Location of a primary meristem in a shoot tip *Acer pseudoplatanus*.

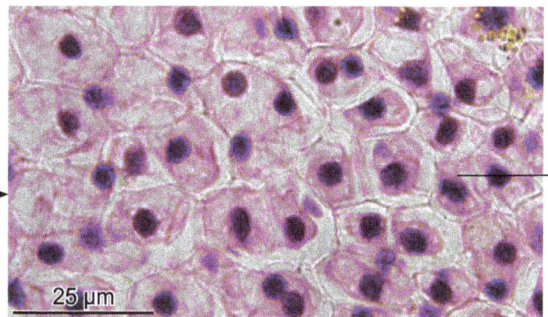

4.9 Microscopic aspect of embryonic cells in the primary meristem in a shoot of *Larix decidua*.

Embryonic cells in root tips, primary meristem

4.10 Macroscopic aspect of the location of a primary meristem in a root tip of *Allium ursinum*.

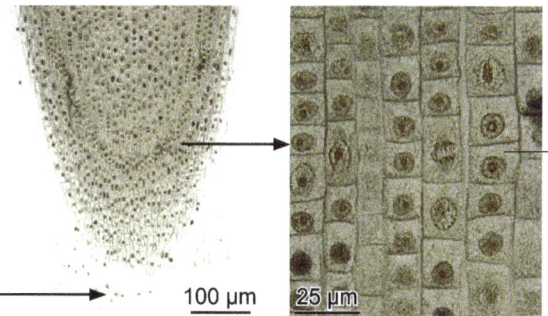

4.11 Embryonic cells in the primary meristem in a root of *Allium ursinum*.

Juvenile cells in secondary and tertiary meristems

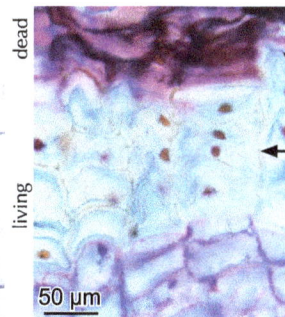

4.12 Cells with nuclei in the tertiary meristem (phellogen) of *Paeonia suffruticosa*.

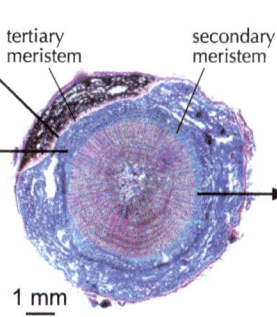

4.13 Location of the secondary and tertiary meristems in *Pinus mugo*.

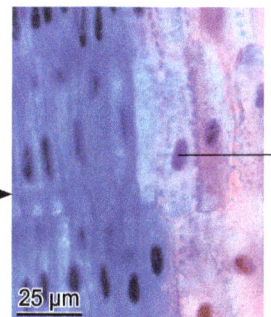

4.14 Juvenile cells with nuclei in secondary meristem and adjacent xylem/phloem in *Viscum album*.

4.3 The cuticula – Protection against dehydration

Cells exposed to the air, mostly epidermis cells, protect internal plant cells from dehydration. The cuticle at the external surface of the epidermis is an effective transpiration protection layer. Leaf surfaces with flat and compact cuticles are glossy and those with rippled cuticles are matte.

Cuticles mainly consist of pectin and cutin. The soluble extract polymerizes after the full development of the organs. Cuticles occur on leaves and young stems without periderm in all taxonomic units of vascular plants. They are absent in roots.

Cuticles are generally absent in water plants, they are thin in plants growing in shadowy conditions, and thick in plants at dry sites. The structure of the surfaces is homogeneous, layered or even granular. Chemically related to cutin are waxes, suberin and sporopollenin.

Macroscopic aspect

4.15 Leaves with a glossy surface have a thick and flat cuticula, like leaves of *Magnolia grandiflora*.

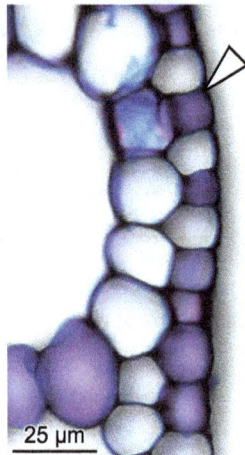

25 µm

4.16 Stem of a water plant (*Potamogeton coloratus*) without cuticula.

Thin and thick cuticles

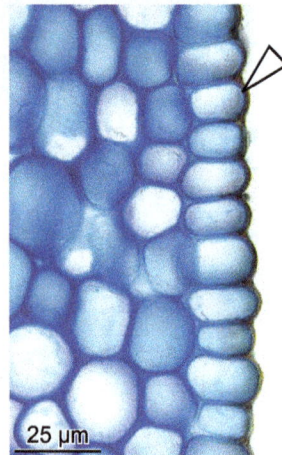

25 µm

4.17 Stem of *Ipomaea tricolor* on a shadowy site, with a thin cuticula.

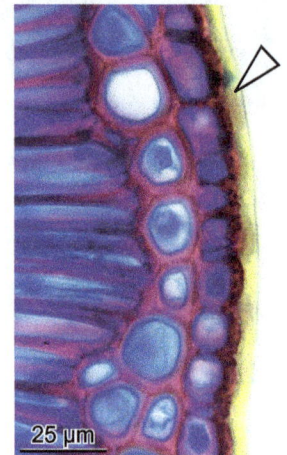

25 µm

4.18 Leaf with a thick, unstructured cuticula of *Zamia* sp. on a very dry site.

Rippled cuticles

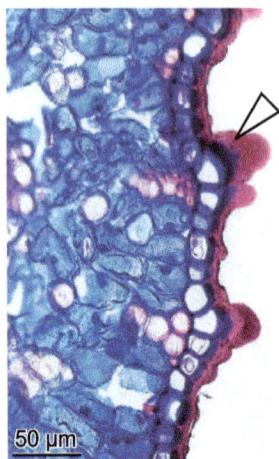

50 µm

4.19 Rippled cuticula of an annual twig of *Ephedra viridis*.

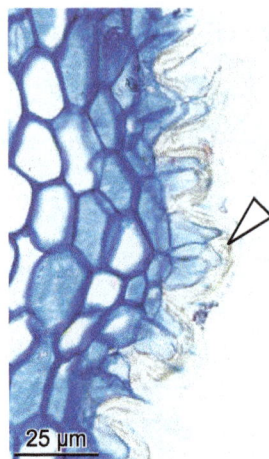

25 µm

4.20 Thick, rippled cuticula on the underside of a leaf of *Buxus sempervirens* on a dry site.

Structured cuticulae

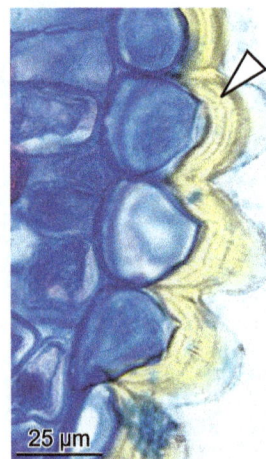

25 µm

4.21 Layered cuticula on a young stem of the mistletoe *Viscum album*.

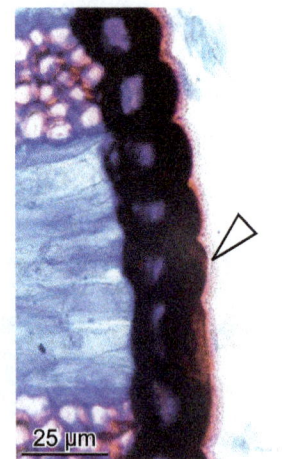

25 µm

4.22 Granulated cuticula on a leaf of *Welwitschia mirabilis*.

4.4 Epidermis – The skin of plants

The epidermis covers the products of primary meristems of most plants. Examples of stems and leaves from ferns, conifers, monocotyledonous and dicotyledonous plants are shown. Epidermis cells form a uniseriate layer of generally isodiametric cells at the periphery of primary plant bodies. Epidermis cells protect internal tissues from dehydration. Local cell wall expansion and cell division form bulliform cells or a variety of trichoms and hairs with special functions.

The epidermis of green plants is punctuated with stomata, allowing gas exchange between the atmosphere and the plant tissue. Since epidermis cell walls are transparent, photosynthesis is possible in all cells below it even when the vacuoles are filled with red-stained anthocyanins (pigments).

Macroscopic aspect of terrestrial plants

4.23 All leaves and young twigs of perennial plants are covered with an epidermis, like here in *Magnolia grandiflora*.

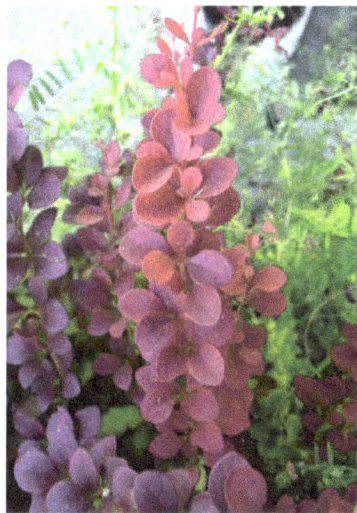

4.24 Red leaves appear red because the epidermis cells are filled with the red pigment anthocyanin, like here in *Berberis* sp.

Epidermis cells in a red leaf

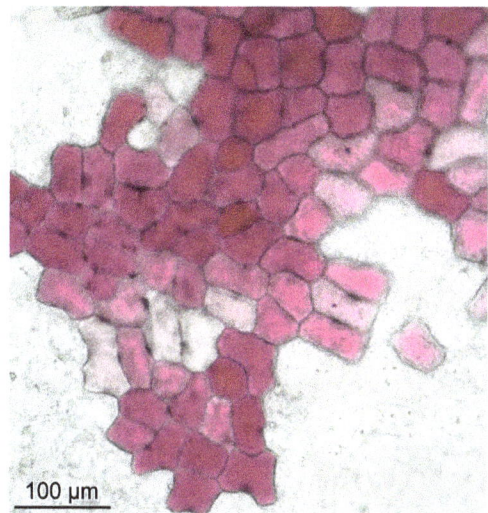

4.25 Vacuoles are filled with the red pigment anthocyanin in the epidermis of a fruit of *Euonymus europaeus*.

Anatomy of a thin-walled epidermis

Water plants

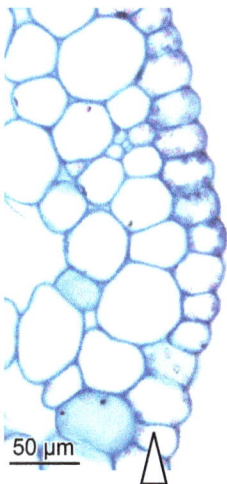

4.26 Very thin-walled epidermis in the water plant *Elodea canadensis*.

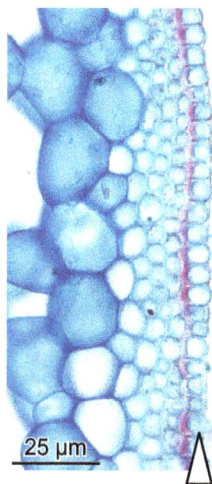

4.27 Partially lignified epidermis in the swamp plant *Scheuchzeria palustris*.

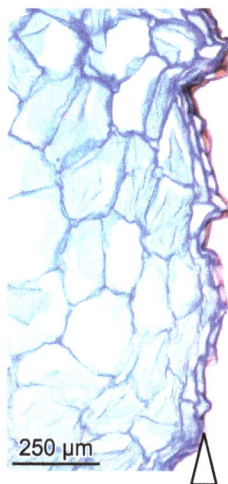

4.28 Externally lignified epidermis of the herb *Adoxa moschatellina* on a wet site.

Terrestrial plants

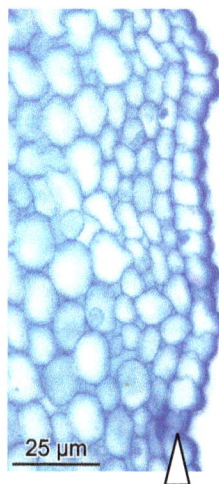

4.29 Externally thick-walled, unlignified epidermis. Young shoot of *Asparagus* sp. on a dry site.

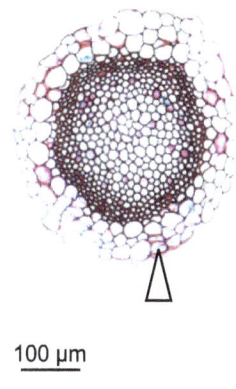

4.30 Very thin-walled epidermis in a *Sphagnum* sp.

Anatomy of a thick-walled epidermis

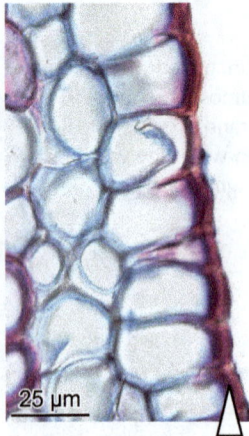

4.31 Externally thick-walled and lignified epidermis of the herb *Psilotum nudum* on a wet site.

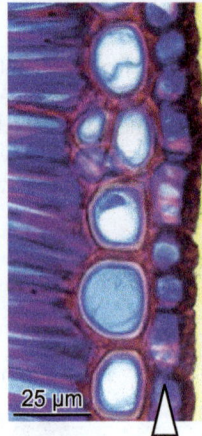

4.32 Very thick-walled epidermis in *Zamia* sp.

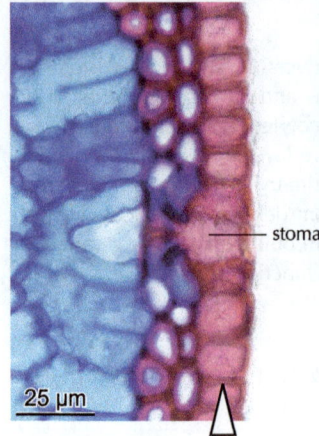

4.33 Thick-walled, lignified epidermis on a needle of the conifer *Pinus nigra*.

Stomata

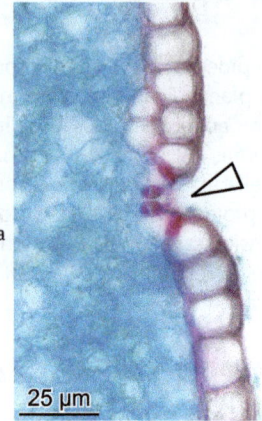

4.34 Sunken stoma in the epidermis of *Elymus farctus*.

Anatomy of elongated epidermis cells

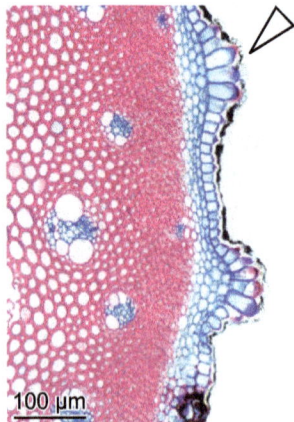

4.35 Locally enlarged epidermis cells in *Asparagus scoparius*.

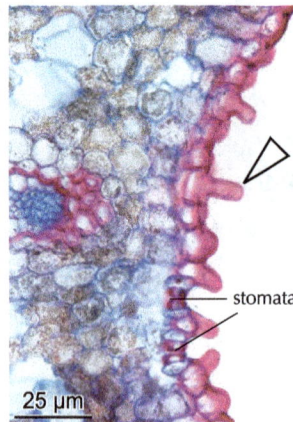

4.36 Bulliform lignified epidermis on a shoot of the monocotyledonous *Carex glareosa*.

Glandulous excretion elements

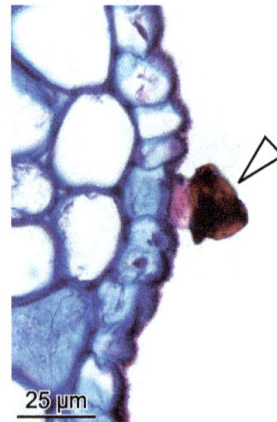

4.37 Excretion in *Lysimachia vulgaris*.

4.38 Excretion in *Salvia pratensis*.

Trichoms

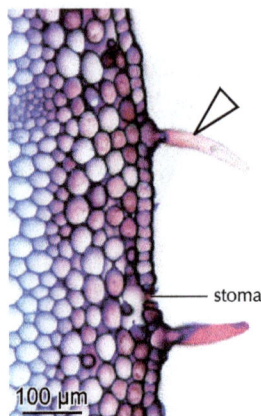

4.39 Unicellular hairs in *Lilium martagon*.

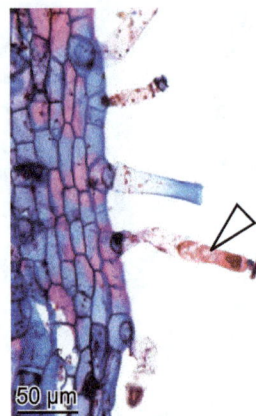

4.40 Bicellular hairs in *Scrophularia peregrina*.

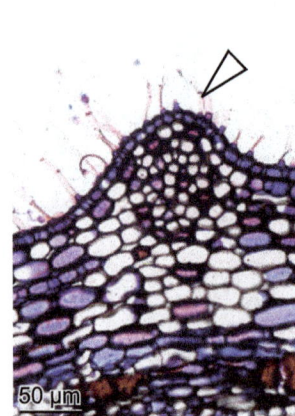

4.41 Multicellular hairs in *Ptilostemon chamaepeuce*.

Rhizoids

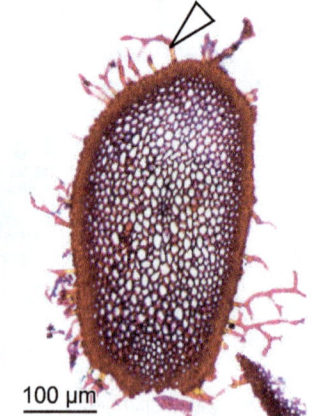

4.42 Stem of the moss *Thuidium tamariscinum*.

4.5 Collenchyma – Local peripheral stability

Collenchyma functions as a stabilizing element in edges and ridges of herbaceous plant stems of various genera in dicotyledonous and monocotyledonous plants. They are a product of primary meristems and occur in the cortex of stems. Collenchyma cells are similar to parenchyma cells but are normally longer and have pointed axial ends. Characteristic are the partially thickened primary cell walls. The walls contain cellulose, and a large amount of pectin, which is indicated by blue to purple staining with Astrablue/Safranin. The appearance largely varies: some cells are only slightly and some intensively thickened (lamellar collenchyma) and some only in the angles (angular collenchyma). At least some cells contain protoplasts and nuclei.

Collenchyma occurs mainly in the cortex of dicotyledonous plants

4.43 *Stachys sylvatica*, a Lamiaceae with quadrangular stems.

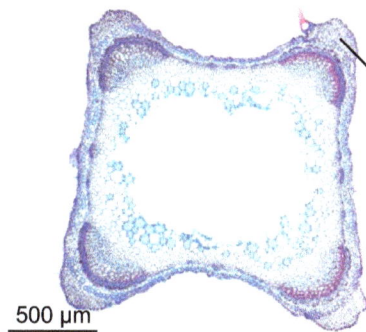

4.44 The quadrangular stem of *Stachys sylvatica* is stabilized by collenchyma in the cortex.

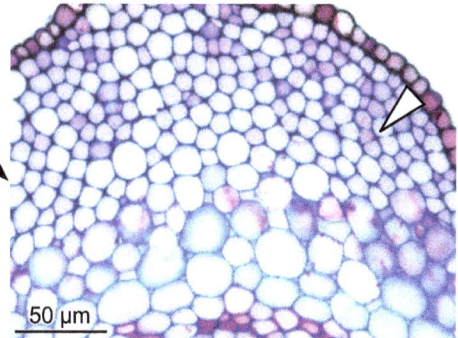

4.45 Cortex with collenchyma, bordered towards the outside by an epidermis and towards the inside by parenchyma cells. Thick edges characterize the collenchyma.

4.46 *Rumex alpinus*, a Polygonaceae with longitudinally ribbed annual shoots.

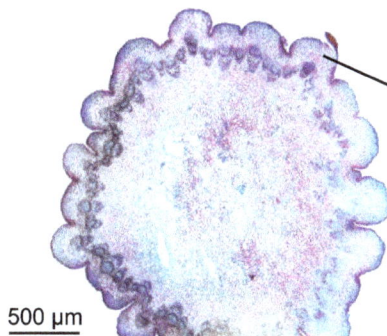

4.47 The ridges in the stems of *Rumex alpinus* are stabilized by collenchyma in the cortex.

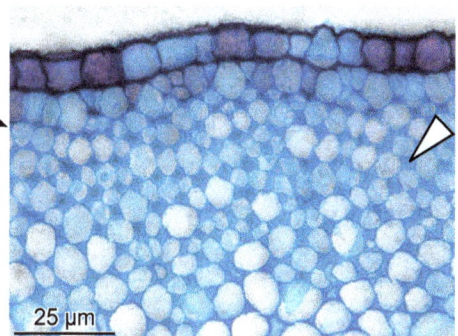

4.48 Living collenchyma cells with intensively thickened angles (angular collenchyma). Most cells contain protoplasts. In a few of them, nuclei are visible.

Collenchyma is rare in monocotyledonous plants

4.49 *Tamus communis*, a monocotyledonous plant, belonging to the Dioscoreaceae, with lightly-ribbed annual liana-like shoots.

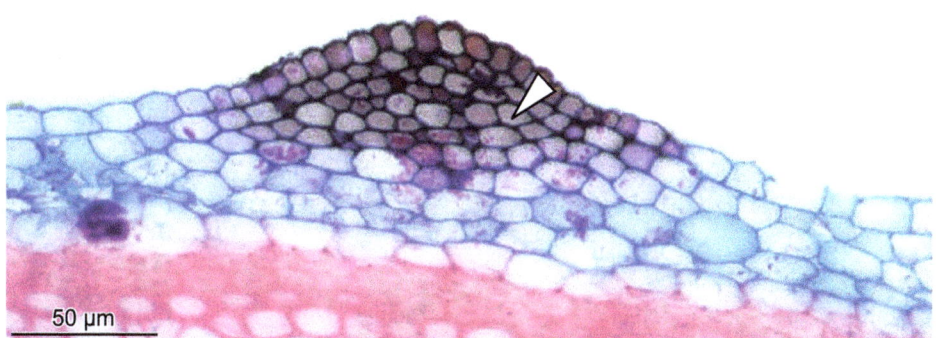

4.50 Morphologically hardly differentiated collenchyma in *Tamus communis*. Only the different reaction to the Astrablue/Safranin staining (purple coloring) indicates the collenchyma.

4.6 Parenchyma cells – Storage and repair

Humans would not be able to exist without plant parenchyma cells because their cell contents, especially carbohydrates, are essential components of our diet. Parenchyma cells are present in all plants. They are mostly small isodiametric or slightly elongated cells without pointed ends. Cell walls of parenchyma cells are thin- or thin- to tick-walled, lignified or unlignified and perforated with simple pits. Parenchyma cells primarily function as storage cells. Living parenchyma cells are totipotent. They have the potential to regenerate new cell types or entire plants under suitable environmental conditions. They play an essential role in regeneration processes after injury (see Chapter 10).

Parenchyma cells are part of stems, roots, leaves, flowers, fruits and seeds. Abundant amounts of parenchyma cells occur in thickened belowground organs. The life span of parenchyma cells of perennial plants is normally very long. Ray-parenchyma cells can live for more than 100 years in the sapwood of conifers or more than 200 years in dwarf shrubs of the arctic.

Parenchyma cells occur in all terrestrial plants

4.51 Potatoes (*Solanum tuberosum*) consist mainly of parenchyma cells. They are able to produce new shoots.

4.52 Approx. 5% of the xylem in trees are parenchyma cells. Parenchyma cells in the phloem can change their mode of behavior and form new shoots.

4.53 Scar on the stem of *Acer pseudoplatanus*. Living parenchyma cells repair wounds and protect the stem against destructive organisms.

Shape of parenchyma cells

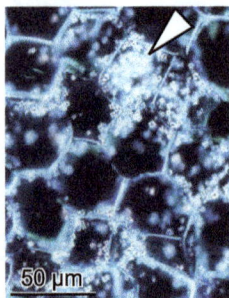

4.54 Parenchyma cells are mostly filled with starch grains in the rhizome of *Anemone nemorosa*, polarized light.

4.55 Isodimetric parenchyma cells in the stem of the moss *Polytrichum commune*.

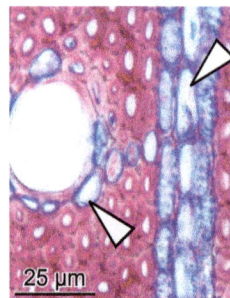

4.56 Parenchyma cells in the xylem of *Sonchus leptophyllus*. Axially sectioned cells are round, radially sectioned cells are elongated.

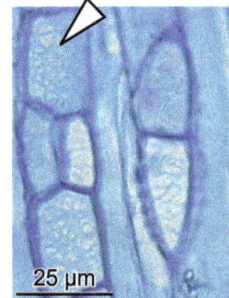

4.57 Simple pits are characteristic for parenchyma cells. Ray cells in the bark of *Paeonia suffruticosa*.

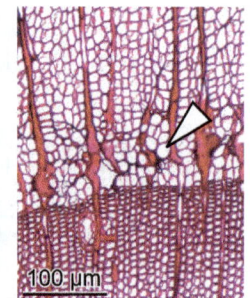

4.58 Irregularly formed parenchyma cells, the callus cells, appear after wounding in *Picea abies*.

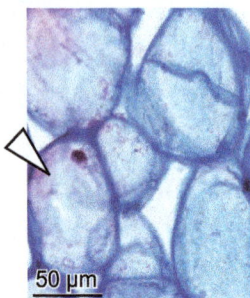

4.59 Water-storing parenchyma cells in the succulent *Sedum acre*.

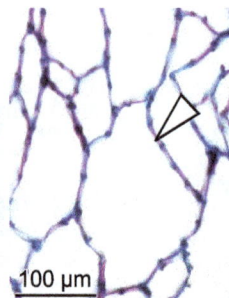

4.60 Very small parenchyma cells between air-conducting spaces in the pith of swamp plant *Scirpus radicans*.

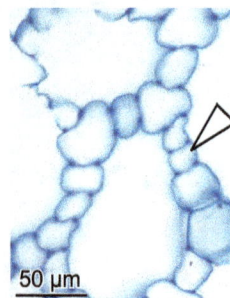

4.61 Parenchyma cells surround air canals in the cortex of the swamp plant *Orontium aquaticum*.

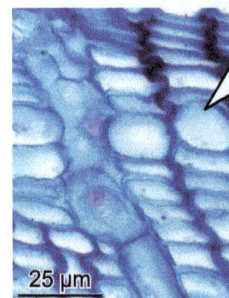

4.62 Parenchyma cells alternate with sieve cells in the bark of conifers such as *Abies alba*.

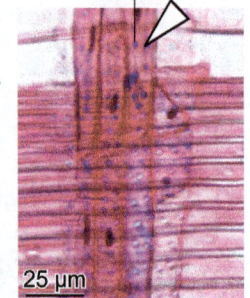

nucleus

4.63 Long-living parenchymatic ray cells with nuclei (blue) in *Picea abies*.

4.7 Fibers and tracheids – Stabilisation and water conduction

Many wooden products, e.g. beams, boards, firewood and paper, are substantial basics of our daily life. The main function of fibers is stabilization, while tracheids are stabilizing and conduct water.

Fibers and tracheids are long cells with elongated, pointed tips. Tips are a result of post-cambial axial elongation. Fibers or tracheids occur in all growth forms and on all sites in spore plants (ferns, horsetails), in conifers, in monocotyledonous and dicotyledonous plants. They are part of annual and perennial stems of roots and leaves. Fibers have more or less thick, lignified cell walls. Characteristic is the presence of secondary walls. Pits in fibers are mostly simple or slightly bordered. Pits in tracheids are bordered (see Chapter 5.2, Fig. 5.19). Fibers occur in the xylem, phloem and cortex, tracheids only in the xylem.

Use of fibers

4.64 Fibers are omnipresent in human life in the form of wood products.

Length of fibers

250 μm

4.65 Short fibers (100–200 μm) in shrubs, such as *Buxus sempervirens*.

250 μm

4.66 Long fibers (500–>1000 μm) in trees, such as *Fagus sylvatica*.

250 μm

4.67 Long (1,000–3,000 μm), de-lignified tracheids (used for pulp) in *Pinus sylvestris*.

Fibers occur in all vascular plants

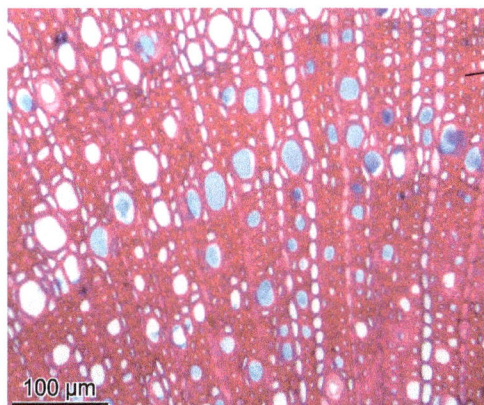

fibers

100 μm

4.68 Fibers in stems of dicotyledonous dwarf shrubs like *Sibbaldia procumbens*.

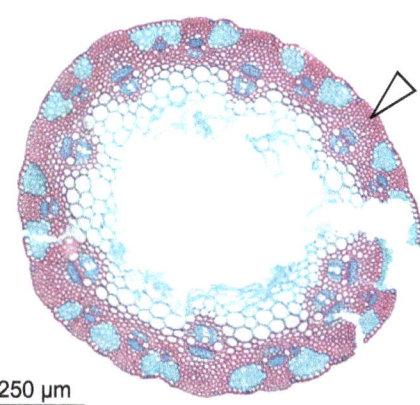

250 μm

4.69 Fibers in culms and leaves of monocotyledonous herbs like *Festuca alpina* (top) or *Festuca erecta*.

Fibers occur in all parts of plants

4.70 Fibers occur in stems, roots and leaves. *Castanea sativa.*

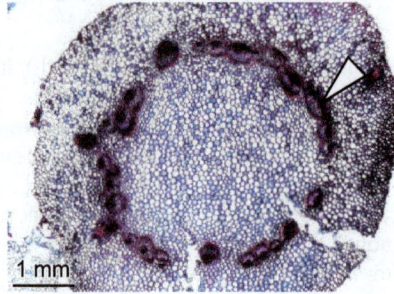

4.71 Fibers in a shoot of *Ranunculus lanuginosus.*

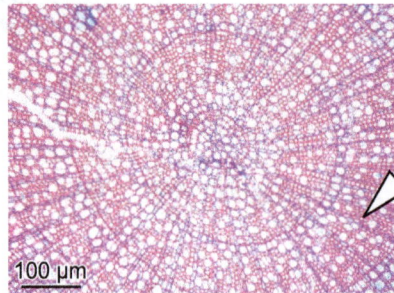

4.72 Fibers in a root of *Ledum palustre.*

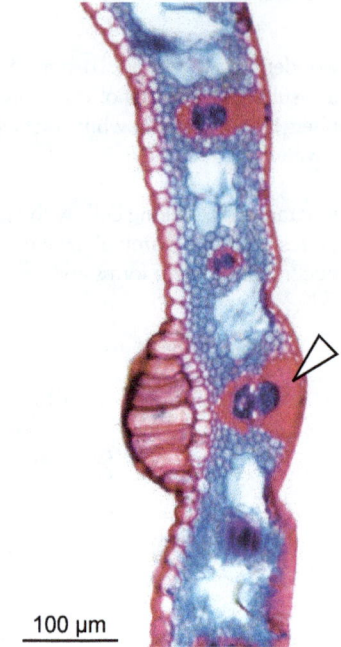

4.73 Fibers in a leaf of *Carex sempervirens.*

Fibers occur in all parts of stems

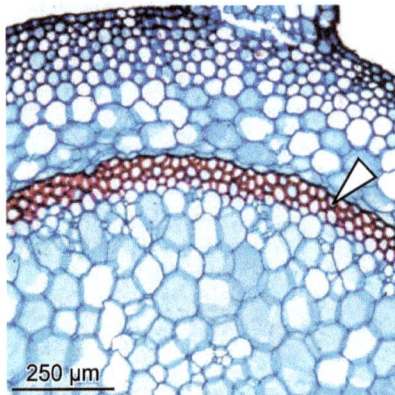

4.74 Fibers in a shoot (cortex) of *Cucumis sativus.*

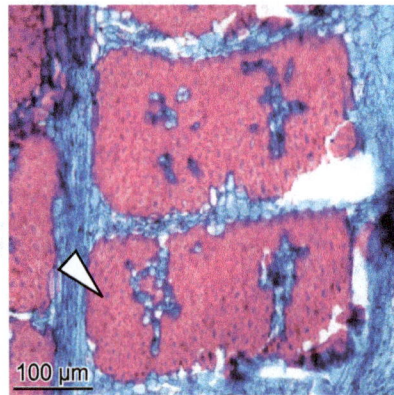

4.75 Fibers in the bark of *Chrysothamnus parryi.*

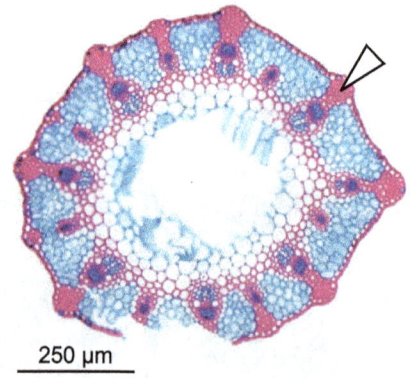

4.76 Fibers in a culm of *Sesleria coerulea.*

Various cell-wall thicknesses of fibers and tracheids

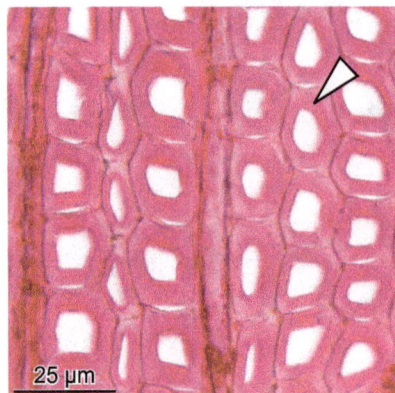

4.77 Thick-walled tracheids in *Larix decidua.*

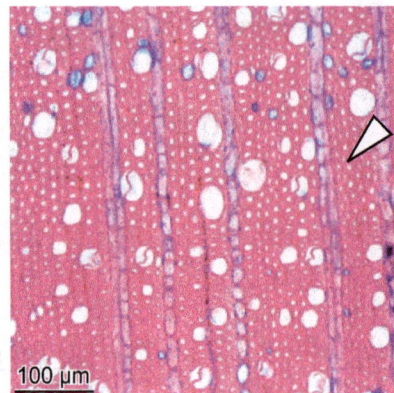

4.78 Thick-walled fibers in *Eryobotrya japonica.*

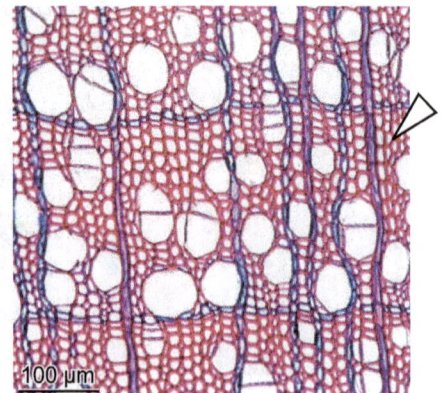

4.79 Thin-walled fibers in *Salix foetida.*

4.8 Sclereids in the bark – Extraordinary cell-wall thickening

We notice sclereids when eating pears: The granules remaining between your teeth are sclereids. Sclereids are absent in the xylem but frequent in the phloem, cortex and pith of trees and herbs. They also occur in fruits and nut shells, and rarely in leaves. Characteristic for sclereids are the irregularly formed cells with thick, secondary walls, distinct simple pits and often distinct growth layers. A special case are star-like sclereids in the aerenchyma of the water lily (*Nuphar* sp). Sclereids in the bark normally occur in the non-conducting (adult) phloem and in rays. Sclereids originate from parenchyma cells with a short-term accelerated growth of secondary walls.

Sclereids in stems

4.80 Hard bark of *Fagus sylvatica*.

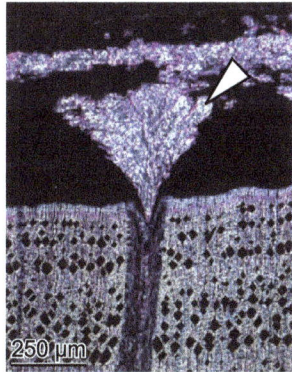

4.81 Many sclereids in a ray and the cortex in *Fagus sylvatica*, polarized light.

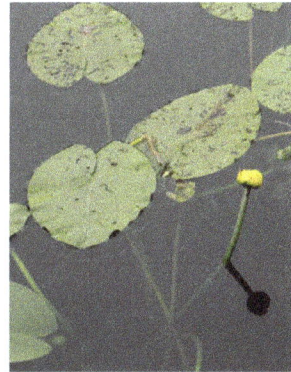

4.82 Flexible and soft culms and a petiole of the water lily *Nuphar lutea*.

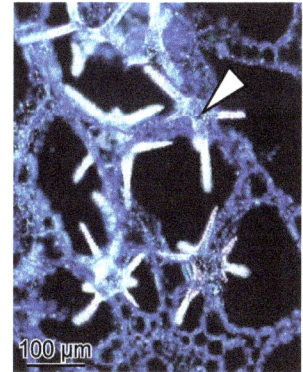

4.83 Star-like sclereids in air-conducting channels (aerenchyma) in *Nuphar lutea*, polarized light.

Sclereids in fruits

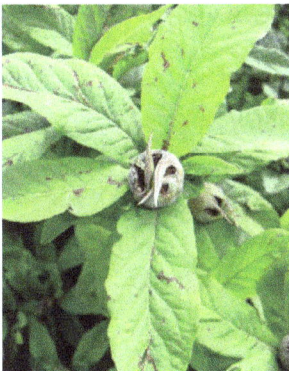

4.84 Fruit of *Mespilus germanica*.

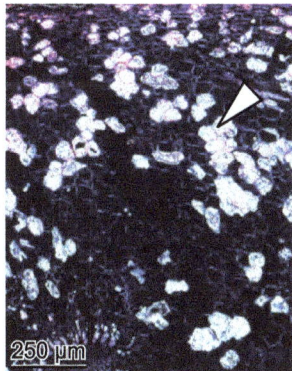

4.85 Sclereids in the peripheral part of the fruit of *Mespilus germanica*, polarized light).

4.86 Fruits of the hazelnut (*Corylus avellana*).

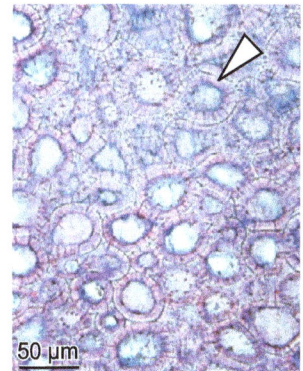

4.87 Sclereids in the shell of a hazelnut.

Single and groups of sclereids

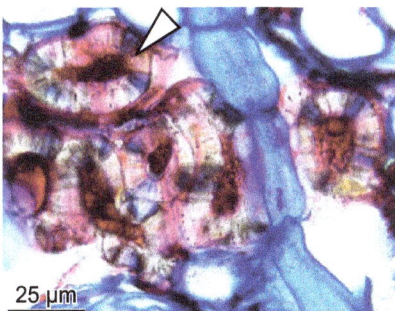

4.88 Group of sclereids in the bark of *Picea abies*.

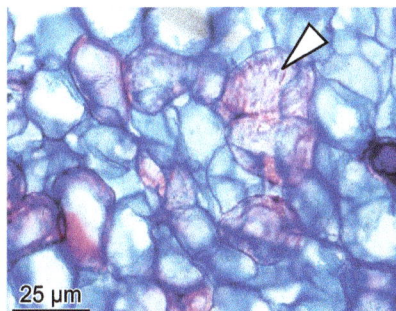

4.89 Isolated sclereids in the fruit of *Mespilus germanica*.

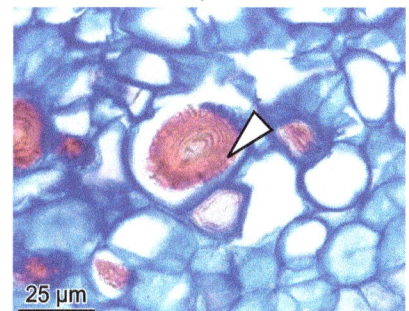

4.90 Isolated sclereids with distinct layers in a leaf of *Welwitschia mirabilis*.

4.9 Vessels – Water conduction

Plant life on terrestrial sites would not exist without the water conducting vessels. However, it is generally difficult to see vessels with bare eyes, because their diameter is usually below the resolution of human eyes. Vessels consist of vessel elements. Fully developed vessel elements are dead and more or less elongated. They are composed of lignified cell walls, perforation plates at their distal ends and pits at the longitudinal walls. Lignification of the walls of vessel elements prevents cell collapse.

Vessels occur in the xylem of most spore plants (ferns, horsetails, lycopods) and most monocotyledons and dicotyledons, except in conifers and few others where they are replaced by tracheids. Vessels occur in all growth forms. Diameters are usually large

in lianas (>200–500 μm), smaller in trees (50–200 μm) and small in herbs (20–50 μm). The length of vessel elements varies from <100 μm in small herbs to >1,500 μm in trees. Vessel elements are normally short in ring-porous and long in diffuse-porous tree species.

Vessels conduct water mainly in axial direction from the root to the leaves. Lateral perforations (bordered pits) also indicate a lateral water transport. Vessels exist in the xylem of all plant organs in various dimensions. Vessels are normally larger in roots than in stems, and even smaller in leaves.

For a discussion of cell-wall structures see Chapter 5.1.

Macroscopic aspect

4.91 Vessels in the earlywood of ring-porous species (e.g. *Quercus* sp.) are visible to the bare eye.

Anatomy of vessel elements

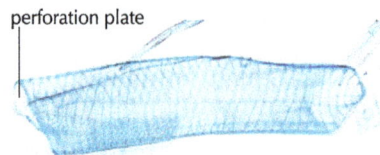

perforation plate

4.92 Long vessel element (500 μm) with helical thickenings in *Tilia* sp.

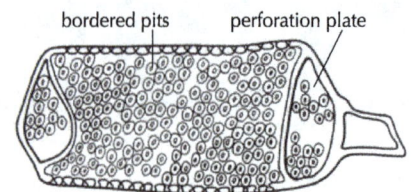

bordered pits   perforation plate

4.93 Short and wide vessel in the earlywood of *Fraxinus excelsior*. Reprinted from Greguss 1945.

Vessels occur in all vascular plants

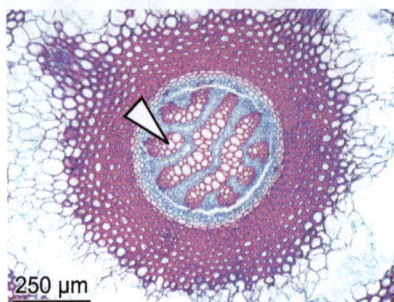

250 μm

4.94 Lycopods: clubmoss *Lycopodium clavatum*.

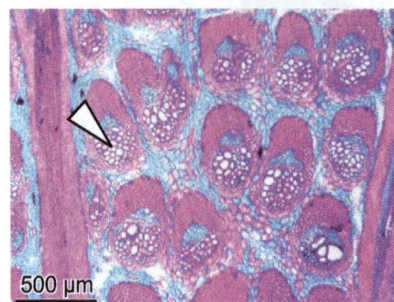

500 μm

4.95 Monocotyledons: stem of palm *Phoenix dactylifera*.

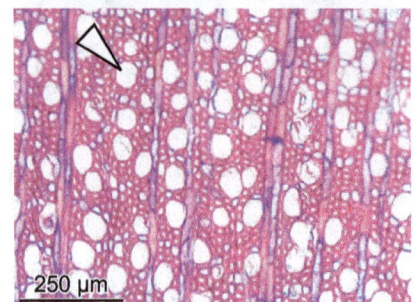

250 μm

4.96 Dicotyledons: stem of apple tree *Malus domestica*.

Vessels occur in all growth forms

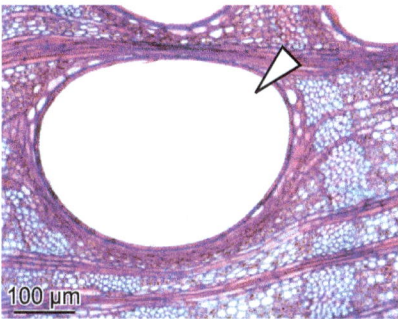

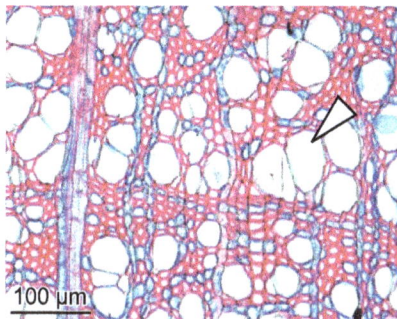

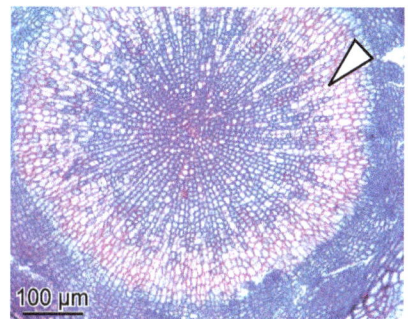

4.97 Very large vessel diameters (550 µm) in lianas *Celastrus* sp. (top) and *Thunbergia* sp. (bottom).

4.98 Large vessel diameter (80 µm) in the dicotyledonous tree *Fagus sylvatica*.

4.99 Very small vessel diameters (20 µm) in the 5 cm-tall annual herb *Erophila verna*.

Vessels occur in most parts of plants

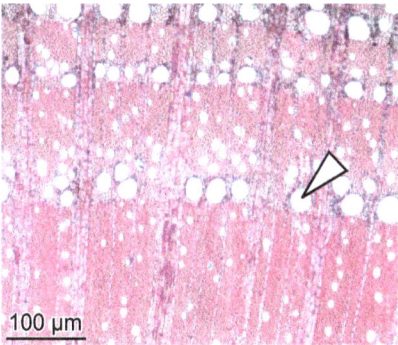

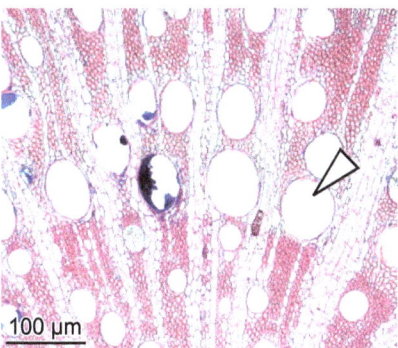

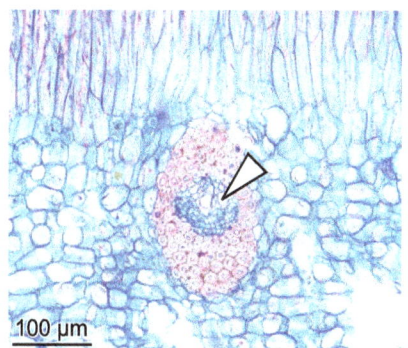

4.100 Diameter of earlywood vessels in the stem (top) is 70 µm, in the root (bottom) 250 µm in *Prunus amygdalus*.

4.101 Small vessel diameters (20 µm) in a vascular bundle of a thick leaf of *Clusea rosea*.

4.10 Cork cells – Defense against organisms, heat and cold

Cork is a perfect insulation material. We use it to seal wine bottles and to insulate walls and floors. Most of the industrially used cork is a product of cork oak (*Quercus suber*). Plants insulate their stems against extreme environmental conditions like intense radiation, blasting by sand and ice crystals, fire or flooding. Cork cells (phellem) occur outside of the phloem and cortex of most taxa and growth forms in conifers and dicotyledons. They are rare in monocotyledonous plants, and occur there only in trees. Cork cell walls consist of suberin, which is difficult to decompose for many fungi. Cork cells of any form can be thin- or thick-walled.

Cork formation occurs usually on the outside of stems. However, a few species produce cork rhythmically within the xylem. Small, long-lived herbs compartmentalize their center by forming cork internally within the stem.

Cork cells in stems of seed plants of all growth forms

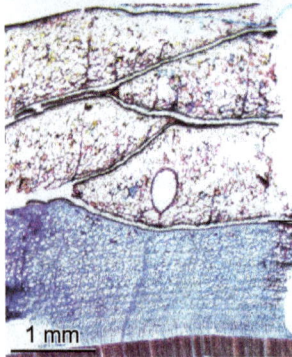

4.102 Conifers: Layered bark of *Pinus sylvestris*.

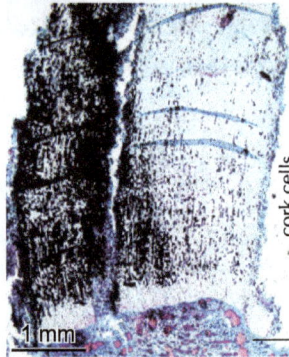

4.103 Dicotyledonous plants: Layered bark of the tree *Quercus suber*.

4.104 Dicotyledonous plants: Layered bark of a small cushion of the alpine *Saxifraga oppositifolia*.

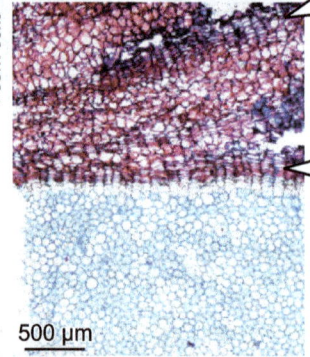

4.105 Monocotyledonous plants: Cork cells in the tree *Dracaena serrulata*.

Thin- and thick-walled cork cells

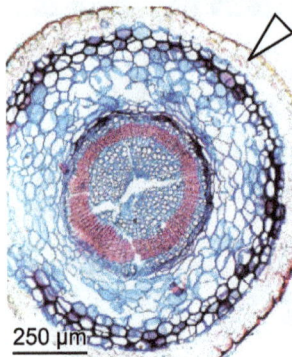

Cork cells within the xylem

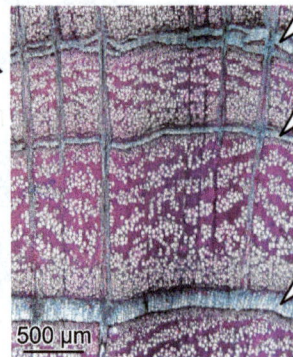

4.106 A small mantle of thin-walled cork cells surrounds the water-storing cortex of the succulent *Sedum acre*.

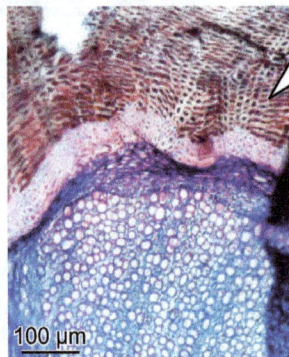

4.107 A large zone of thick-walled, rectangular cork cells surrounds the soft, unlignified xylem of the alpine cushion plant *Saxifraga caesia*.

4.108 Tangential cork layers between compartments of xylem in *Artemisia tridentata*.

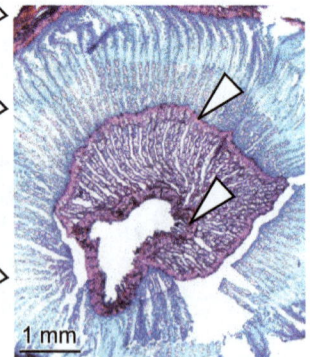

4.109 Cork layers in a stem separate the living from the dead xylem in the center of the stem of *Scorzonera virgata*.

4.11 Sieve cells, sieve tubes and companion cells
– Conduction of assimilates

Sieve elements are part of the bast. Neolithic settlers knew sieve cells and bast fibers very well and used the bast of several plants to braid baskets and tissues. Today, papier mâché ("chewed paper") is probably the only remaining useful product.

Sieve cells and sieve tubes are a major part of the phloem, which mainly conducts assimilates from leaves to parenchyma cells. Sieve elements are accompanied by parenchyma cells and companion cells. Sieve cells have sieve fields on their lateral sides, while sieve tubes have plates at their distal ends and

on their lateral walls. However, the anatomical differentiation in cross sections is generally difficult. Sieve tubes and sieve cells are long, thin-walled and unlignified. Adult sieve elements do not contain nuclei. The metabolism of the sieve elements is maintained by the nuclei in the companion cells. Companion cells are smaller and always adjacent to sieve tubes.

Sieve cells occur in all vascular plants from ferns to dicotyledons. Sieve cells and sieve tubes often collapse following their death. More information is given in Chapter 6.2.8.

Macroscopic aspect of the phloem

4.110 Neolithic tissues made of *Tilia* bast. Photo: Amt für Städtebau Unterwasserarchäologie, Univ. Zürich.

4.111 Stripped bark of *Populus* sp. Bast fibers dismantled from the xylem in the cambial zone.

4.112 Large layered phloem between the xylem and cork in *Juglans regia*.

Anatomy of sieve elements and companion cells

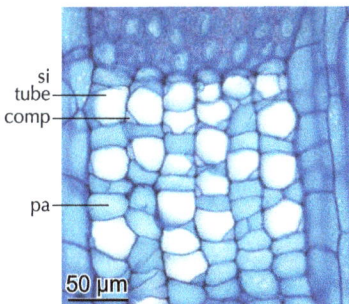

4.113 Phloem of the tree *Adansonia digitata*, consisting of sieve tubes, companion cells and parenchyma cells.

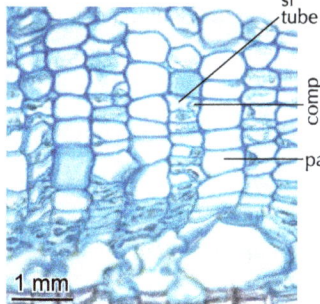

4.114 Phloem of the shrub *Lonicera alpigena*, consisting of sieve tubes, companion cells and parenchyma cells.

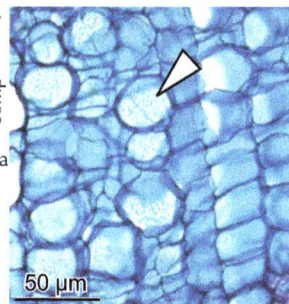

4.115 Sieve plates of sieve tubes in *Adansonia digitata*.

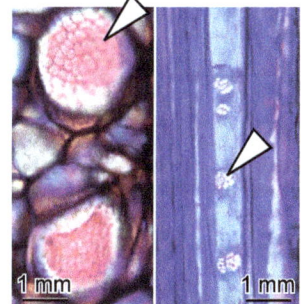

4.116 Sieve elements of the dicotyledonous herb *Bryonia dioeca* (left) with sieve plates at distal ends, and of the conifer *Larix decidua* with lateral sieve plates (right).

Sieve elements in ferns, conifers, dicotyledonous and monocotyledonous plants

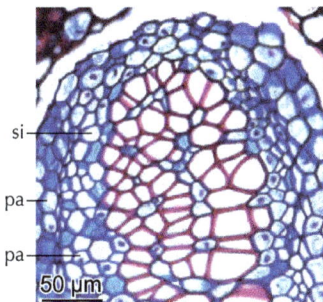

4.117 Sieve cells in the phloem of the fern *Cryptogramma crispa*.

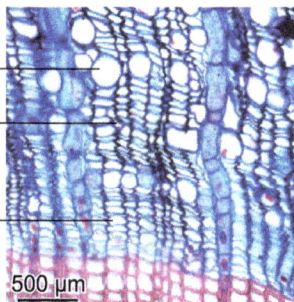

4.118 Sieve cells in the phloem of the conifer *Picea abies*.

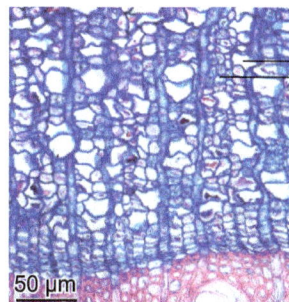

4.119 Sieve tube in the phloem of the dicotyledonous *Quercus ilex*.

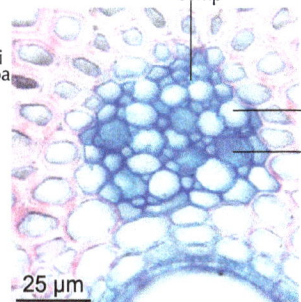

4.120 Sieve tubes in a vascular bundle of the monocotyledonous *Arundo donax*.

4.12 Secretory cells – Defense

Today, the best known products of secretory cells are resin condensates, e.g. colophony, amber, incense, and oils in orange skins. The resins are excreted from resin ducts in trees. Fifty years ago, latex as a product of secretory canals was intensively used as gum for tires and chewing gum.

Plant secretory cells are formed either externally, from epidermal tissues, or internally, by primary or secondary meristems. Described in this section are mainly internal secretory cells. They sporadically occur in the whole taxonomic system of terrestrial plants from ferns to dicotyledonous plants. Secretory cells are thin-walled, unlignified cells. Single cells occur in the xylem of a few trees and produce oil. Laticifers (latex ducts) form long, uni- or multicellular tubes. Very frequent are ducts that are surrounded by resin-producing, long-lived epithelial and parenchyma cells. Many slime-producing secretory cells are anatomically identical to normal parenchyma cells.

Secretory cells occur around ducts in the pith, xylem, phloem and cortex of stems, roots, leaves and fruits.

Resin use in art and religion

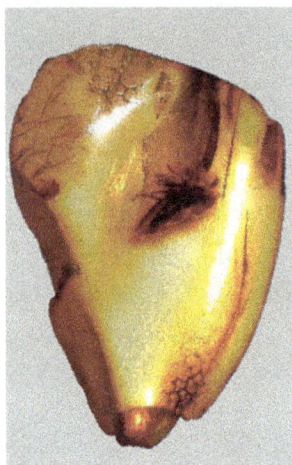

4.121 Amber is a fossilised, condensed resin, produced from various conifers 100 million years ago. It has been regarded as a magic "stone" since Neolithic times. Photo: L. Monshausen.

4.122 Frankincense is a product of resin ducts in the bark of the desert tree *Boswellia sacra*. Incense smoke is used for many ritual acts in a number of different religions. Photo: S. Fleckney.

Essential oils

4.123 Ducts in the skin of oranges (*Citrus sinensis*) produce essential oils.

Latex

4.124 Caoutchouc harvesting on a tree of *Hevea brasiliensis*.

Resin ducts in the xylem of conifers

4.125 Resin in the sapwood of the conifer *Pinus sylvestris*.

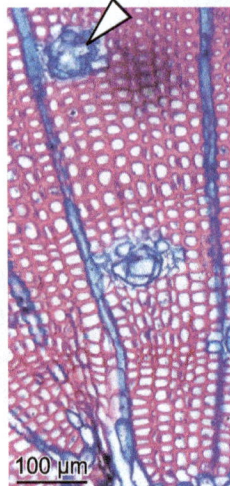

4.126 Resin ducts in the xylem of the conifer *Pinus mugo*.

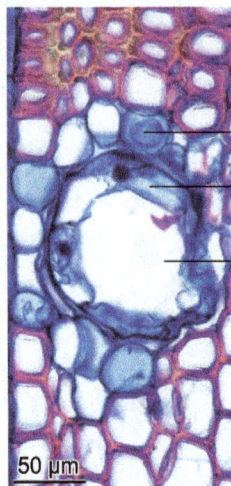

4.127 Living epithelial and parenchyma cells around a resin duct in *Pinus sylvestris*.

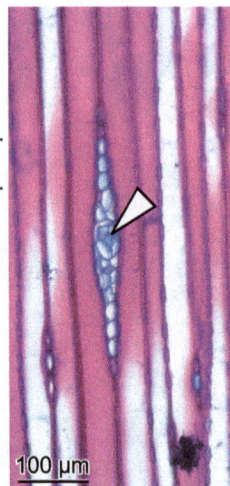

4.128 Resin duct in a ray in *Pinus sylvestris*, tangential section.

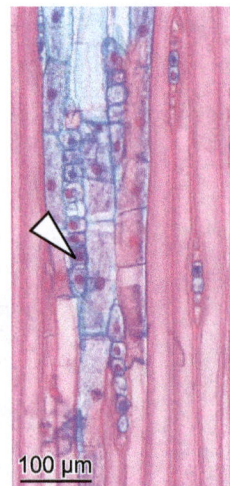

4.129 Living epithelial and parenchyma cells around a resin duct of *Pinus sylvestris*, radial section.

Anatomy of ducts in stems of dicotyledonous plants

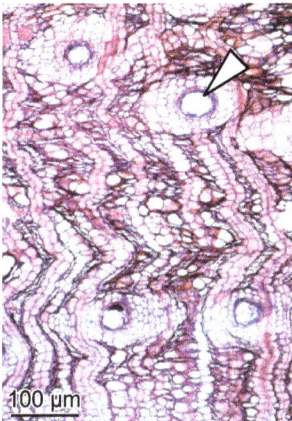

4.130 Resin ducts in the phloem of the incense tree *Boswellia sacra*.

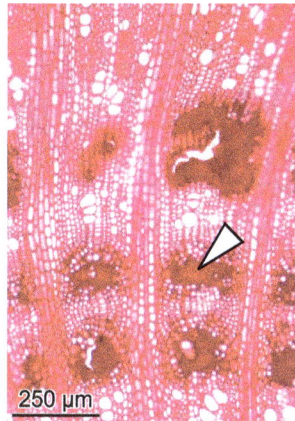

4.131 Traumatic "resin" ducts in the xylem of an almond tree (*Prunus amygdalus*).

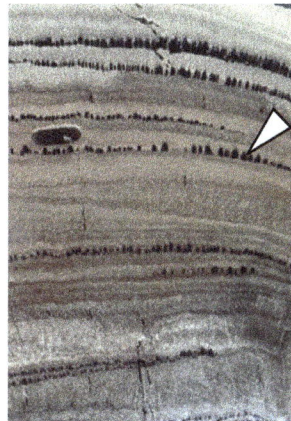

4.132 Macroscopic aspect of traumatic "resin" ducts (kino veins) in the xylem of *Eucalyptus obliqua*. Photo: P. Majewski.

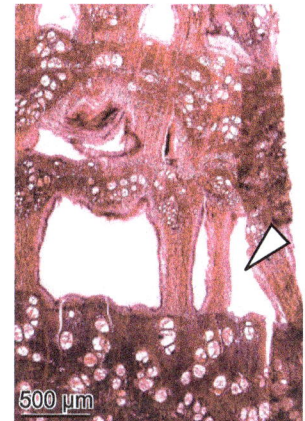

4.133 Microscopic aspect of traumatic "resin" ducts (kino vein) in the xylem of a *Eucalyptus* sp.

Anatomy of ducts in various parts of plants

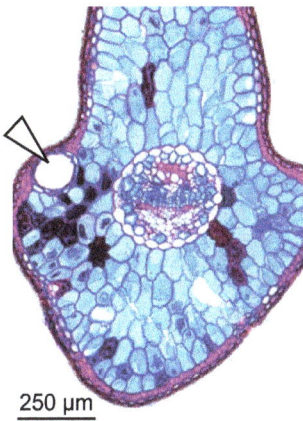

4.134 Resin duct in a needle of the conifer *Picea abies*.

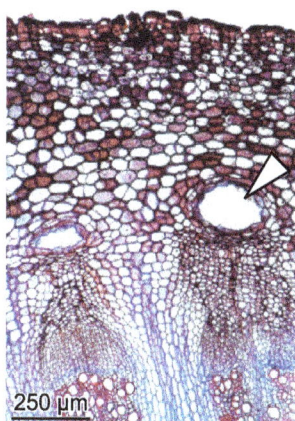

4.135 Duct in the cortex of the bark of the herb *Petasites paradoxus*.

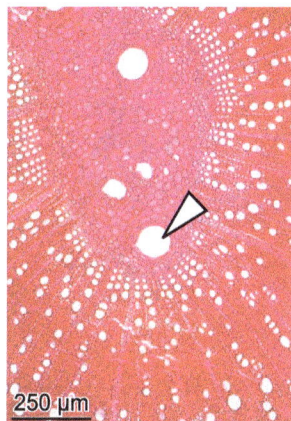

4.136 Ducts in the pith of a twig of *Grewia villosa*.

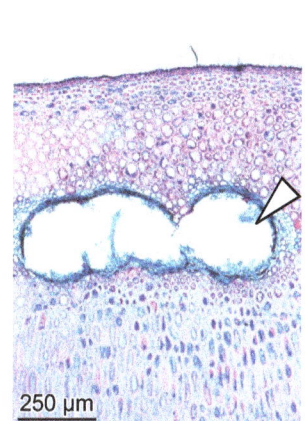

4.137 Ducts in the shell of a hazelnut (*Corylus avellana*).

Anatomy of oil cells

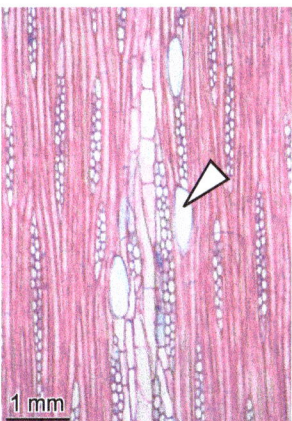

4.138 Enlarged oil cells in the xylem of the tree *Phoebe nanmu*, tangential section. It was the preferred wood for the construction of the Forbidden City in Beijing.

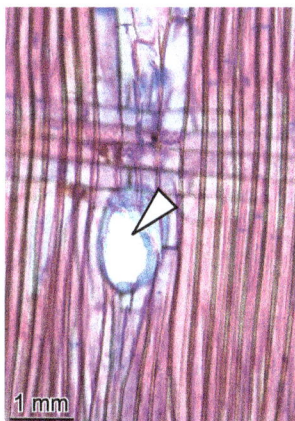

4.139 Enlarged oil cells in the xylem of *Phoebe nanmu*, radial section.

Anatomy of laticifers

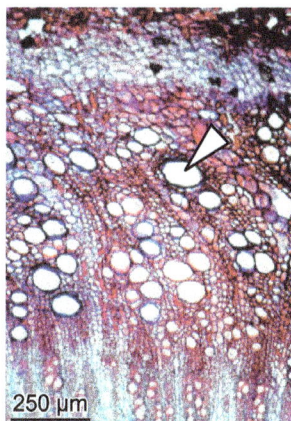

4.140 Latex-producing laticifers in the phloem and the cortex of the shrub *Euphorbia armena*.

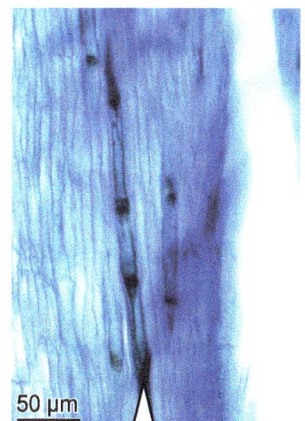

4.141 A single laticifer, consisting of several secretory elements in *Scrophularia dentata*.

4.13 Intercellulars and aerenchyma – Air circulation within the plant

The very light-weighted shoots of reed (*Phragmites communis*) and many other grass-like shore plants have been used for the construction of boats, and the insulation of roofs and floors. Hay from wet meadows in the European Alps is used as bedding in cattle stables. These practical uses are based on the hollow stems, and the presence of aerenchyma in the shoots. Intercellulars of any form occur mainly in wetland plants. They guarantee the gas exchange from the stomata in the leaves to all cells within the plant. Large intercellulars are defined as aerenchyma. Hollow shoots and aerenchyma in water plants allow

them to float. Small intercellulars occur mainly in leaves, pith and cortex of plants in all taxonomic units of vascular plants, and aerenchyma with particular structures occur in plants of wet environments, e.g. in swamps and lakes (helophytes and hydrophytes).

Intercellulars are small spaces between round cells. Aerenchyma consists of parenchyma cells surrounding the intercellulars. They form nets, sponge-like tissues in thick stems, star-like groups, channels, irregular and radial spaces and lacunas.

Use of wetland plants

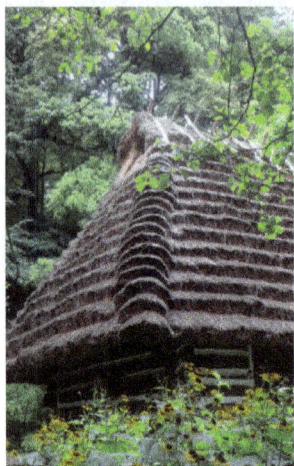

4.142 Roofing with reed (*Phragmites communis*).

4.143 Stacks of dry, stiff, nutrient-poor grasses and sedges in a wet montane meadow. The material is used instead of cereal straw for bedding in cattle stables. Photo: M. Küchler.

Anatomy of "air bags"

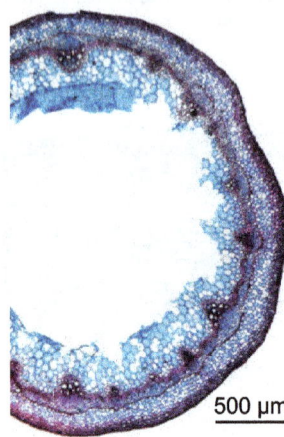

4.144 Hollow pith as the result of extreme stem expansion in *Polygonum amphibium*.

Aerenchyma

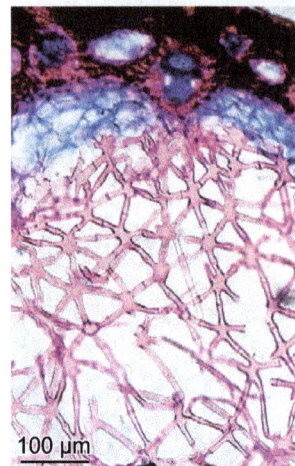

4.145 Stellate combined cells form an air-filled shoot center in *Juncus conglomeratus*.

Aerenchyma

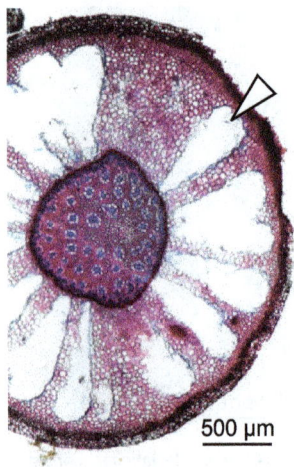

4.146 Air-filled radial spaces in the cortex of a rhizome of *Juncus articus*.

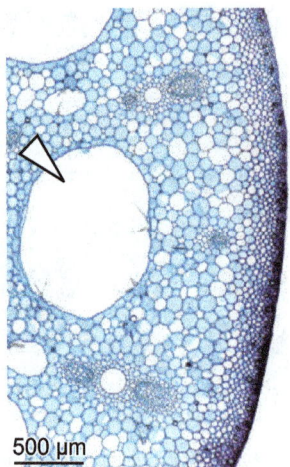

4.147 Large air tubes conduct air from the leaf to the root in *Nymphaea alba*.

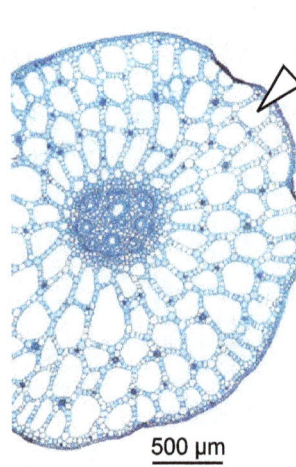

4.148 Small air tubes conduct air from the leaf to the root in *Potamogeton natans*.

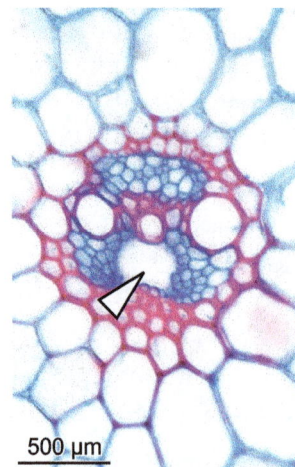

4.149 Small lacunas within vascular bundles conduct air from the leaf to the roots in *Dactylis glomerata*.

Cell Wall: Structure and Components

5.1 Principal cell-wall structure – Form and stability

The cell wall of seed plants principally consists of several layers: the middle lamella, the primary, secondary and tertiary wall, a concept presented by Evert 2006. Secondary walls consist of macro-fibrils. They are visible under the light microscope. However, the micro-fibrils are only visible in longitudinal sections of compression wood in conifers (see Chapter 10.4.1, Fig. 10.105) and under the electron microscope. In reality, there is a great chemical and morphological diversity in cell-wall structures (Evert 2006).

The middle lamella and the primary walls are optically difficult to distinguish. Primary walls are thin and consist of irregularly distributed micro-fibrils, which are embedded in a matrix of hemicellulose and pectin. Therefore they disappear in polarized light. The links between the micro-fibrils are not permanent and can be dissolved by enzymes. Therefore, cell-wall expansion is possible.

The secondary walls are normally thick and lignified. They start to form after cell expansion is completed. Characteristic at electron-microscopic scale are the helically oriented macro-fibrils consisting of cellulose and hemicellulose. They are embedded in lignin. The direction of the macro-fibrils changes within the wall but they always remain helically oriented. Due to the crystalline structure of the cellulose, secondary walls are anisotropic and are visible under polarized light. The links between the macro-fibrils are permanent; therefore the size of the cell remains stable after its formation. Secondary walls provide mechanical support against gravity and turgor pressure.

The tertiary wall is normally thin, and often covered with warts. Micro-fibril orientation is random and they are therefore not visible under polarized light. Layered cellulosic cell walls occur in all taxonomic units of vascular plants, in all habitats, from the desert to the water.

Principal structure of cell walls

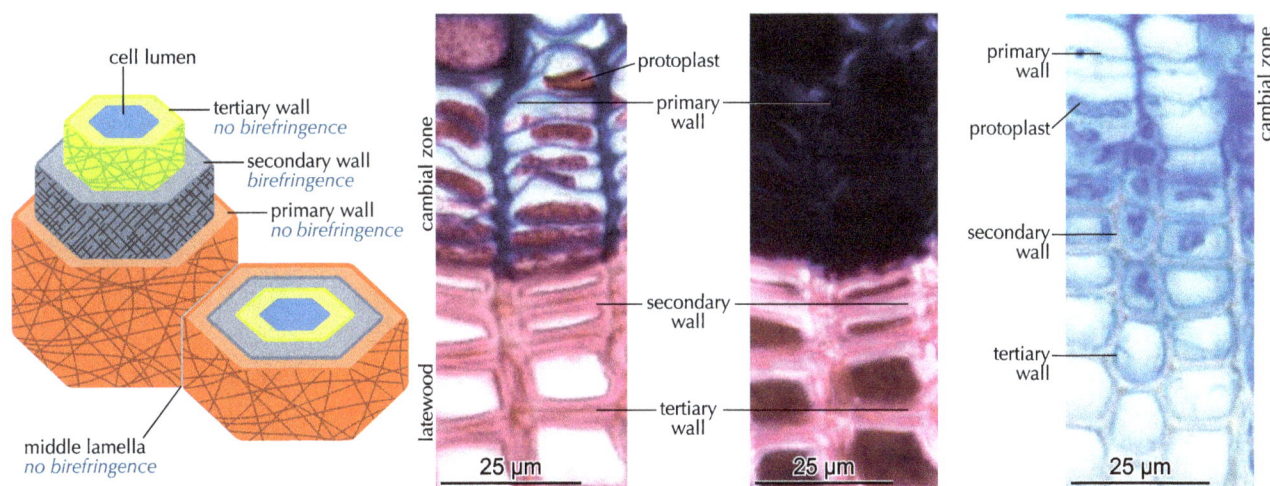

5.1 Cell walls of vascular plants consist of the middle lamella, the primary, secondary and tertiary wall.

5.2 Latewood (red) and cambium (blue) of *Larix decidua* shown in transmitting (left) and polarized light (right). Cell walls in the cambial zone are not lignified and consist only of primary walls. They therefore disappear in polarized light.

5.3 Latewood of *Picea abies* in formation. Thin tertiary walls (blue) appear immediately after the formation of secondary walls (light).

Primary walls

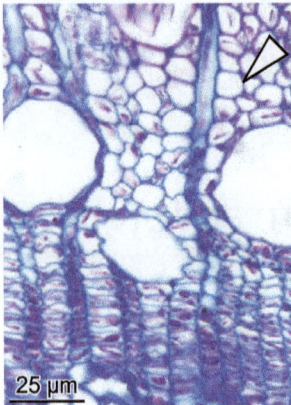

25 µm

50 µm

5.4 Primary walls without lignification in the cambium area of a stem of *Corylus avellana*. All cells, except vessels, contain protoplasts.

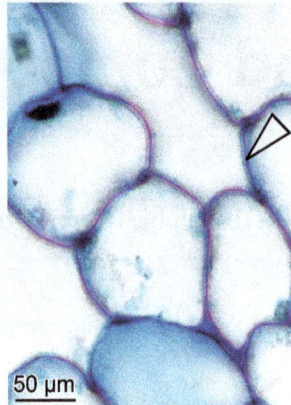

5.5 Fully developed primary walls and rudimentary walls in all cells of the water plant *Ranunculus trichophyllus*.

Secondary walls

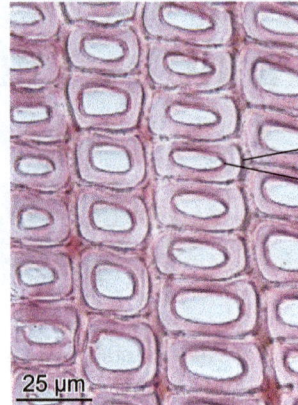

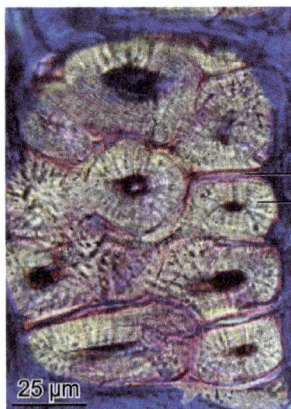

secondary wall primary wall

25 µm

secondary wall primary wall

25 µm

5.6 Fully developed cell walls with primary, secondary and tertiary walls of tracheids in the latewood of the conifer *Pinus elliottii*.

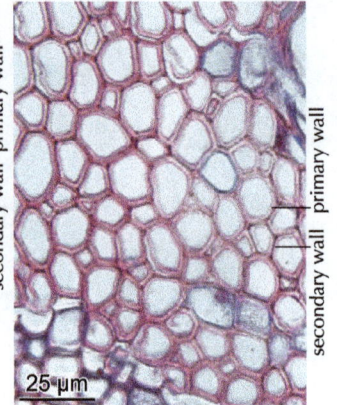

5.7 Fully developed cell walls with primary and secondary walls of fibers in the xylem of *Juglans regia*. Tertiary walls are only visible in the parenchyma cells.

Secondary walls

secondary wall primary wall

25 µm

secondary wall primary wall

25 µm

5.8 Multilayered cell walls of sclereids in the phloem of *Picea abies* in normal and polarized light.

Tertiary walls

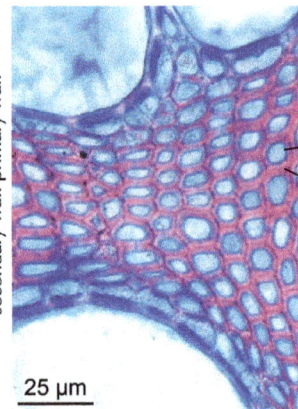

25 µm

secondary wall tertiary wall

25 µm

5.9 Fully developed cell walls with primary, secondary and tertiary walls of fibers in the xylem of *Vitis vinifera*. Tertiary walls are not lignified.

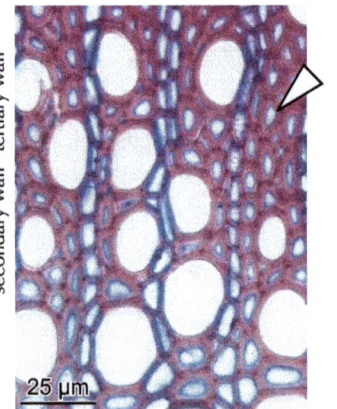

5.10 Fibers with thin-walled, unlignified tertiary walls (blue) in *Stachys sylvatica*.

Cell walls of ferns and fungi

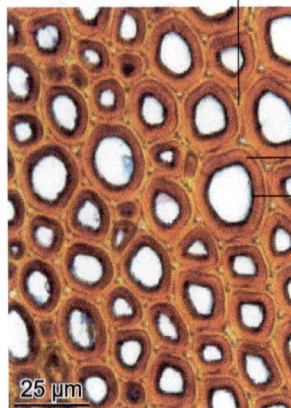

primary wall

tertiary wall secondary wall

25 µm

25 µm

25 µm

25 µm

5.11 Cell walls of fern fibers (*Lygodium* sp.) are constructed the same as in seed plants. Visible are middle lamella, primary, secondary and tertiary walls.

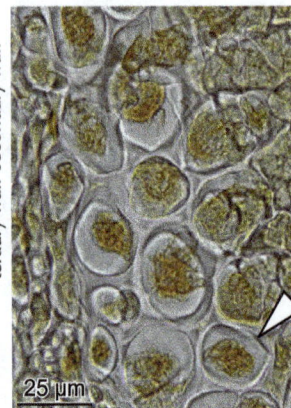

5.12 The unstructured cell walls of marine algae (*Fucus serratus*) consist of cellulose and glycoproteins.

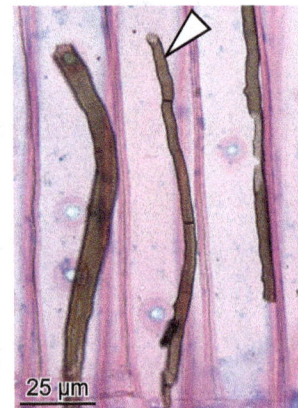

5.13 The unstructured cell walls of the blue-stain fungus *Grosmannia clavigera* consist of chitin.

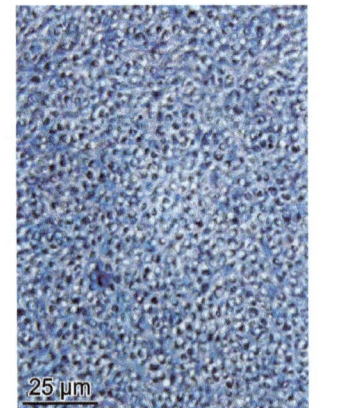

5.14 Unstructured cell walls of fungi in the lichen *Usnea barbata*.

5.2 Pits – Lateral contact between cells

Cells communicate with each other. Two neighboring cells are anatomically connected by pits. Through their channels, cells communicate physiologically. There are two principal structures; simple pits and bordered pits. However, many transition forms exist.

Simple pits

Simple pits occur in all vascular plants. They form a channel through the secondary wall of neighboring cells. The unlignified middle lamella in the pit blocks the channel but is perforated by plasmodesmata (not visible with light microscopes), through which sugars, amino acids, hormones and nutrients can flow. Simple pits of different forms occur in axial parenchyma cells, horizontal parenchymatic ray cells, rays and fibers. If simple pits connect rays and vessels they are called vessel-ray pits.

The form and size of the opening varies. Pits in parenchyma cells are round, and those in fibers are often slit-like. They can be small (<2 µm) or large (>4 µm) or horizontally or axially enlarged.

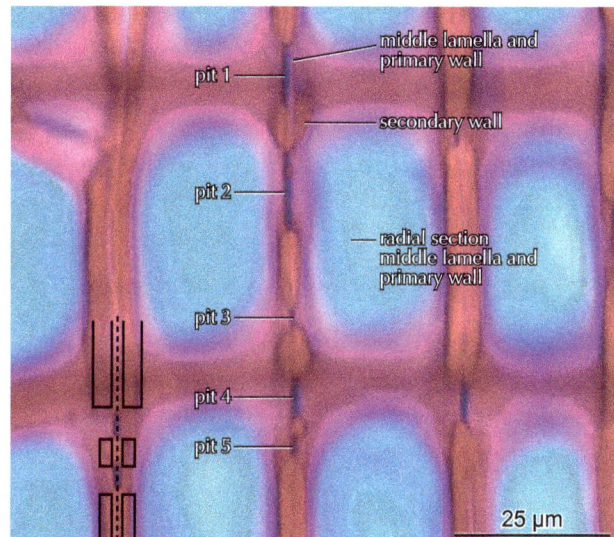

5.15 Structure of simple pits in rays of *Pinus sylvestris*. The middle lamellae and primary walls are not lignified (blue) in young cells.

Structure of simple pits

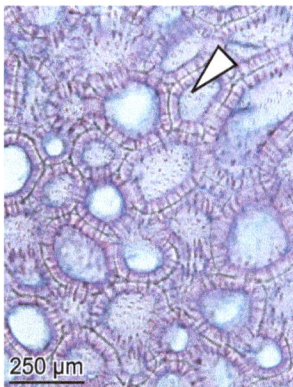

5.16 Cross section of simple pits in sclereids of hazelnut shells (*Corylus avellana*).

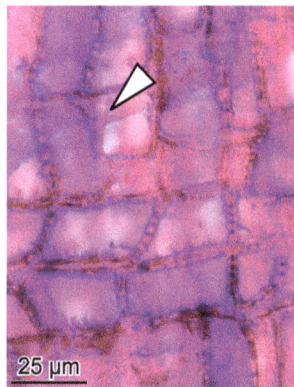

5.17 Cross section of simple pits on ray cells in *Ephedra viridis*.

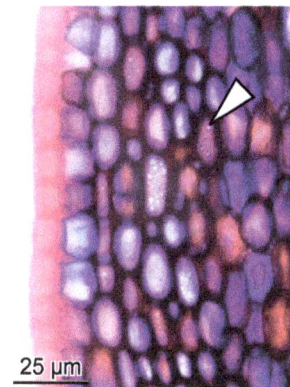

5.18 Top view of round simple pits in cortex-parenchyma cells of *Rosa canina*.

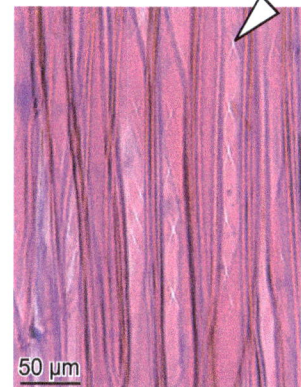

5.19 Radial view of slit-like simple pits in fibers of *Magnolia acuminata*.

Size and form of simple pits

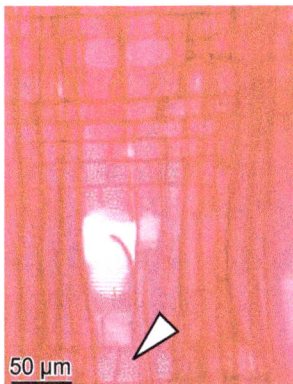

5.20 Small vessel-ray pits in the xylem of *Betula aetnensis*.

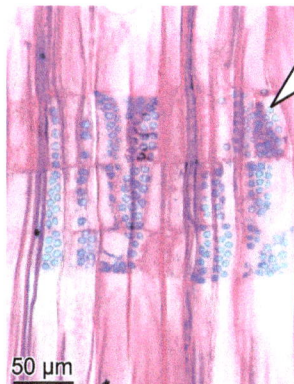

5.21 Large round vessel-ray pits in the xylem of *Salix arctica*.

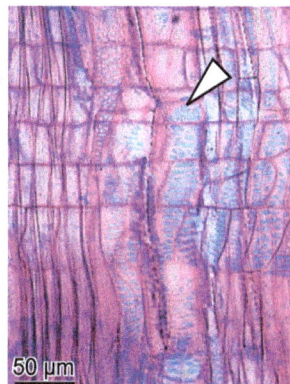

5.22 Scalariform vessel-ray pits in the xylem of *Euphorbia calyptrata*.

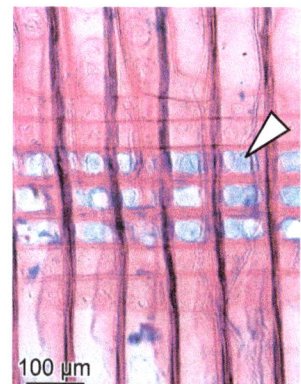

5.23 Large, fenestrate ray pits in the xylem of *Pinus sibirica*.

Bordered pits

Bordered pits occur in vessels, and axial and radial tracheids. The difference to simple pits is that secondary walls arch over the pit channel and that the middle lamellae and the primary walls are thickened (torus). Tori in pits of living cells are unlignified and laterally flexible. Bordered pits with unlignified tori act as valves. If pressure in one cell decreases, e.g. by injury, the torus gets pressed to the wall with the higher pressure (healthy cell). If bordered pits connect vessels they are called intervessel pits. A special case are intervessel pits with warts on their openings, the so-called **vestured pits**.

The form of the outer borders, the size, and the position in a cell wall varies. They can be small (<2 μm) or large (>4 μm), with round or angular outlines, or horizontally enlarged (**scalariform pits**) or net-like (**reticulate pits**). Pits are mostly alternating, rarely they are arranged opposite. Bordered pits in tracheids of conifers are in one, two or multiple rows. All these characteristics are taxonomically relevant.

Structure and location of bordered pits

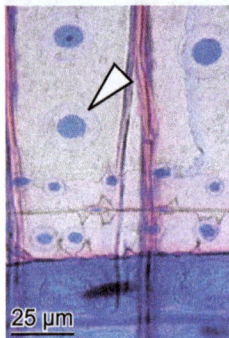

5.24 Ray tracheid in *Pinus sylvestris*, with unlignified tori.

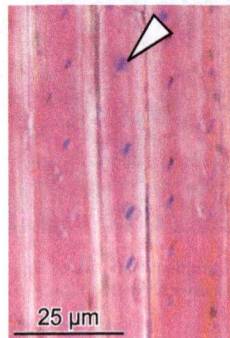

5.25 Tracheid in the dwarf shrub *Sarcococca hookeriana*.

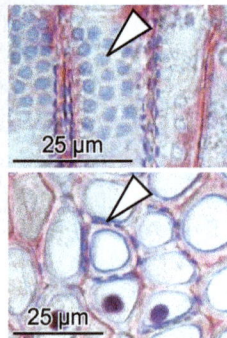

5.26 Vessels with pits in the mistletoe *Viscum album*, radial (top) and cross (bottom) section.

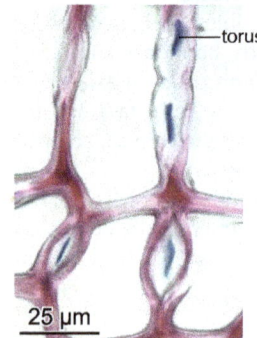

5.27 Tracheids in *Picea excelsa*, cross section, with unlignified tori.

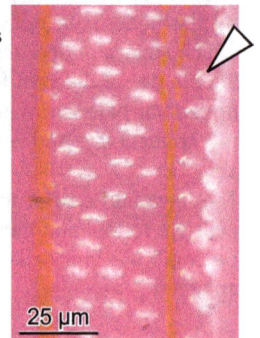

5.28 Vessel in *Sinapidendron frutescens*, with vestured pits.

Form and arrangement of bordered pits in deciduous plants

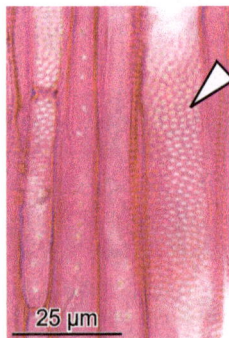

5.29 Small in *Betula nana*.

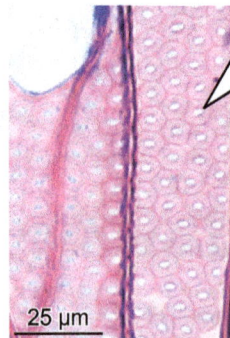

5.30 Large in *Salix alba*.

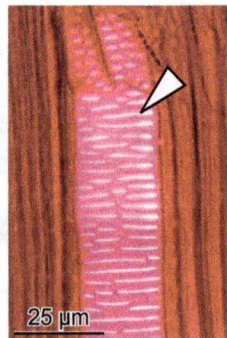

5.31 Scalariform in *Ribes alpinum*.

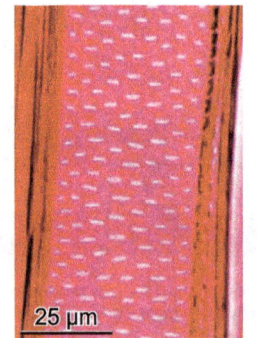

5.32 Alternate in *Reseda suffruticosa*.

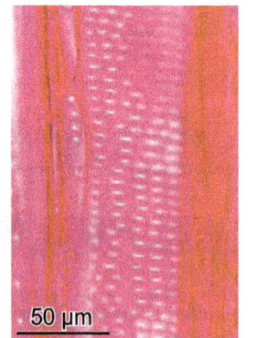

5.33 Opposite in *Platanus* sp.

Arrangement of bordered pits in tracheids of conifers

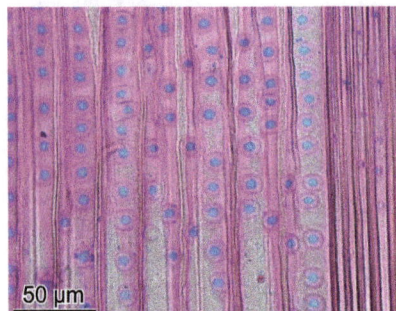

5.34 One row in *Pinus banksiana*.

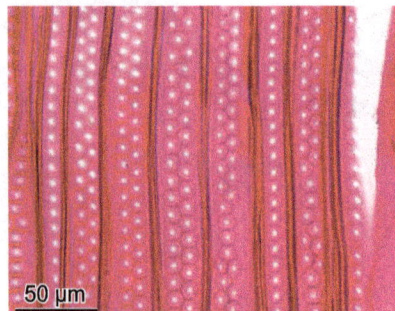

5.35 Two rows in *Araucaria angustifolia*.

5.36 Multiple rows in the carboniferous *Dadoxylon* sp.

5.3 Perforation plates – Axial contact between vessels

Vessels are perforated at their axial ends. The perforation plates are normally oblique positioned at their radial walls. Therefore they can be observed on radial sections.

Three perforation-plate types exist principally:

Simple perforation plates are characterized by their large, round to oval opening. The vast majority of species have simple perforations.

Foraminate perforation plates with several round openings occur only in the family of Ephedraceae, which stands taxonomically between conifers and dicotyledonous plants.

Scalariform perforation plates are characterized by their oblique position and horizontal bars. The number of bars varies

from one to more than 30. The thickness of the bars and the size of the plates are also variable. The occurrence of scalariform perforation plates is species-specific. They mainly occur in larger plants from shrubs to trees. They are rare in tropical environments and in small plants. Hardly any herb has scalariform perforation plates.

Aberrant scalariform perforation plates have been observed in a few species of small plants. They occur mainly in the family of Asteraceae.

It has been suggested that foraminate perforation plates stand at the beginning of an evolutionary process of trees, which ends with simple perforation plates (Bailey 1944).

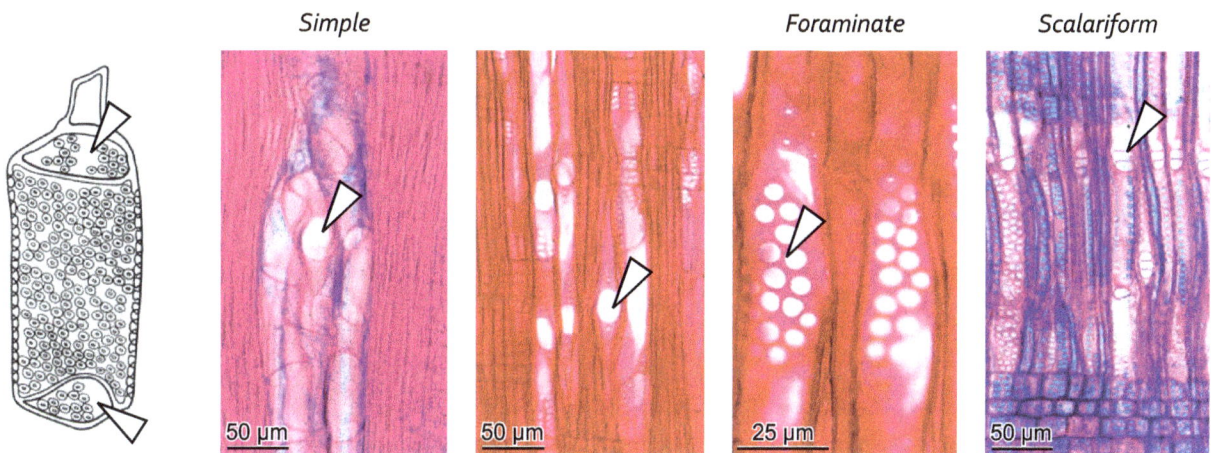

	Simple		Foraminate	Scalariform

5.37 *Fraxinus excelsior.* Reprinted from Greguss 1945.

5.38 *Carduus macrocephalus*

5.39 *Salix helvetica*

5.40 *Ephedra distachya* ssp. *helvetica*

5.41 <10 bars, *Paeonia fruticans.*

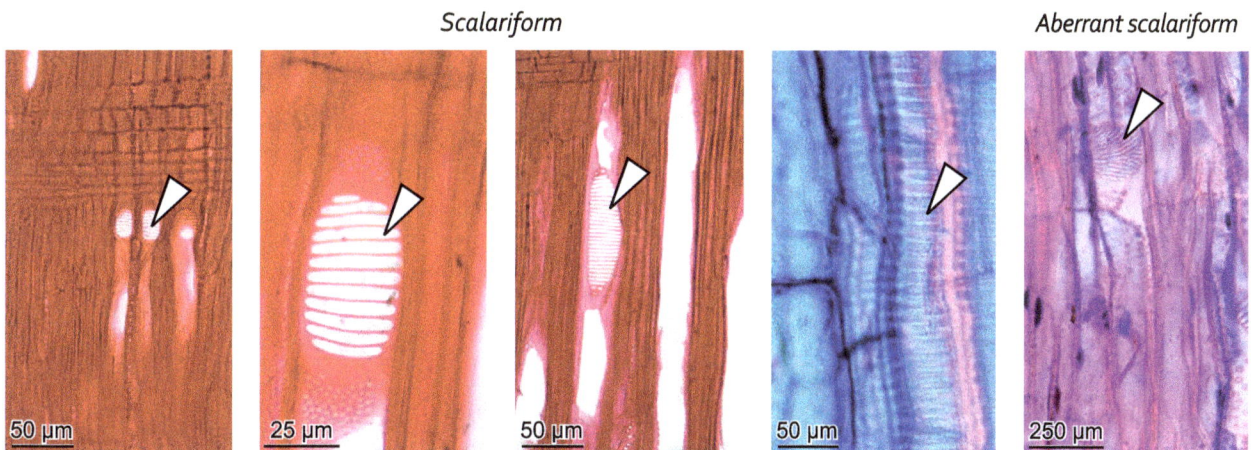

		Scalariform		Aberrant scalariform

5.42 <10 bars, *Buxus sempervirens.*

5.43 10–20 bars, *Betula humilis.*

5.44 >20 bars, *Viburnum opulus.*

5.45 >20 bars, *Menyanthes trifoliata.*

5.46 *Bidens tripartita*

5.4 Helical thickenings – Special wall thickenings

The secondary wall in fibers, tracheids and vessels of many species occurs as annular rings, spirals or nets. Those structures are absent in parenchyma cells of the xylem. The principal function of spiral thickenings might be stabilization but their occurrence in the plant body indicates also an ontogenetic and phylogenetic component. Helical thickenings have occurred in the xylem of all taxonomic units of vascular plants since their move to the land 350 million years ago.

Annular thickenings occur exclusively in the protoxylem and metaxylem of vascular bundles. The rings represent an early ontogenetic form of secondary walls.

Helical thickenings of various thickness occur in the mature xylem. Helical thickenings are of great taxonomic value because they appear only in specific taxa:
◦ In the tracheids of some genera of conifers, e.g. *Taxus* and *Pseudotsuga*.
◦ In the tracheids and vessels of dicotyledonous plants. In some species they are thick-walled and easily visible, in others thin-walled and difficult to recognize. Variations occur also in the angle of the spiral and the density within a vessel.

Helical thickenings are not to be confused with helical cavities in the secondary walls of compression wood in conifers, or with reticulate pits in some dicotyledonous plants.

Helical thickenings in the proto- and metaxylem

5.47 Cross and radial section of a species of *Rosa*. Helical thickenings occur immediately around the pith.

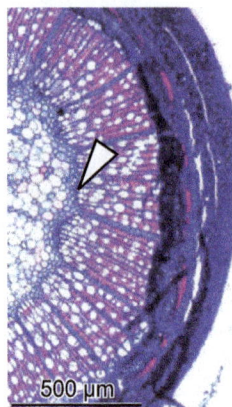

5.48 Cross section of a species of *Rosa*. Helical thickenings occur in vessels at the initial point of vascular bundles.

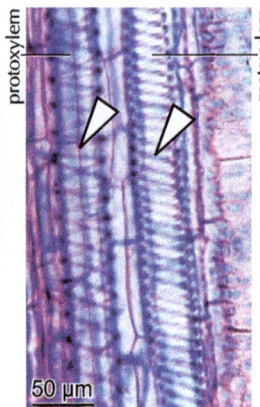

5.49 Helical thickenings in vessels of *Colocynthis vulgaris*. The gap between the spiral bands in the first vessel (protoxylem?) is wider than that in the second vessel (metaxylem).

Helical thickenings in tracheids and vessels

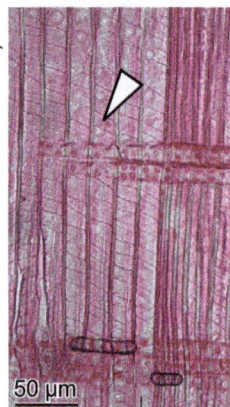

5.50 Helical thickenings in tracheids in the xylem of the conifer *Taxus baccata*.

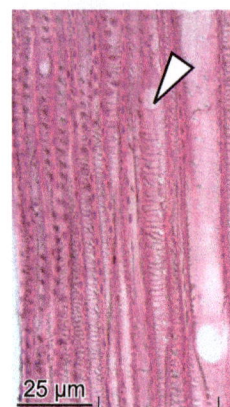

5.51 Helical thickenings in tracheids and vessels of the dicotyledonous dwarf shrub *Digitalis obscura*.

Helical thickenings in vessels of dicotyledonous plants

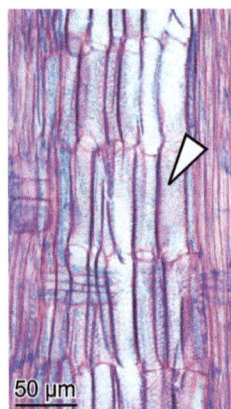

5.52 Densely positioned helical thickenings with a flat angle in *Adenocarpus viscosus*.

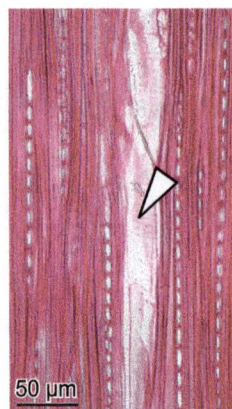

5.53 Thin helical thickenings in vessels in *Tilia platyphyllos*.

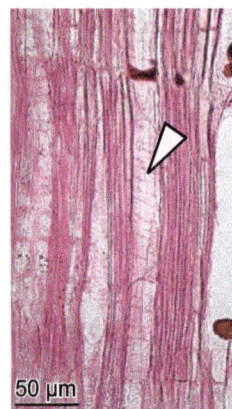

5.54 Thin helical thickenings in vessels in *Lonicera xylosteum*.

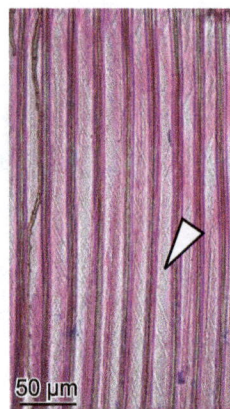

Do not confuse

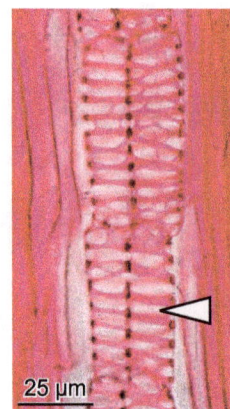

5.55 Helical cavities in compression-wood tracheids of the conifer *Pinus sylvestris*.

5.56 Reticulate pits with very wide openings in *Aeonium urbicum*.

5.5 Tyloses – Permanent interruption of water flow

Tyloses occur in vessels. They represent an essential feature in compartmentalized parts of stems at the boundary between heartwood and sapwood, living and dead parts of branches, or in injured parts of the xylem. Tyloses block the axial water transport and defend living tissues against pathogens.

Tyloses are irregularly formed cell walls with various wall thicknesses inside vessels. Their origins are specially formed cell wall layers inside secondary walls in neighboring parenchyma cells which expand balloon-like through pit openings into the vessels. Such cells are called contact cells. Tyloses are generally unstructured and unlignified. However, tyloses with simple pits also exist. Tyloses in older tissues are often lignified and impregnated with phenols. Nuclei often migrate into the tyloses. Tyloses occur in vessels of the primary and secondary xylem of trees and herbs.

Tyloses in a defense zone

5.57 Tyloses occur at the heartwood-sapwood boundary of oaks (*Quercus* sp.).

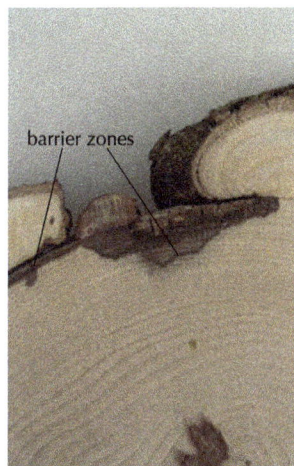

5.58 Tyloses inside an injured part of a stem of *Acer pseudoplatanus*.

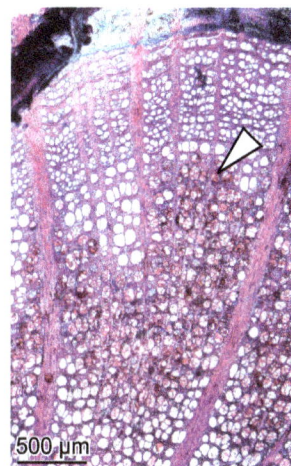

5.59 Tyloses in a defense zone (barrier zone) in a root of *Fagus sylvatica*.

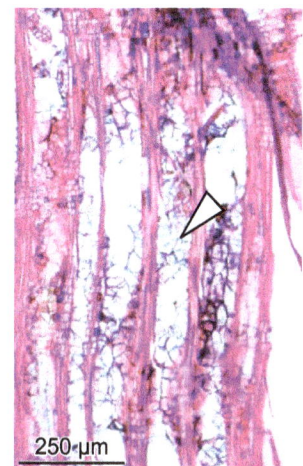

5.60 Tyloses in vessels of a defense zone block the water conductance in *Fagus sylvatica*.

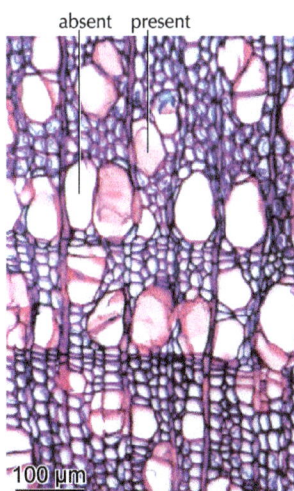

Anatomy of tyloses — Nuclei in tyloses

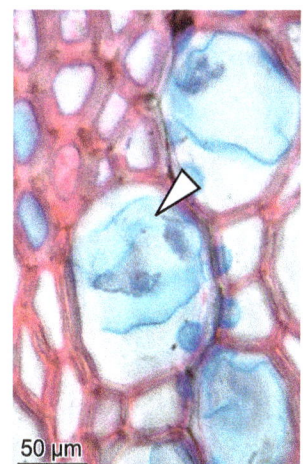

5.61 Selective occurrence of thin-walled, lignified tyloses in vessels of the arctic dwarf shrub *Salix arctica*.

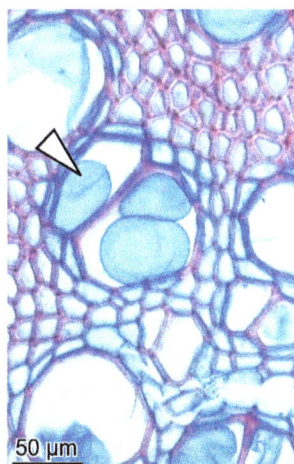

5.62 Tyloses in vessels of the nettle *Urtica dioica*.

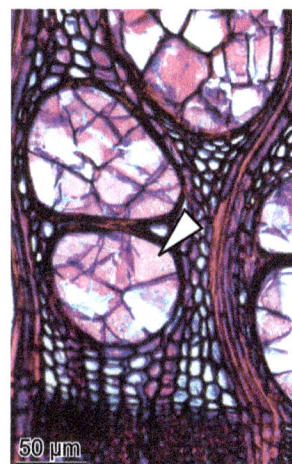

5.63 Tyloses with simple pits in *Maclura pomifera*.

5.64 Tyloses in an earlywood vessel with nuclei in the climber *Hedera helix*.

5.6 Cell contents – Everything inside the cell wall

Living adult cells of vascular plants principally contain a ***proto-plast*** and vacuoles. The protoplast is the living unit of a cell. It is separated by the tonoplast (semipermeable thin biomembrane) from the vacuole. The protoplast contains a liquid (cytoplasm) in which various organelles such as the nucleus with nucleolus, plastids (starch grains, chloroplasts, chromoplasts, leucoplasts), mitochondria, the endoplasmatic reticulum, Golgi vesicles and ribosomes occur. The vacuole contains ergastic substances such as crystals, oil bodies and various types of phenols. Most organelles contain genetic information.

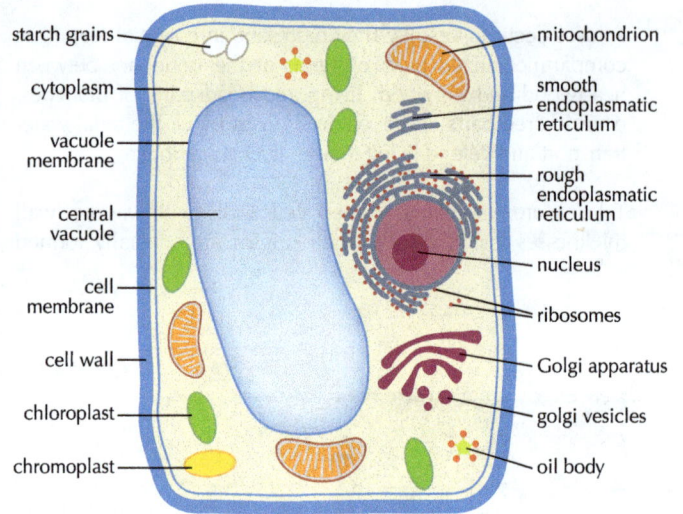

5.65 Schematic representation of an adult cell. Visible by light microscopy are nuclei, starch grains, chloroplasts, chromoplasts, crystals, oil bodies and phenols.

5.6.1 Nuclei in protoplasts – Metabolic centers of the plant cells

The cell nuclei are the basic structural and functional unit of organisms. They occur in all cell types and all organs of plants, however, their life span greatly varies.

Nuclei in meristems

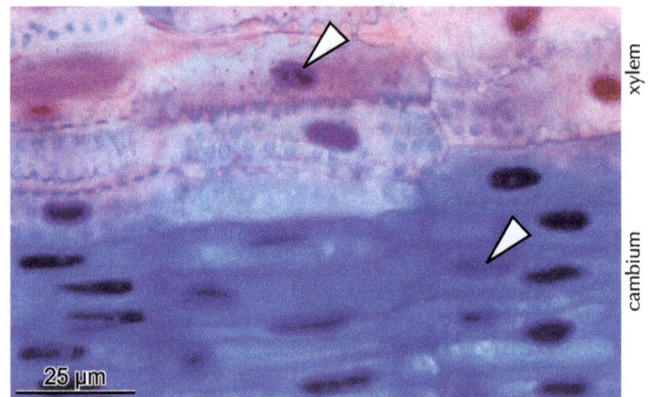

5.66 Nuclei in a secondary cambium and rays of the dwarf shrub *Viscum album*, radial section.

Nuclei in parenchyma cells

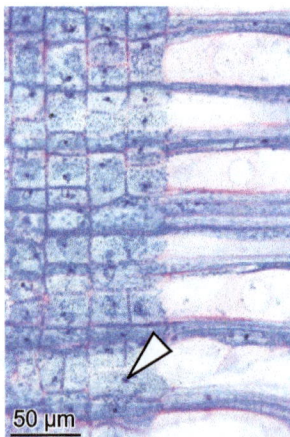

Nuclei in tracheids and fibers

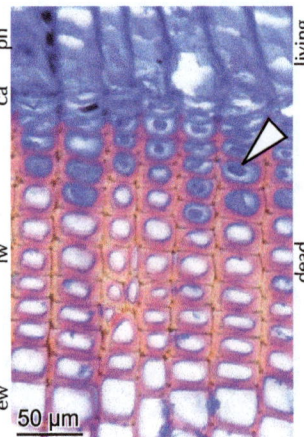

5.67 Small, round, 80-year-old nuclei in ray cells of the arctic dwarf shrub *Rhododendron lapponicum*.

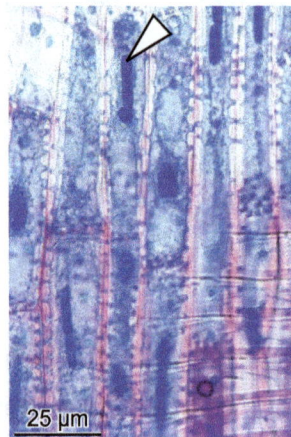

5.68 Radially elongated nuclei in young ray cells of the conifer *Abies alba*.

5.69 Protoplast and nuclei in latewood tracheids of the conifer *Pinus sylvestris*, cross section. Only the most recently formed tracheids are living, they contain protoplasts.

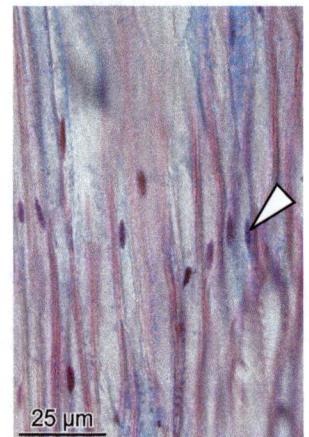

5.70 Nuclei with pointed axial ends in fibers of the herb *Anthemis palestina* in the desert, radial section.

5.6.2 Plastids – Green, yellow and white bodies

All living cells contain plastids. Chloroplasts are green-pigmented plastids. They contain the green photosynthetically active chlorophyll and the yellow carotenoids. Chromoplasts are yellow-pigmented plastids and contain carotenoid pigments. Leucoplasts are non-pigmented plastids.

Chloroplasts occur in all green parts of plants. They represent a substantial part in leaves (10–200 chloroplasts per cell) but also occur in parenchymatic parts of stems and fruits. Disk-like bodies, the thylacoids, are only visible in electron-microscopic magnifications.

Chromoplasts occur mainly in yellow- and orange-pigmented parts of plants such as in flowers, fruits and roots. Chromoplasts are also a product of aging chloroplasts in leaves.

Leucoplasts do not contain pigments. Some types can develop into chloroplasts or chromoplasts under the influence of light.

Chloroplasts make the world green

5.71 Green leaves contain chloroplasts. *Hamamelis virginiana*.

5.72 Green parts of stems contain chloroplasts. *Clematis alpina*.

5.73 Chloroplasts in the needle of *Pinus nigra*.

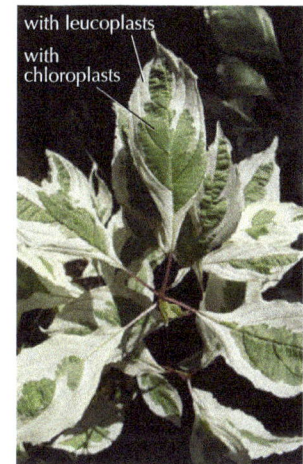

5.74 Chloroplasts and leucoplasts in a variegated leaf of *Cornus sericea*.

Chromoplasts make the world colorful

5.75 Aging leaves of *Hosta* sp. lose their chlorophyll, and yellow carotenoids now characterize the aspect.

5.76 Yellow flowers of *Hypericum* sp. contain chromoplasts.

5.77 Orange carrots (*Daucus carota*) contain chromoplasts.

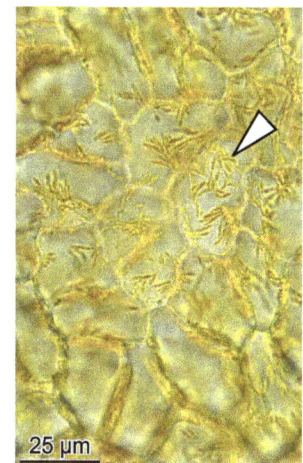

5.78 Chromoplasts in the petal of *Lonicera tatarica*.

5.6.3 Starch grains – Stored energy

Starch is a carbohydrate and plays a fundamental role for life on earth. Without starch from potatoes, manihot or cereals such as wheat, corn, rice and many others, human existence is unimaginable. Starch stores the energy of polysaccharides in a non-osmotic efficient way in parenchyma cells in all parts of all vascular plants. When sugars are re-synthesized to starch, the grains are called amyloplasts.

Starch grains are easy to recognize under the microscope in polarized light by the characteristic "Maltese cross", or by staining with potassium iodide. Form and size of the grains vary. In most cases they are more or less globular with a central point (hilum). Layers are the result of alternating deposits of different polysaccharides.

Starch as the base for (human) life

5.79 Flour from cereals.

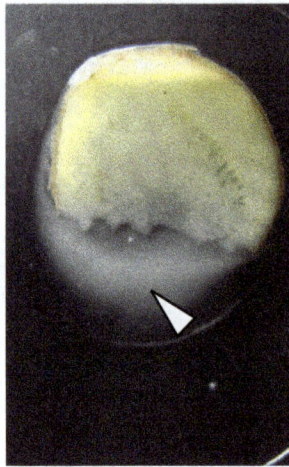

5.80 Section of a potato (*Solanum tuberosum*) with starch flowing out.

Starch grains in tubers of potatos

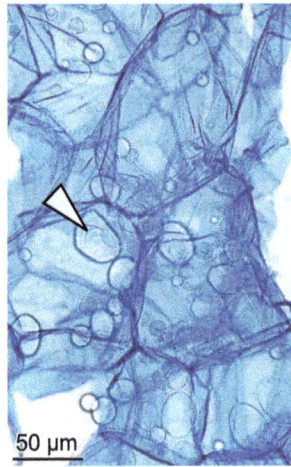

5.81 Large starch grains in *Solanum tuberosum*.

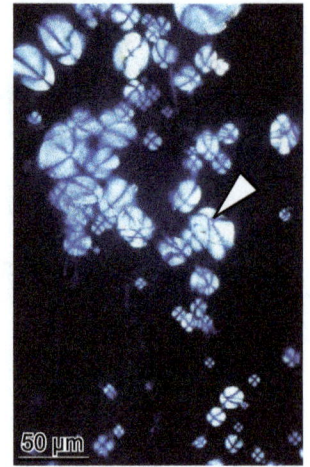

5.82 Starch in *Solanum tuberosum* with the characteristic "Maltese crosses" seen in polarized light.

Starch grains in shoots

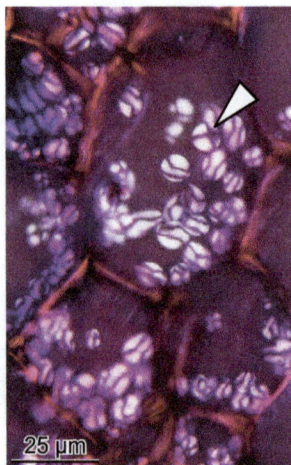

5.83 Starch in the pith of a rhizome of the herb *Tellima grandiflora*, polarized light.

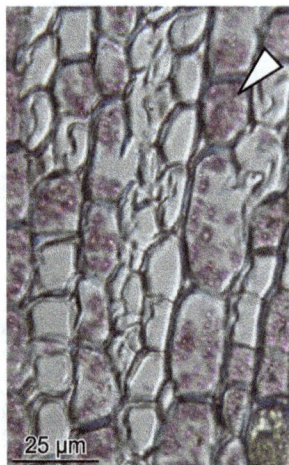

5.84 Small starch grains in rays and axial parenchyma cells in the xylem of a twig of *Fraxinus excelsior*, cross section.

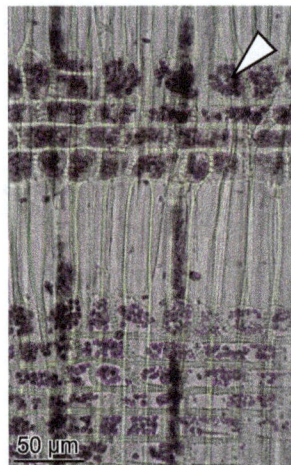

5.85 Starch stained with potassium iodide in ray cells of the wood of *Abies alba*, radial section.

Starch grains in fruits

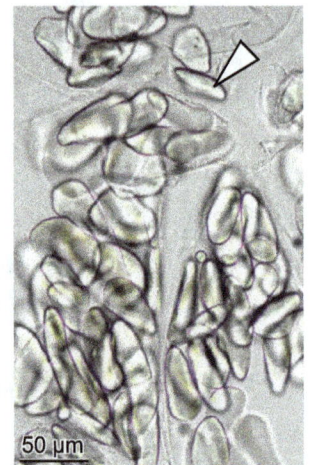

5.86 Starch in the soft fruit of a banana (*Musa × paradisiaca*).

5.6.4 Crystals in vacuoles – Regulators and metabolic waste

Crystals are excreted from protoplasts and deposited in vacuoles. Most crystals in cells are calcium oxalates. They have two major functions. Calcium is an essentially required element for plant growth, therefore calcium oxalate crystals often occur in meristematic tissues where calcium ions regulate the transport of organic molecules. Calcium oxalate is an end product of metabolic processes (metabolic waste), therefore calcium oxalates are often deposited in vacuoles of adult cells in all parts of plants, where they remain until cell death. Some plants form special cells, so-called crystal idioblasts, where crystals are deposited.

Calcium oxalate crystallizes in the form of prismatic crystals of various shapes: as druses, raphides and irregular small grains. The distribution and shape of calcium oxalates in tissues is a valuable taxonomic feature.

Crystals in meristematic tissues

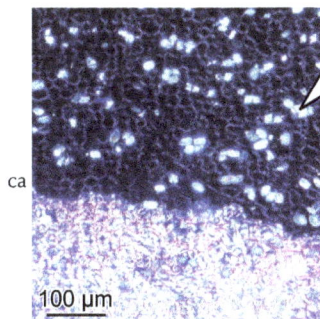

5.87 Crystals in the meristematic tissue of a bud of *Acer pseudoplatanus*, polarized light.

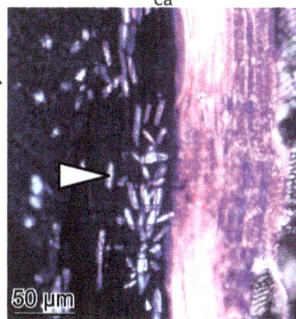

5.88 Elongated crystals (styloids) in the cambial zone of the herb *Hippocrepis comosa*, polarized light.

Crystal forms

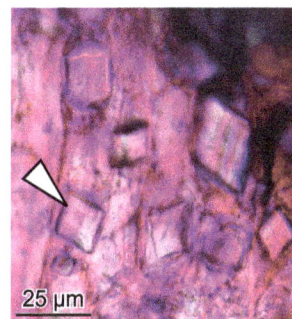

5.89 Prismatic crystals in a xylem ray of *Fagus sylvatica*, polarized light.

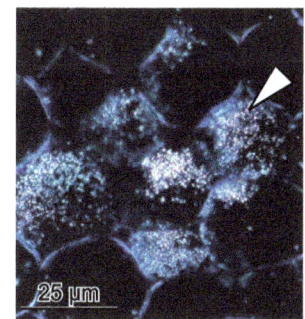

5.90 Crystal sand in cortex cells of the tropical *Piper nigrum*, polarized light.

Crystal forms

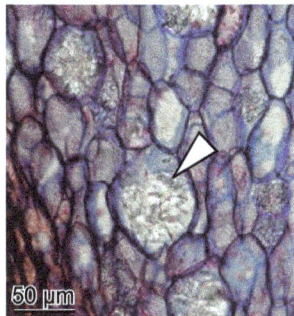

5.91 Crystal druses in expanded cells of the cortex of the alpine herb *Astrantia major*, polarized light.

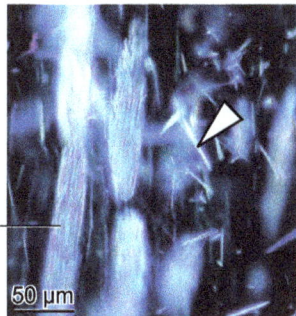

5.92 Elongated crystals (raphides) in idioblasts in the cortex of the liana-like dwarf shrub *Rubia tibetica*, polarized light.

5.93 Elongated and layered crystal in the rhizome of *Iris sibirica*, polarized light.

Crystals in chambers

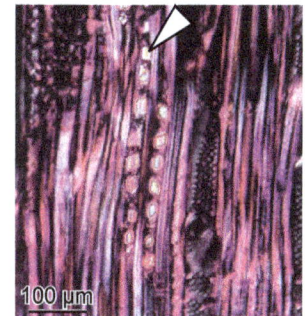

5.94 Prismatic crystals in a large idioblast (radial section) of the small shrub *Neochamaelea pulverulenta*, polarized light.

Crystal arrangement in bark

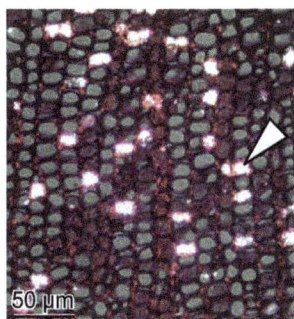

5.95 Irregularly dispersed crystal druses in the phloem of the shrub *Buxus sempervirens*, polarized light.

5.96 Tangentially arranged crystals in the phloem of the alpine shrub *Ribes alpinum*, polarized light.

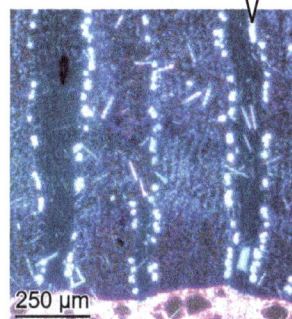

5.97 Radially arranged crystals along large rays in the phloem of the climber *Parthenocissus inserta*, polarized light.

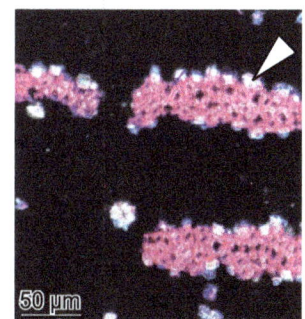

5.98 Prismatic crystals, arranged around fiber groups in the phloem of *Salix purpurea*, polarized light.

5.6.5 Stained substances within the stem – Defense

Stained substances occur in the phloem, xylem and fruits of all vascular plants. Most obvious are discolorations in stems with heartwood, injuries or subfossil wood.

The term indicates the chemical and anatomical heterogeneity of the amorphic substances. Different terms are used, e.g. organic compounds, organic extractives, tannins (polyphenols), gums, resins and oils. Characteristic for all of them are the colors, which range without staining from yellow to red and dark brown, and with Astrablue/Safranin staining from blue to red and black.

A selective variety of different stained substances are presented, which occur in the xylem in vessels, axial and radial parenchyma cells, fibers and cells in the phloem. The producing cell structures are discussed in Chapter 4.12.

Macroscopic aspect of dark-stained substances

5.99 Dark-stained heartwood in various Australian wood species.

5.100 Dark-stained zones in biological defense zones (compartmentalization) of *Fagus sylvatica*.

5.101 Dark-stained heartwood of a subfossil, waterlogged *Quercus* sp.

Stained substances in vessels

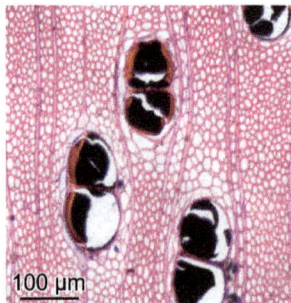

5.102 Black-stained substances in vessels of *Acacia longifolia* (Astrablue/Safranin-stained).

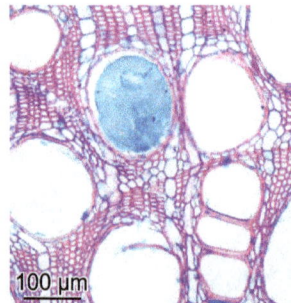

5.103 Dark-blue-stained substances in vessels and parenchyma cells in the heartwood of *Juglans regia* (Astrablue/Safranin-stained).

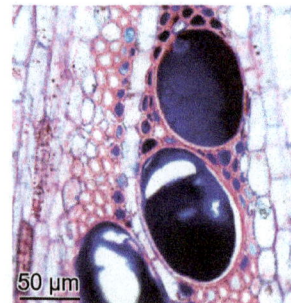

5.104 Dark-blue-stained substances in vessels of a *Paliurus spina-christi* (Astrablue/Safranin-stained).

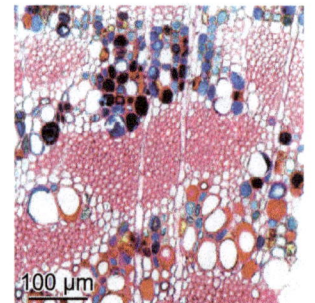

5.105 Black-, red-, blue- and yellow-stained substances in vessels and vascular tracheids in *Spartocytisus supranubius* (Astrablue/Safranin-stained).

Stained substances in parenchyma cells

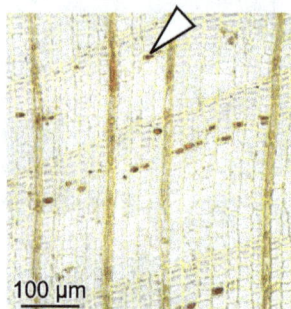

5.106 Brownish-stained substances in parenchyma cells in the heartwood of *Juniperus communis* (unstained).

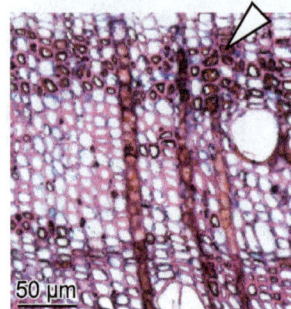

5.107 Dark-red-stained substances in axial and radial parenchyma cells in a chestnut-blight-affected *Castanea sativa* (Astrablue/Safranin-stained).

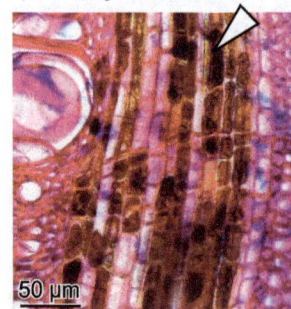

5.108 Brown-stained substances in a large ray of a dead part of the vine *Vitis vinifera* (Astrablue/Safranin-stained).

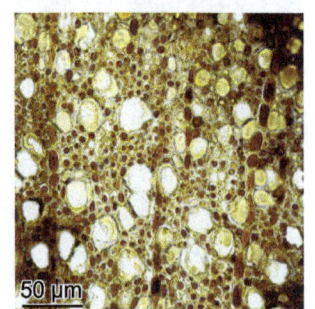

5.109 Brown-stained substances in parenchyma cells and fibers in the heartwood of the dwarf shrub *Arctostaphylos uva-ursi* (unstained).

Stained substances in fibers and vessels

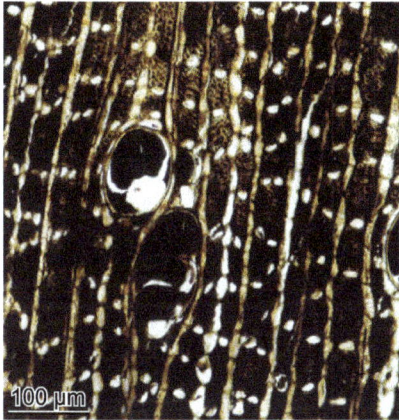

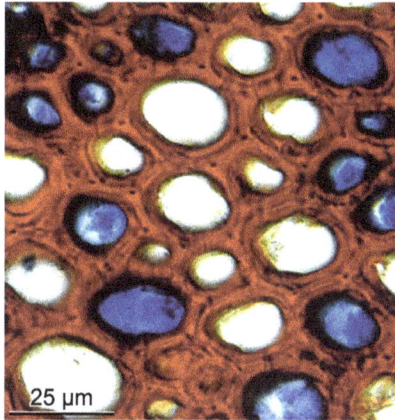

5.110 Brown-stained substances in vessels and fibers in the heartwood of ebony *Diospyros* sp. (unstained). Parenchyma cells do not contain dark-stained substances.

5.111 Red-stained substances in parenchyma cell walls and blue-stained substances in cell lumina of a rhizome of the fern *Osmunda regalis* (Astrablue/Safranin-stained).

Stained substances in decaying wood

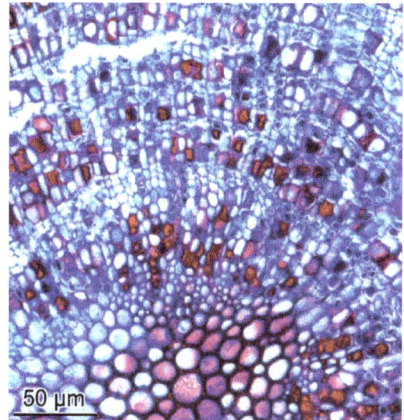

5.112 Red-stained substances in vessels of a delignified, decaying stem of the dwarf shrub *Arctostaphylos alpina* (Astrablue/Safranin-stained).

Stained substances in subfossil wood

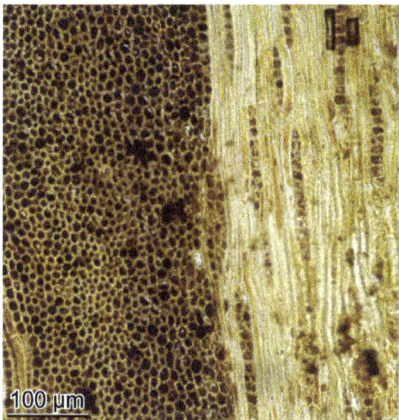

5.113 Brown-stained substances in ray cells of the black heartwood of a subfossil, waterlogged *Quercus* sp. (unstained). Fibers are slightly impregnated by yellowish substances.

Stained substances in cells of the phloem

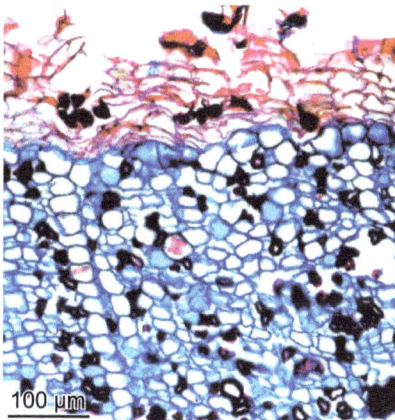

5.114 Black-stained substances in idioblasts in the phloem of the desert shrub *Krameria grayi* (Astrablue/Safranin-stained).

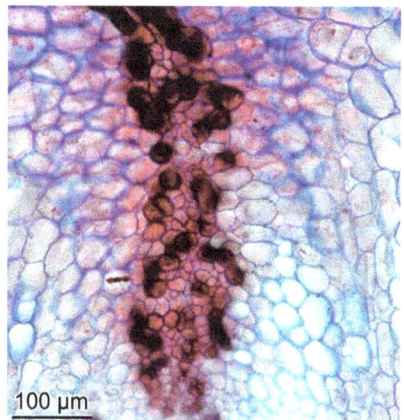

5.115 Dark-red-stained substances in parenchyma cells of the phloem of the herb *Scorzonera graminifolia* (Astrablue/Safranin-stained).

Stained substances in cells of the phloem

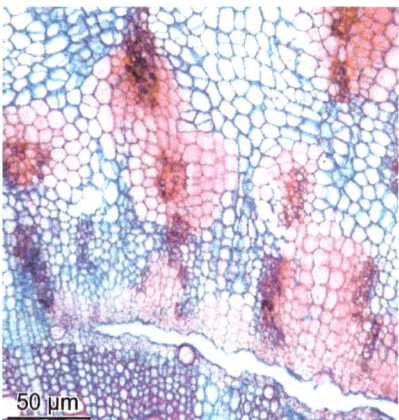

5.116 Red-stained substances around groups of sieve tubes in *Cichorium intybus* (Astrablue/Safranin-stained).

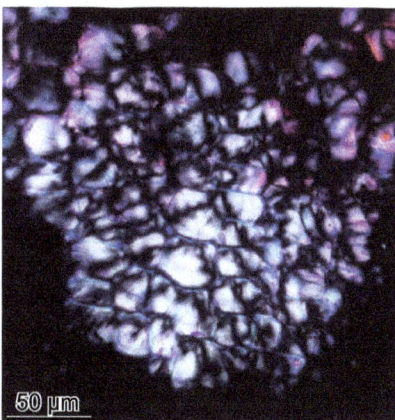

5.117 Unknown substances in the bark of the herb *Symphytum creticum*, which reflect in polarized light.

Stained substances in / around ducts

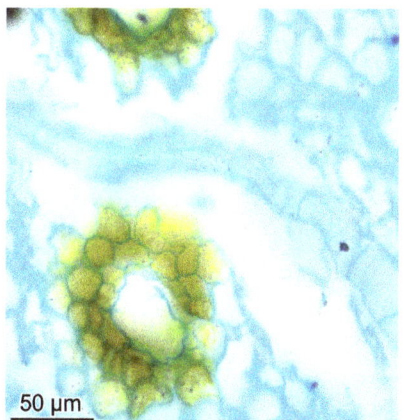

5.118 Yellow-stained substances, probably oil, around ducts in the phloem of the herb *Heracleum pinnatum* (Astrablue/Safranin-stained).

6

Meristems: A Comprehensive Study

Definitions

Meristems initiate longitudinal growth on tips of shoots and roots of plant bodies. Meristematic tissues consist of living cells, which produce new cells.

Primary meristems on shoot tips (apical meristems) are embryonic tissues, which originate from seeds. They produce the epidermis, the cortex, the leaves and the pith.

Secondary meristems originate from primary meristems and produce the xylem and phloem. The whole conducting system is called stele. The arrangement of vascular bundles within the central strand defines the type of stele. Protostele: one vascular bundle (mosses); plectostele, polystele: several vascular bundles in the center (lycopods); eustele: concentrically arranged isolated or laterally connected vascular bundles in a ring (most dicotyledons). The term stele is used here only as an anatomical characteristic, not in relation to evolutionary stages.

Tertiary meristems originate from parenchymatic tissues, which are located within xylem, phloem and cortex.

Structural variation in meristematic products will be discussed in the following chapters.

Production rates

Production rates and proportions between shoot length, diameters and number of cells within the pith, xylem, phloem, cortex and phellem vary greatly. Most obvious are differences in the xylem (product of the cambium) and the product of the primary and secondary meristems (cortex, phloem, phellem).

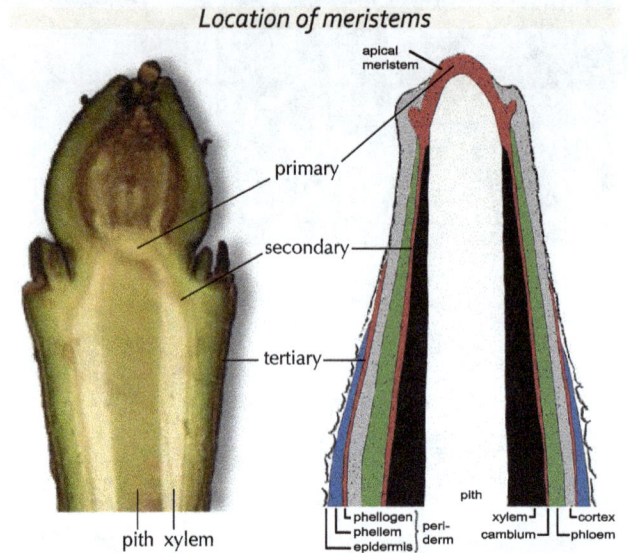

Location of meristems

6.1 Left: Primary, secondary and tertiary meristems in a twig of *Fraxinus excelsior*. Right: Schematic representation of primary, secondary and tertiary meristems. Reprinted from Schweingruber *et al.* 2008.

Long and short shoots

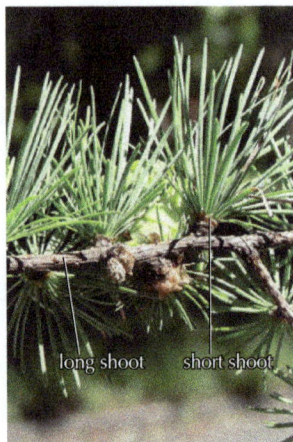

6.2 Short shoots on a long shoot of the conifer *Larix decidua*.

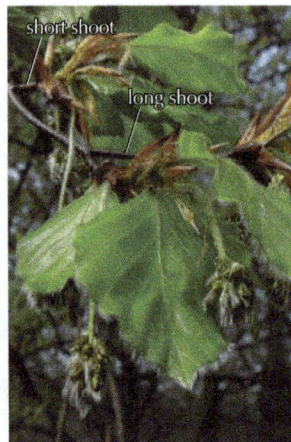

6.3 Short and long shoots in *Fagus sylvatica*.

Proportions of xylem, phloem, cortex and phellem

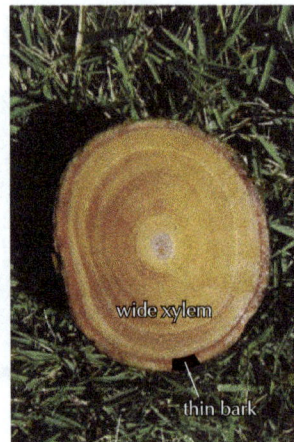

6.4 Xylem to bark proportion of 10:1 in the mangrove *Rhizophora mangle*.

6.5 Xylem to bark proportion of 2:1 in a young stem of *Quercus suber* (cork oak).

Small and large pith

cortex
phloem
xylem

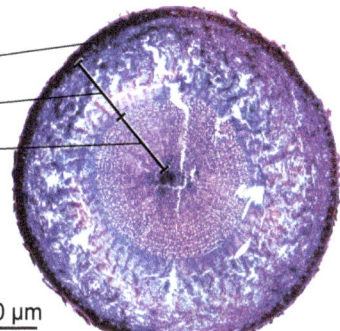

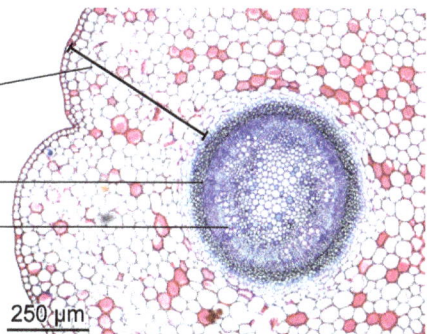

pith

eustele

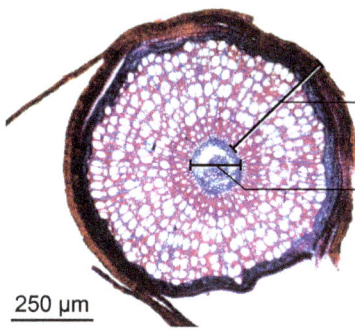

6.6 Herb with a small pith. Pith to xylem and bark proportion of 1:6 in *Schistophyllidium bifurcum*.

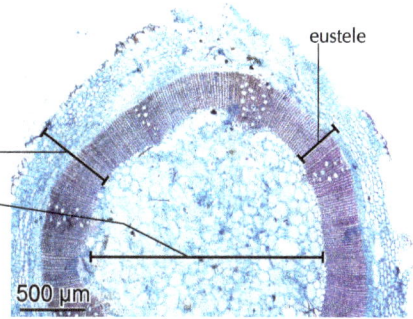

6.7 Herb with a large pith. Pith to xylem and bark proportion of 1:0.3 in *Impatiens macroptera*.

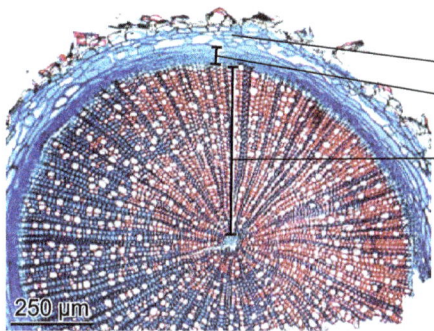

Small and large phloem

cortex
phloem
xylem

6.8 Herb with a small phloem. Phloem to xylem proportion of 1:10 in *Linum bienne*.

6.9 Herb with a large phloem. Phloem to xylem proportion of 1:1 in *Draba cachartena*.

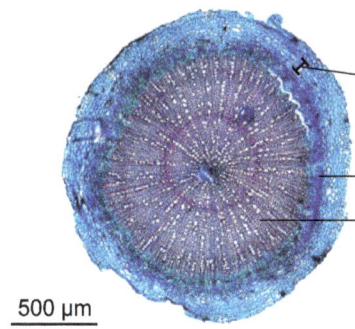

Small and large cortex

cortex

phloem
xylem

6.10 Herb with a small cortex. Cortex to phloem proportion of 1:1 in *Bupleurum bladensis*.

6.11 Herb with a large cortex. Cortex to phloem proportion of 8:1 in *Honkenia peploides*.

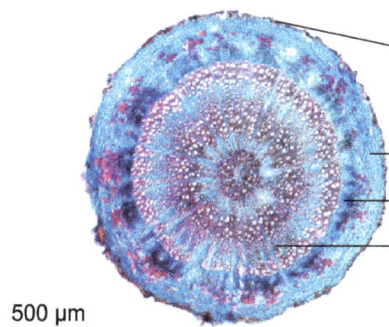

Small and large phellem

phellem

cortex
phloem

xylem

6.12 Herb with a small phellem. Phellem to phloem and cortex proportion of 1:5 in *Thesium arvense*.

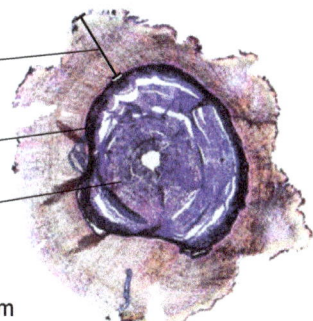

6.13 Herb with a large phellem. Phellem to phloem and cortex proportion of 4:1 in *Saxifraga oppositifolia*.

6.1 Primary meristems in apical zones – Initials of longitudinal and radial growth

6.1.1 Macroscopic aspect of primary meristems in apical shoots and roots – Grow higher, grow deeper

The origin of primary meristems is in the seed and all meristematic derivates in apical zones are also primary meristems. As soon as the seed germinates, the germs divide into a root and a shoot. Apical meristems occur on roots and shoots, on the primary as well as on all adventitious shoots and roots. In dormant as well as in active periods apical meristems in roots are not protected by buds but often wrapped in a mantel of hyphae (mycorrhiza).

Apical meristems occur in mosses and in all vascular annual and perennial plants.

Apical meristems in shoots first form stems with leaves. These meristems often change their mode and also form flowers and fruits (see also Chapter 10, Fig. 10.1).

Apical meristems on shoots and roots

6.14 *Castanea sativa* seedling with a primary shoot and primary root.

6.15 *Carpinus betulus* seedling with primary apical meristems.

6.16 *Lonicera xylosteum* sapling with apical meristems on shoots and roots.

6.17 Adult grass *Festuca rupestris* with apical meristems in the root zone.

6.18 Moss *Polytrichum commune* with apical meristems on shoots. Rhizoides are covered by mycorrhiza.

Apical meristems on adventitious shoots   Apical meristems in buds

6.19 Injured stem of *Taxus baccata* with adventitious shoots, which contain apical meristems.

6.20 Adventitious shoot on a *Fagus sylvatica* stem.

6.21 Terminal shoot of *Acer pseudoplatanus* with a terminal bud and two adventitious buds.

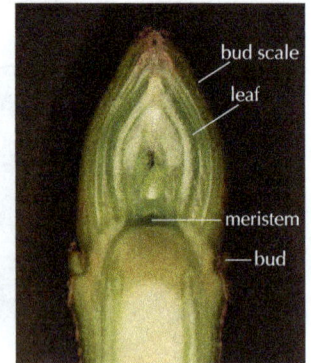

6.22 Terminal shoot of *Acer pseudoplatanus* with an apical meristem wrapped in undeveloped leaves and bud scales.

6.1.2 Apical shoot dynamics – Long and short shoots – Grow fast, grow slow

The aspect of plants is determined by the position of buds, the formation of long and short shoots, and the growth and death dynamic of apical meristems in shoots. The past activity of apical meristems on shoots can be determined by **bud scale scars**. They are overgrown wounds of deleted bud scales after leaf flushing. The terms short and long shoot are vaguely defined.

Internodes between bud scale scars indicate an extreme **variability of longitudinal growth**. The long distances between bud scale scars, e.g. in long shoots, make it easy to **determine the age** of twigs. Short distances, or indiscernible bud scale scars at the outside of shoots (e.g. short shoots) hinder macroscopic age determination of shoots. However, microscopic age determination is possible with remaining **pith bridges** in the position of bud scale scars. **Ring counting** in short shoots is normally not reliable.

Macroscopic aspect of bud scale scars and short shoots

6.23 Bud scales in *Picea abies*.

6.24 Short shoots in *Pinus mugo*.

6.25 Bud scale scars in *Pinus mugo*.

6.26 Bud scale scar in *Quercus robur*.

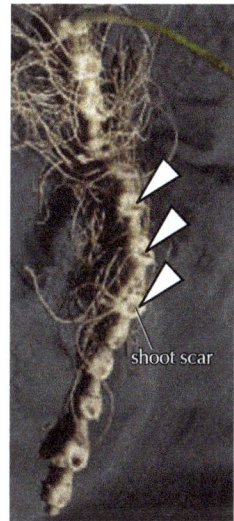

6.27 Rhizome with short shoots and shoot scars in *Polygonum officinale*.

External and internal delimitations of short shoots

6.28 Short shoots in *Larix decidua*.

6.29 Longitudinal section of a short shoot with annual bridges in the pith of *Larix decidua*.

6.30 Cross section of *Larix decidua* short shoot. Annual rings are absent.

6.31 Shoots of the arctic dwarf shrub *Cassiope tetragona*.

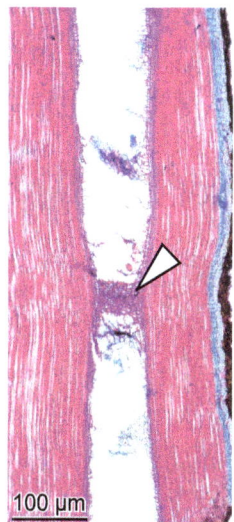

6.32 Annual latewood bridge in the latewood of a shoot of *Cassiope tetragona*.

6.1.3 Shoot death and metamorphosis – The end of longitudinal growth: Twigs must die

Normal shoot formation happens in a period of two to four weeks (e.g. in *Quercus*) or lasts for the whole vegetation period (e.g. in *Populus*). However, the life span of shoots varies between one and more than 1,000 years.

Very soon after the formation of twigs, the *self pruning* process starts. The majority of twigs die and drop after a few years. Only a few dominating shoots remain on the plant for the whole lifetime on the individual. The survivors develop into branches and stems and form the crowns of trees, shrubs and herbs. The crown form is a result of selective death of twigs and branches. The principal stem is the winner of an extensive programmed dying process.

Twig shedding, also called twig abscission or cladaptosis, occurs principally in three forms:
a) Twigs dry out, get affected by fungi and drop even due to slight mechanical disturbances. This is the most common type of twig shedding.

b) The shedding zone is not anatomically visible, but the strength is reduced near the base of the twig in *Salix*. The twig drops before it dries out.
c) The shedding zones are anatomically predetermined in *Quercus* and *Populus*. This shedding mechanism is expressed by dramatic anatomical change between the remaining and discarded part and the breaking zone.

The most common shoot transformation (metamorphosis) is the change from the vegetative to the generative form; from shoot to flower formation.

Very common is the transformation of shoots into thorns. In this case the apical meristem loses its replication capacity and changes its mode to an extensive growth of secondary walls.

Twigs die and drop

6.33 Twigs lose their vitality, die, get affected by fungi and break off. *Corylus avellana*.

6.34 Twigs break near the base at a predetermined mechanical weak zone in *Salix alba*.

Breaking zones compartmentalize

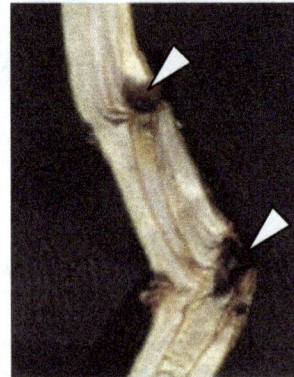

6.35 Compartmentalized wounds of broken twigs in *Fraxinus excelsior*.

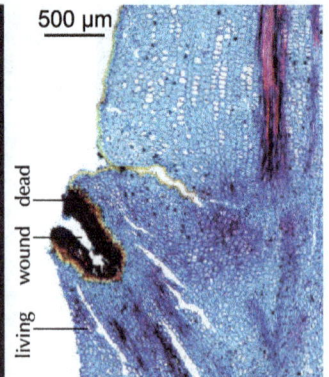

500 µm

dead
wound
living

6.36 Compartmentalized wound in *Viscum album*.

Macroscopic aspect of breaking zones

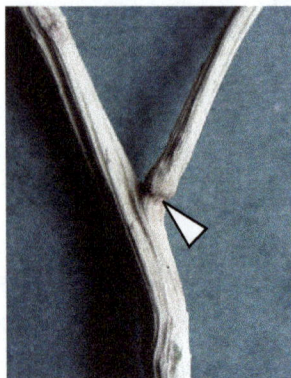

6.37 Predetermined breaking zone on a debarked twig of *Quercus robur*.

6.38 Scar of a broken twig in *Quercus robur*.

6.39 Shed twigs of *Quercus robur*.

6.40 Shed twig of *Gnetum gnemon*.

Microscopic aspect of breaking zones in Quercus robur

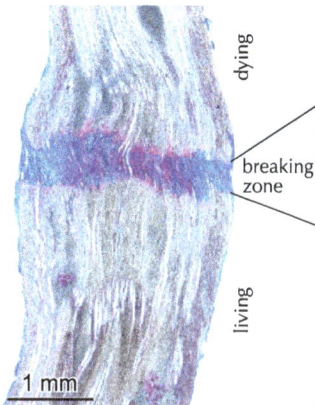

6.41 Longitudinal section through a breaking zone. It is characterized by a poor lignification (blue zone).

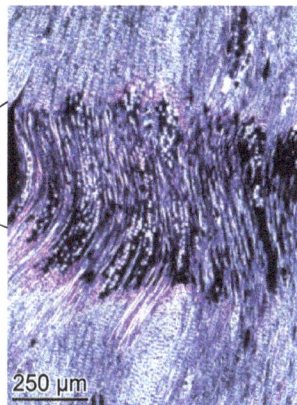

6.42 Breaking zone with numerous crystals, polarized light.

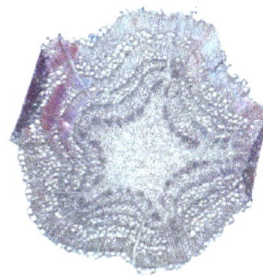

6.43 Anatomical structure below the breaking zone. This structure is typical for oak wood.

6.44 Anatomical structure of the breaking zone. This structure is very different from the basal twig and characterized by the absence of fibrous latewood and small vessels.

Transformations of apical meristem to flowers, fruits and thorns

Flowers on long shoots

Flowers on long shoots, needles on short shoots

6.45 *Carduus macrocephalus*

6.46 *Sempervivum wulfenii*

6.47 *Gentiana utricularia*

6.48 *Pinus mugo*

Flowers and fruits on short shoots

Thorns on long shoots

6.49 *Alnus viridis*

6.50 *Buxus sempervirens*

6.51 *Crataegus monogyna*

6.52 *Crataegus monogyna*

6.1.4 Microscopic aspect of apical meristems of shoots and roots – Towards heaven and earth

The principal differences between apical root and shoot growth are shown below. Differences and similarities between apical meristems in roots and shoots are obvious in microscopic sections. This is here demonstrated on some dicotyledonous and monocotyledonous species.

Omnipotent cells in the center of the shoot are common for the apex of roots and shoots. Bipolarity is also common, which means meristematic cells produce cells towards two axial directions: geocentric and heliocentric. Central cells of the roots produce the root cup, which determines the trajectory and protect the inner central cells. Root cup cells get sloughed off by abrasive soil particles. Central shoot cells primarily produce leafs, but also all parts of flowers.

The major difference between the two types is in the zone behind the tip. Soon after the first cell differentiation, the dicotyledonous plant produces a cambial zone which separates the cortex with the initial leaves (leaf primordial) from the central cylinder.

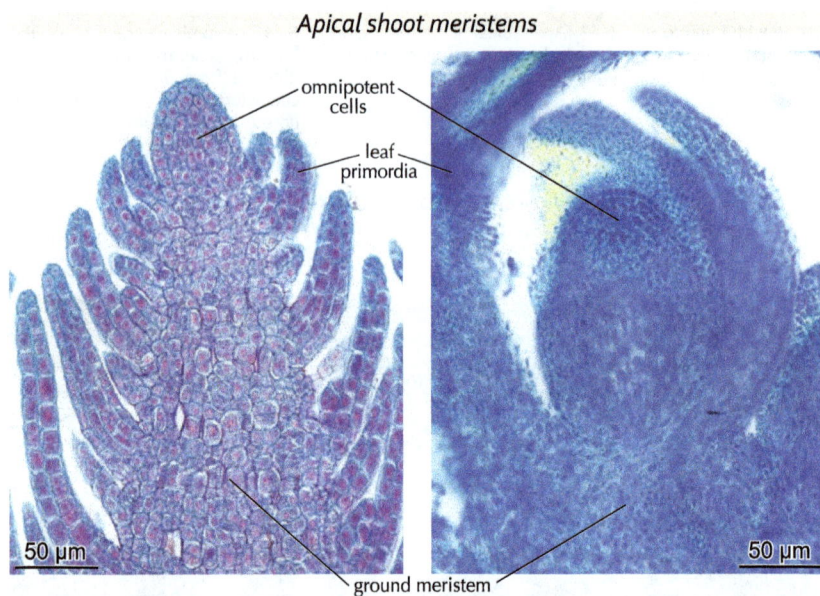

Apical shoot meristems

Development behind the initial zone

6.53 Apical meristem of *Elodea canadensis*, a monocotyledonous water plant. Slide: J. Lieder.

6.54 Apical meristem of *Euphorbia cyparissias*, a dicotyledonous terrestrial plant.

6.55 No cambium in the monocotyledonous water plant *Elodea canadensis*. Slide: J. Lieder.

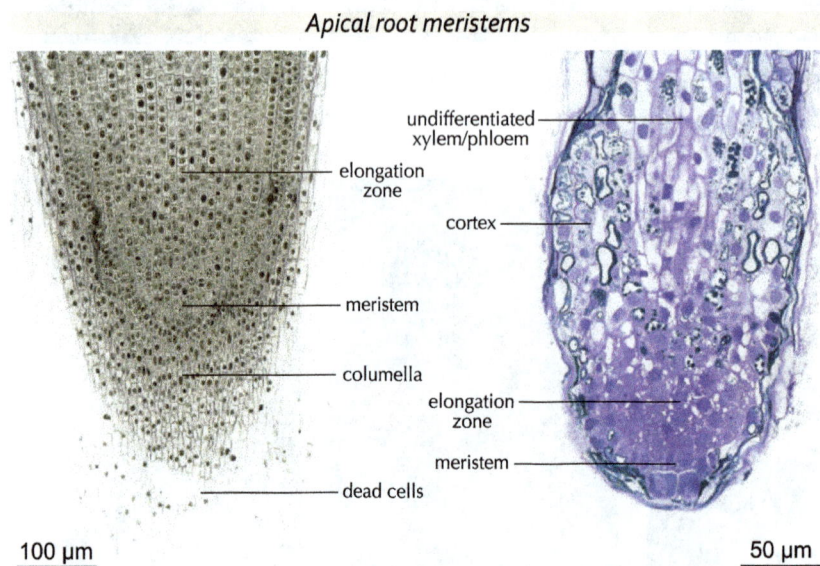

Apical root meristems

6.56 Apical root meristems in the monocotyledonous plant *Allium ursinum*.

6.57 Apical root meristems of an unidentified dicotyledonous species. Slide: S. Egli.

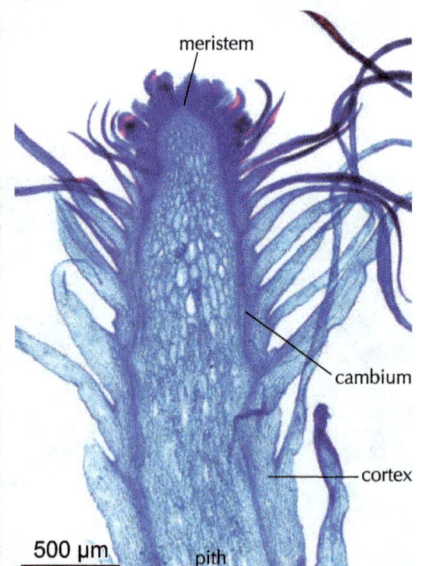

6.58 Cambium present in the dicotyledonous *Euphorbia cyparissias*.

6.1.5 From primary apical meristem to secondary lateral meristems in *shoots* – From longitudinal to radial growth

The transformation from primary to secondary meristems occurs in apical zones of roots and shoots of dicotyledonous plants. Initial apical meristems in herbs are mostly unprotected, while in trees they are mostly protected by bud scales.

The principles of secondary meristem formation are similar in all shoots of plants, however, in the detail there are many differences. The herb *Euphorbia chamaecyparissias* and the trees *Acer pseudoplatanus* and *Fraxinus excelsior* are discussed here.

In shoots, the formation of leaves in the cortex and the central pith are common. In all examined species, xylem and phloem formation starts near the apex and lignification occurs later. In detail: Cells of the central part of shoots of dicotyledonous plants remain in a parenchymatic, undifferentiated state (the pith). Around the primary meristem a ring of collateral vascular bundles is formed, which consists of protoxylem and protophloem. Lignification occurs in *Euphorbia* 10 mm and in *Acer* and *Fraxinus* 2 mm behind the apex. Vessels of the protoxylem and metaxylem are characterized by annular and helical thickenings. Crystals of various forms are very frequent in *Acer* and *Fraxinus* but are almost absent in *Euphorbia*. Crystals play a role in cell wall formation.

Protected in leaf sheath of bud scales – macroscopic aspect

6.59 A mantel of poorly developed leaves wraps the meristematic apex in *Euphorbia cyparissias*.

6.60 Bud scales wrap the meristematic apex in *Acer pseudoplatanus*.

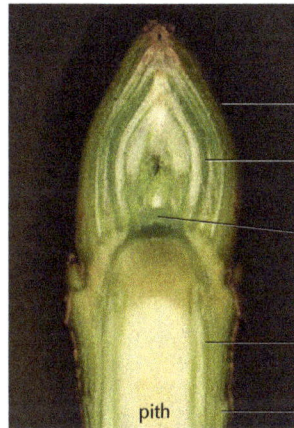

6.61 External bud scales and internal initial leaves protect the meristematic apex in *Acer pseudoplatanus* and *Fraxinus excelsior*.

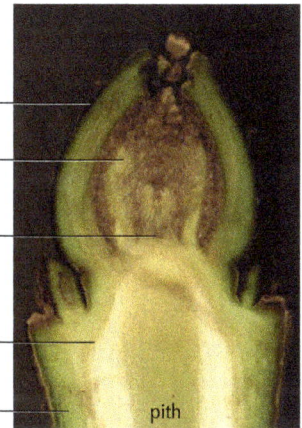

Pith and cortex with initial leaves – product of the primary meristem

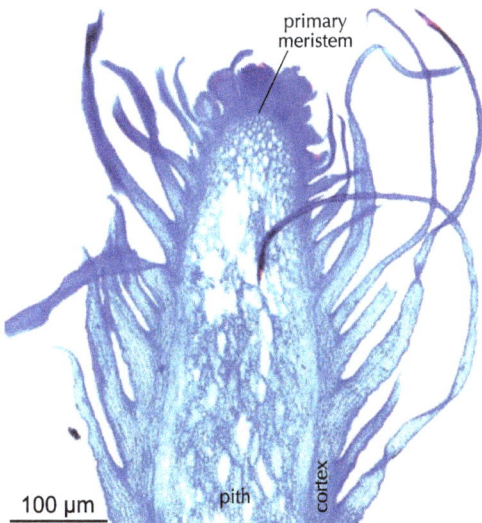

6.62 Longitudinal section of *Euphorbia cyparissias*.

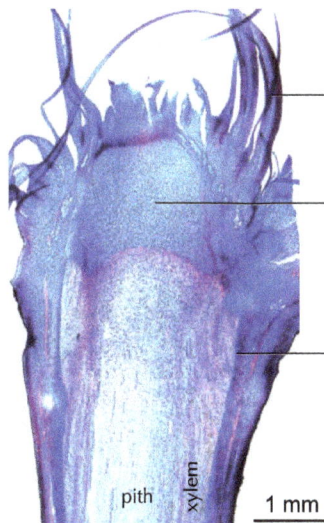

6.63 Longitudinal section of *Acer pseudoplatanus*.

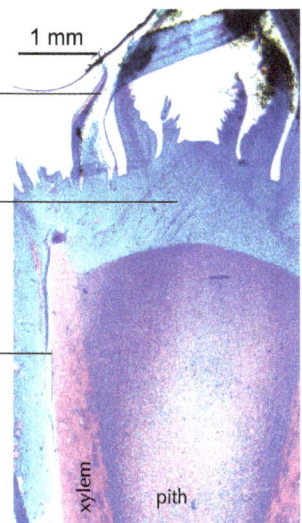

6.64 Longitudinal section of *Fraxinus excelsior*.

Secondary meristem creates initial xylem and phloem

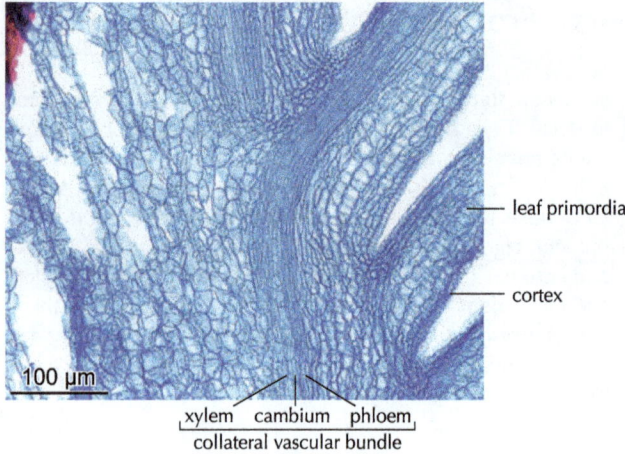

leaf primordia

cortex

100 µm

xylem cambium phloem
collateral vascular bundle

6.65 Longitudinal section of a vascular bundle in the tip of *Euphorbia cyparissias*.

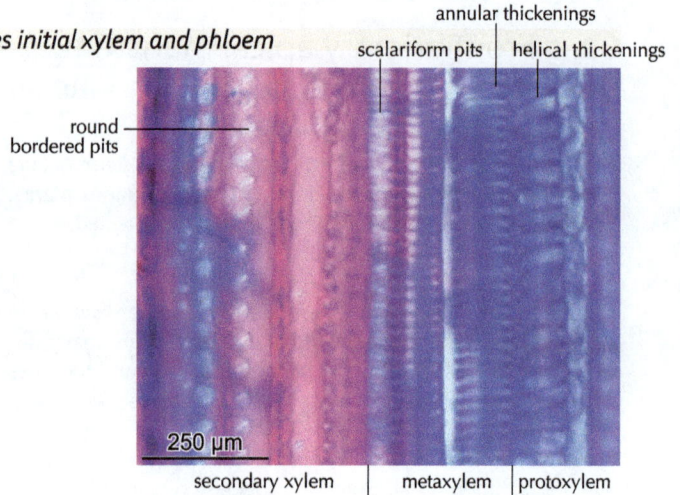

annular thickenings

scalariform pits | helical thickenings

round bordered pits

250 µm

secondary xylem | metaxylem | protoxylem

6.66 The vessel wall structure changes from helical thickenings in the protoxylem to round bordered pits in the secondary xylem of *Cycas revoluta*.

Delayed lignification

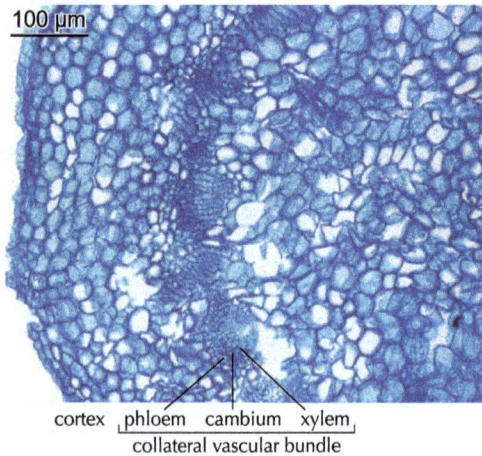

100 µm

cortex phloem cambium xylem
collateral vascular bundle

6.67 Cross section near the tip of a young shoot of *Acer pseudoplatanus*. First vascular bundles are formed but there is no lignification.

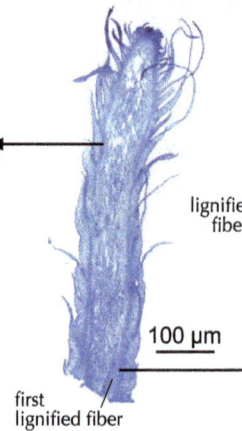

100 µm

first lignified fiber

6.68 Longitudinal section of a young shoot of *Euphorbia cyparissias* with initial vascular bundles.

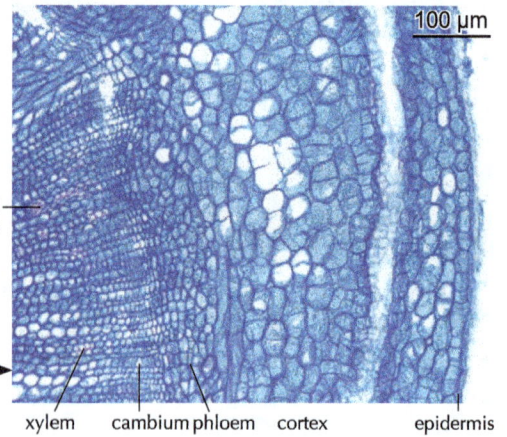

100 µm

lignified fibers

xylem cambium phloem cortex   epidermis

6.69 Cross section of *Euphorbia cyparissias* 10 mm behind the tip. The xylem of the vascular bundles already contains a few lignified fibers.

Cell-wall formation and calcium oxalate crystals

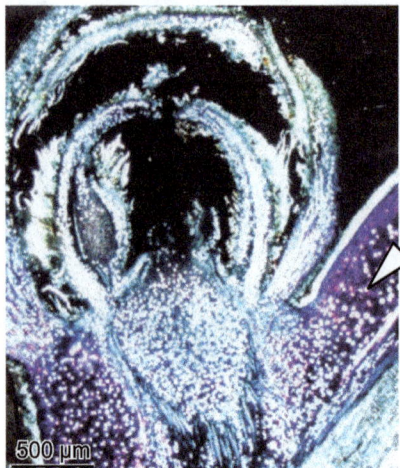

500 µm

6.70 Vegetation point within a bud of *Acer pseudoplatanus*, polarized light.

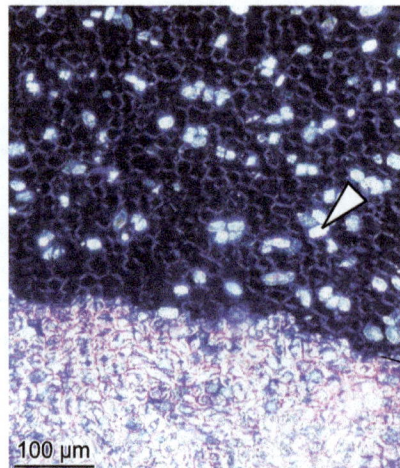

100 µm

cambium

6.71 Cambial zone of *Acer pseudoplatanus*, polarized light.

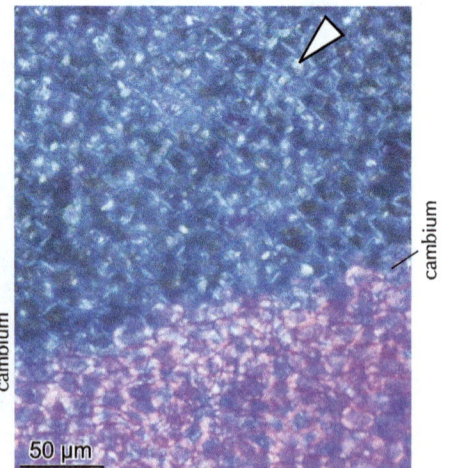

50 µm

cambium

6.72 Cambial zone of *Fraxinus excelsior*, polarized light.

6.1.6 From primary apical meristem to secondary lateral meristems in *roots* – From longitudinal to radial growth

Differentiation between shoot and root takes place in the so-called root collar, the zone between the cotyledons and the root which can be found in herbs, shrubs and trees. Shoots are characterized by a pith, while roots have none.

In contrast to the shoot, apex cells of the root differentiate very soon after their formation, xylem towards the inside and phloem towards the outside. Root apex cells behave like a secondary meristem. Therefore the roots of dicotyledonous plants have no pith. However, the width of the transition zone between the pith-filled shoot and the pith-less root varies between 5 mm and 20 cm.

Root collar – transition zone between root and shoot

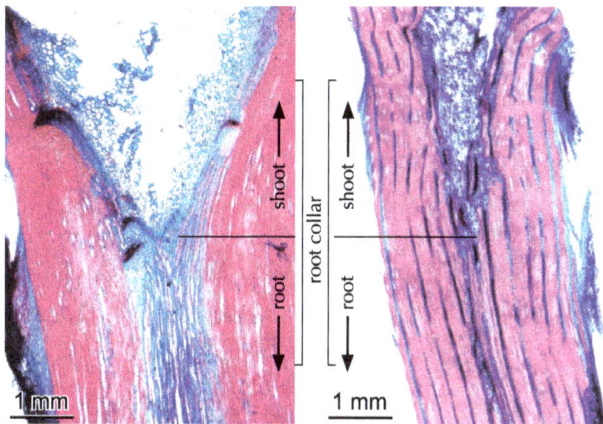

6.73 Root collar of *Tordylium apulum*.

6.74 Root collar of *Chenopodium opulifolium*.

Shoot and root in an annual dicotyledonous herb

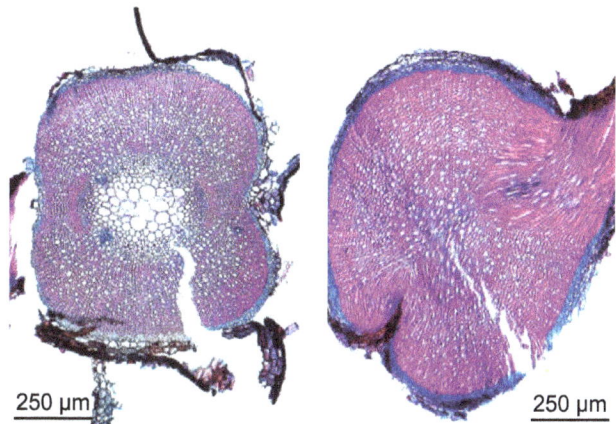

6.75 Shoot, with a pith, of *Euphrasia* sp.

6.76 Root, without a pith, of *Euphrasia* sp.

Shoot and root in a conifer

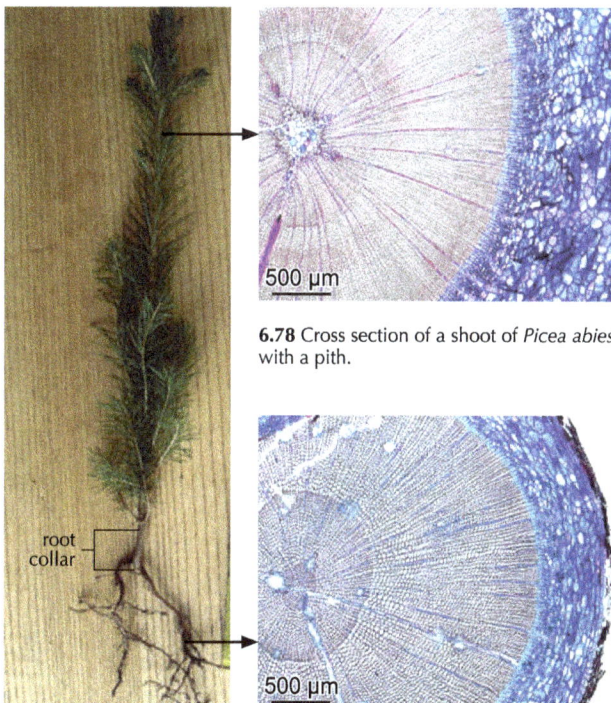

6.78 Cross section of a shoot of *Picea abies* with a pith.

6.77 Sapling of *Picea abies*.

6.79 Cross section of a root, 10 cm below the ground, of *Picea abies* without a pith.

Shoot and root in a dicotyledonous tree

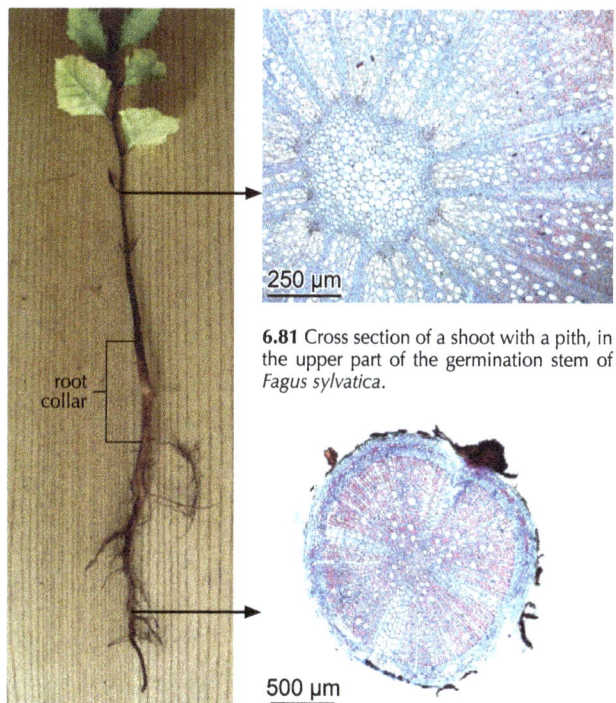

6.81 Cross section of a shoot with a pith, in the upper part of the germination stem of *Fagus sylvatica*.

6.80 A 30 cm-tall sapling of *Fagus sylvatica*.

6.82 Cross section of a root without a pith, 10 cm below the ground, of *Fagus sylvatica*.

6.1.7 From primary apical meristem in shoots to roots in plants without cambium (*monocotyledons*)

The taxonomic and morphological diversity is enormous within the monocotyledons, be it shoots, rhizomes or roots. Dramatic anatomical changes occur along the stem axis. Each section is characterized by typical anatomical structures.

Cells of the apical meristem of shoots and rhizomes differentiate very soon after their formation into parenchyma and isolated closed vascular bundles (no cambium). The bundles in the flower stalk (culm) are collateral, those in the rhizome in general concentric. Cells of the apical meristem of roots form a central vascular cylinder and a cortex. The cylinder is surrounded by an endodermis and a pericycle. The pericycle occasionally initiates lateral roots. Vascular bundles are located in the central cylinder inside an endodermis.

This is shown here for a few species from different families. However, the anatomical diversity is much larger.

Macroscopic aspect of shoots, rhizomes and roots

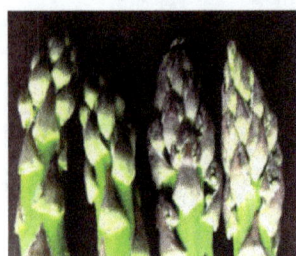

6.83 Young apical meristem in *Asparagus officinalis*.

6.84 Rhizome of *Juncus conglomeratus*.

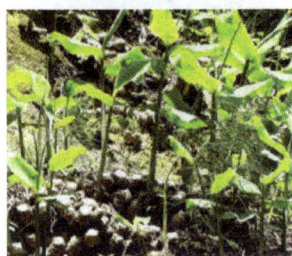

6.85 Rhizomes of *Hedychium gardnerarum*.

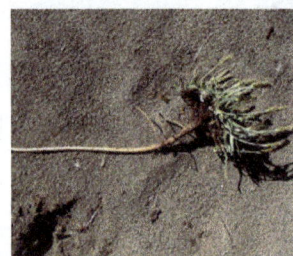

6.86 Polar root of *Plantago maritima*.

Morphological and anatomical stem structure of shoots, rhizomes and roots

Carex pendula, Cyperaceae

6.87 Flower stalk of *Carex pendula*.

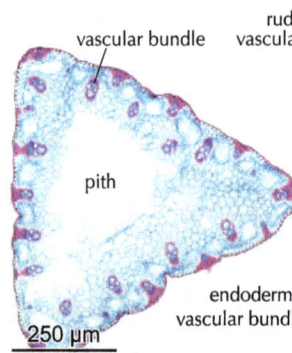

6.88 Cross section of a triangular culm. Vascular bundles are located at the periphery.

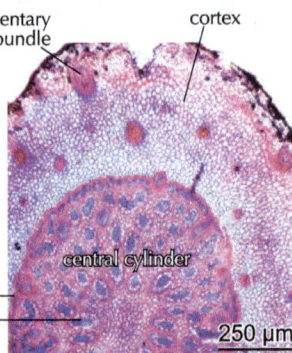

6.89 Cross section of a rhizome. Concentric vascular bundles are located in the central cylinder inside of a thick-walled endodermis.

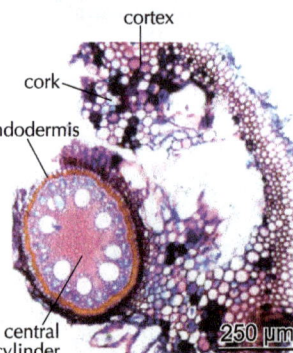

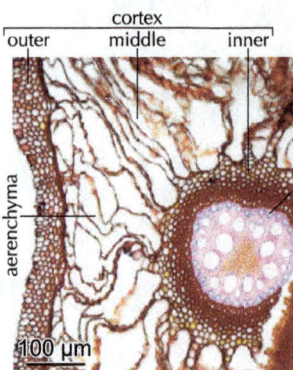

6.90 Cross section of a root. Vascular bundles are located around a thick-walled central fibrillose center.

Juncus inflexus and *conglomeratus*, Juncaceae

6.91 Culms of *Juncus inflexus*.

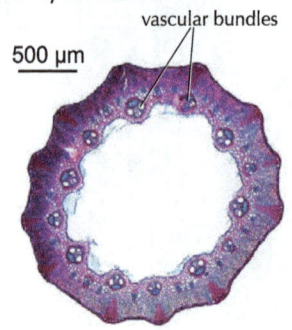

6.92 Cross section of a culm of *Juncus inflexus*. Large and small vascular bundles alternate at the periphery.

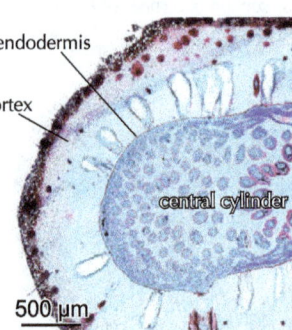

6.93 Cross section of a rhizome of *Juncus conglomeratus*. Vascular bundles are located in the central cylinder inside a thick-walled endodermis.

6.94 Cross section of a root of *Juncus conglomeratus*. Vascular bundles are located around a thick-walled central fibrillose center. The cortex contains large aerenchymatic tissue.

Phoenix canariensis, Palmaceae

6.95 *Phoenix canariensis*

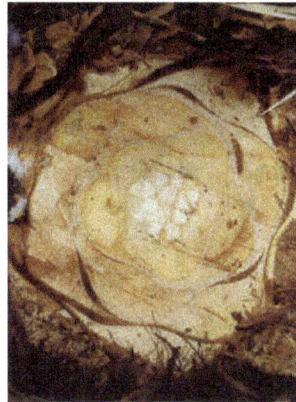

6.96 Cross section of a vegetation point from where palm syrup is harvested.

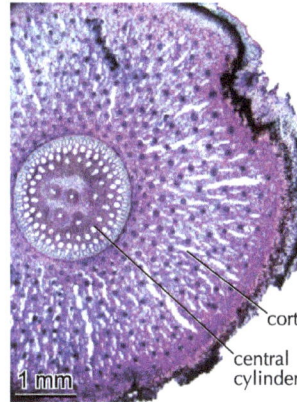

6.97 Cross section of a root. Vascular bundles are located around a thick-walled central fibrillose center.

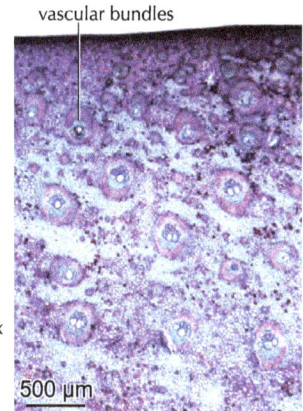

Structure of vascular bundles in shoots, rhizomes and roots

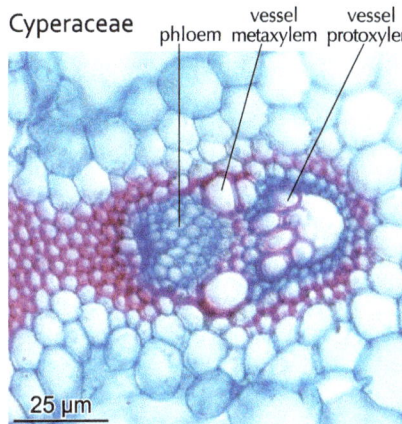

Cyperaceae

6.98 Closed collateral vascular bundle in a shoot of *Carex pilosa*. The xylem consists of a group of protoxylem and a few lateral metaxylem vessels. The phloem consists of sieve tubes and companion cells. The vessels are surrounded by a layer of fibers.

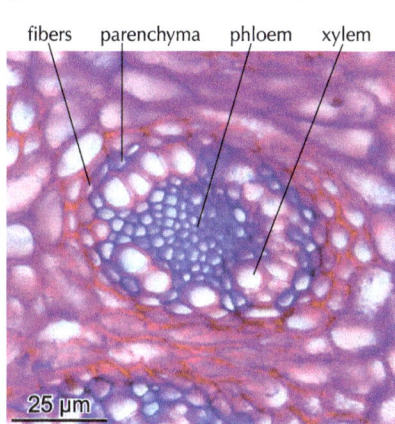

6.99 Concentric vascular bundle in a rhizome of *Carex pilosa*. Vessels surround a central group of sieve tubes. A small sheath of fibers surrounds the vascular bundle.

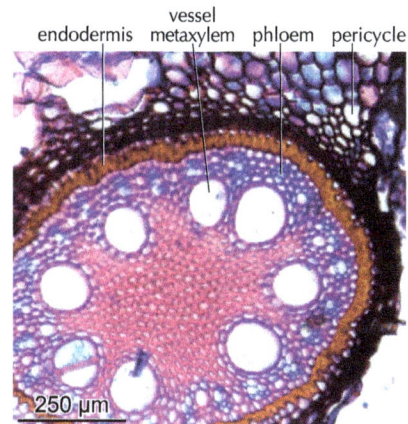

6.100 Closed collateral vascular bundles in a root of *Carex pendula*. The closed collateral vascular bundles are located inside of a thick-walled endodermis.

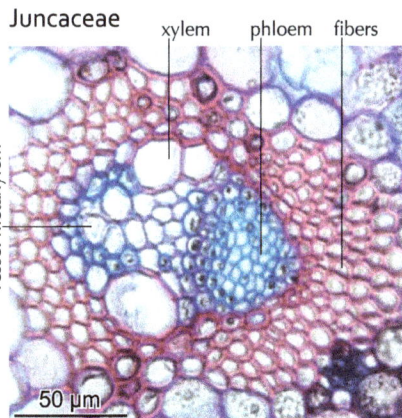

Juncaceae

6.101 Closed collateral vascular bundle in a shoot of *Juncus arcticus*. The xylem consists of a group of protoxylem and a few lateral metaxylem vessels.

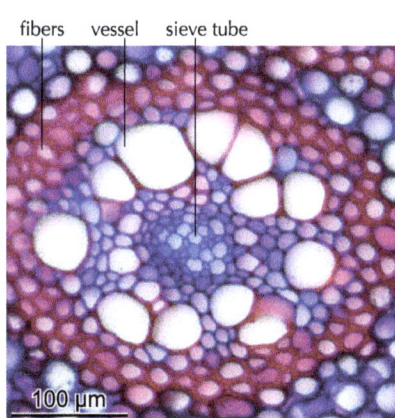

6.102 Concentric vascular bundle of a rhizome of *Juncus arcticus*. Vessels surround a central group of sieve tubes and companion cells. A sheath of thick-walled fibers surrounds the vascular bundle.

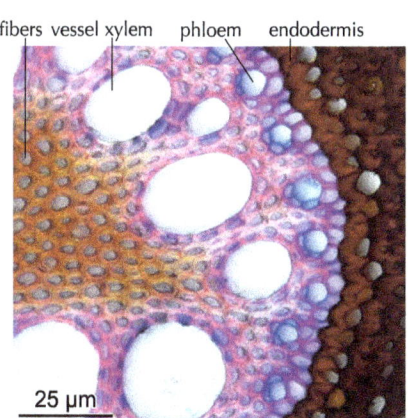

6.103 Separated xylem and phloem inside a thick-walled endodermis in a root of *Juncus conglomeratus*.

6.1.8 From primary apical meristem in shoots to roots in vascular *spore plants*

There is a great taxonomic and morphologic variety within the vascular spore plants, e.g. the lycopods, spikemosses, horsetails and ferns. Little variation occurs in the first three units, however, it is tremendous within the ferns. Major vascular spore plants have no secondary growth but anatomical changes occur along the stem axis. Products of apical meristem in shoots are leaves, often with sporophytes, and rhizomes. The product of geotropic apical meristem is the root.

The plant size and morphological variability is rather small in spikemoss (*Selaginella*), clubmoss (*Lycopodium*) and in horsetails (*Equisetum*). All types form long shoots and rhizomes with thin roots. In contrast, size and morphological variability is extremely large in ferns. All plant parts have concentric vascular bundles with the xylem in the center. Their bundles are surrounded by a cortex. The form varies from round to long oval. The number of vessels is normally high in *Selaginalla*, clubmosses and ferns. It is reduced to a few vessels in horsetails. This section presents an overview. More details are shown in Chapter 7.

Macroscopic aspect of the whole plant

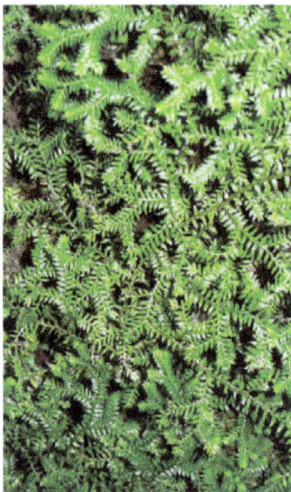

6.104 Perennial prostrate shoots of the spikemoss *Selaginella denticulata*.

6.105 Fertile annual shoots of the clubmoss *Lycopodium clavatum*.

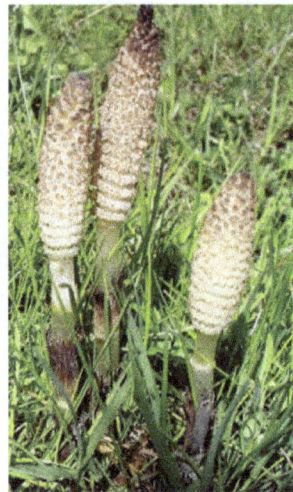

6.106 Fertile annual shoots of the horsetail *Equisetum telmateia*.

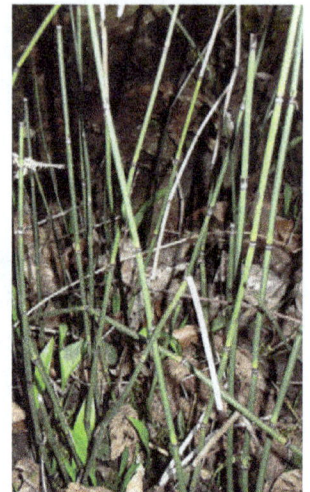

6.107 Sterile annual shoots of the horsetail *Equisetum hiemale*.

6.108 Tree fern *Cyathea cooperi*.

6.109 Climbing fern *Lygodium* sp.

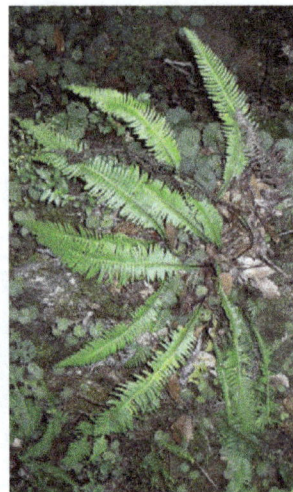

6.110 Hemicryptophytic fern *Blechnum spicant*.

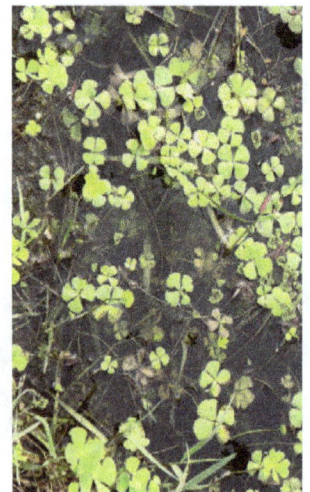

6.111 Hydrophytic fern *Marsilea quadrifolia*.

Anatomical structure of shoots, rhizomes and roots

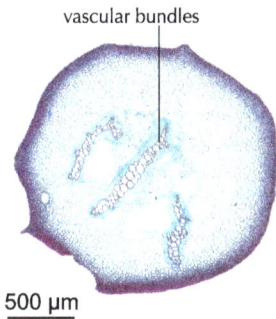

vascular bundles

500 µm

6.112 Shoot of *Selaginella* sp. with three vascular bundles.

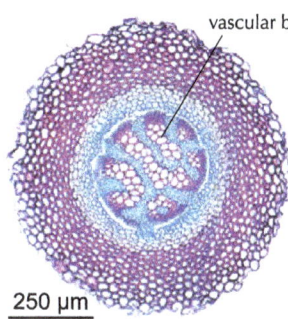

vascular bundles

250 µm

6.113 Shoot of *Lycopodium alpinum*. Irregularly distributed vascular bundles in a central cylinder (stele).

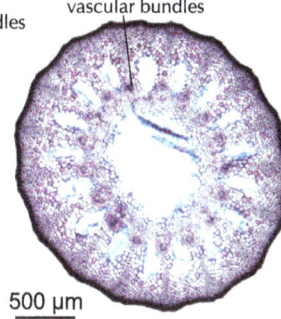

vascular bundles

500 µm

6.114 Shoot of *Equisetum hiemale*. Circular arranged, round vascular bundles.

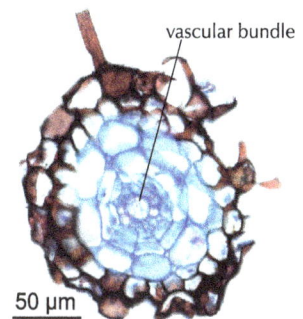

vascular bundle

50 µm

6.115 Root of *Equisetum arvense*. One concentric vascular bundle.

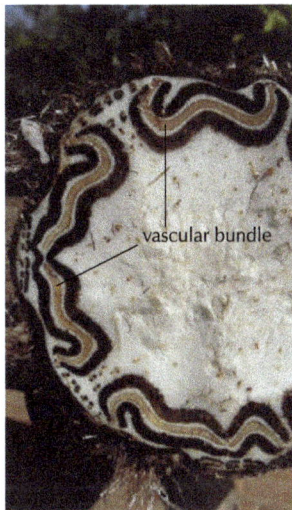

vascular bundle

6.116 Stem cross section of the tree fern *Cyathea cooperi*.

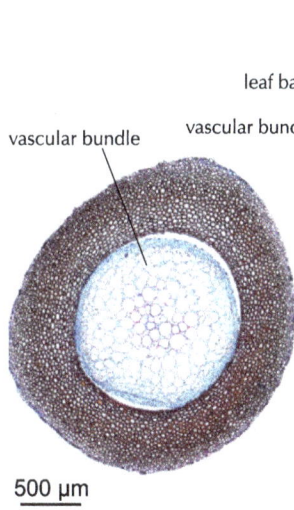

vascular bundle

500 µm

6.117 One central vascular bundle in the liana-like fern *Lygodium* sp.

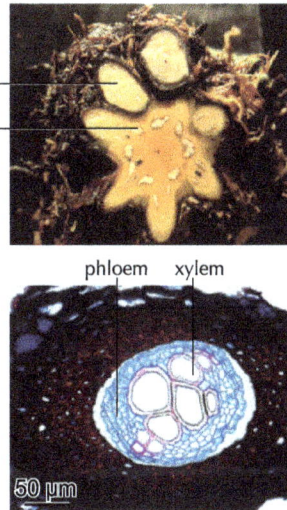

leaf base

vascular bundle

phloem xylem

50 µm

6.118 Fine root of the hemicryptic fern *Dryopteris filix-mas* with a single concentric vascular bundle.

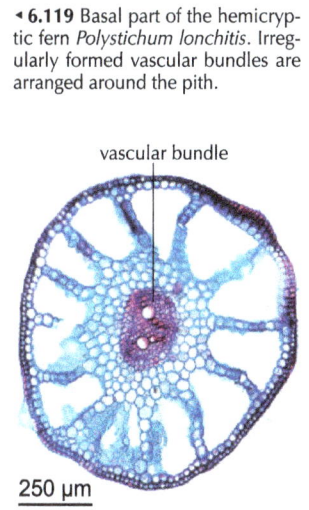

◄ **6.119** Basal part of the hemicryptic fern *Polystichum lonchitis*. Irregularly formed vascular bundles are arranged around the pith.

vascular bundle

250 µm

6.120 Microscopic cross section of a petiole of the hydrophytic fern *Marsilea quadrifolia* with collateral vascular bundles.

Structure of vascular bundles in shoots, rhizomes and roots

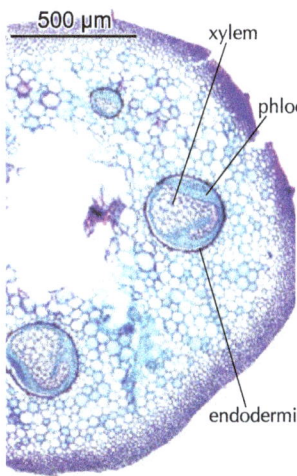

500 µm

xylem

phloem

endodermis

6.121 Round vascular bundle in a leaf of *Dryopteris filix-mas*.

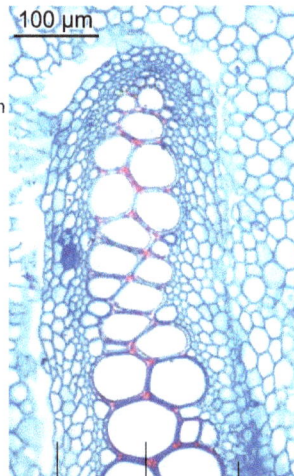

100 µm

endodermis xylem phloem

6.122 Long oval vascular bundle in a shoot of *Selaginella* sp.

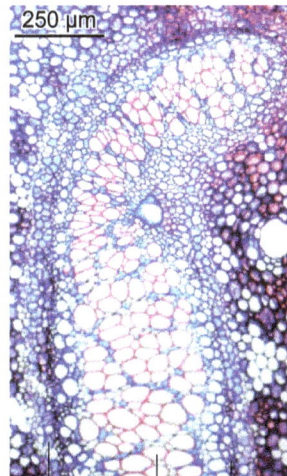

250 µm

endodermis xylem phloem

6.123 Long oval vascular bundle in a stem of the tree fern *Cyathea cooperi*.

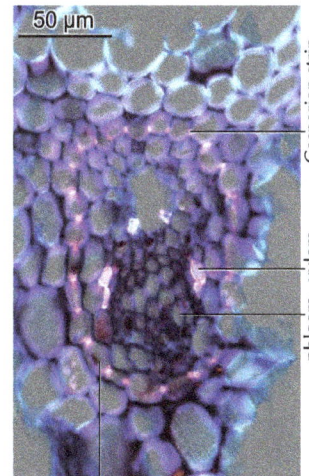

50 µm

Casparian strip

xylem

phloem

endodermis

6.124 Round vascular bundle with reduced xylem in the hemicryptophytic horsetail *Equisetum hiemale*, polarized light.

6.1.9 Pericycle and endodermis – Separation of central cylinder and cortex

Cortex and central cylinder (stele) in roots and rhizomes of monocotyledonous and dicotyledonous plants are separated by a pericycle and an endodermis. The pericycle is the outermost layer of the stele and the endodermis is the innermost cell layer of the cortex. The pericycle is a meristematic relict of the primary root meristem; it keeps its protoplast. Most of the time it is in a dormant state, but if it is in a meristematic mode it produces lateral roots. In young states it also initiates the cork cambium. The endodermis primarily regulates hydrological differences between the central cylinder and the cortex. It maintains the root pressure and protects the central vascular bundles from toxic substances, which occasionally occur in the cortex.

Only optimally developed endodermis and pericycle zones in a few roots are described in textbooks. In reality endodermis and pericycle are often unrecognizable or difficult to distinguish. Also, the anatomy of endodermis varies. Presented here are a few "unproblematic" examples.

In many monocotyledonous plants the cell walls are extremely thick-walled on the inner and lateral sides. Often described but rarely occurring is the endodermis with Casparian strips. The strips form a lignified band of radial and transverse walls.

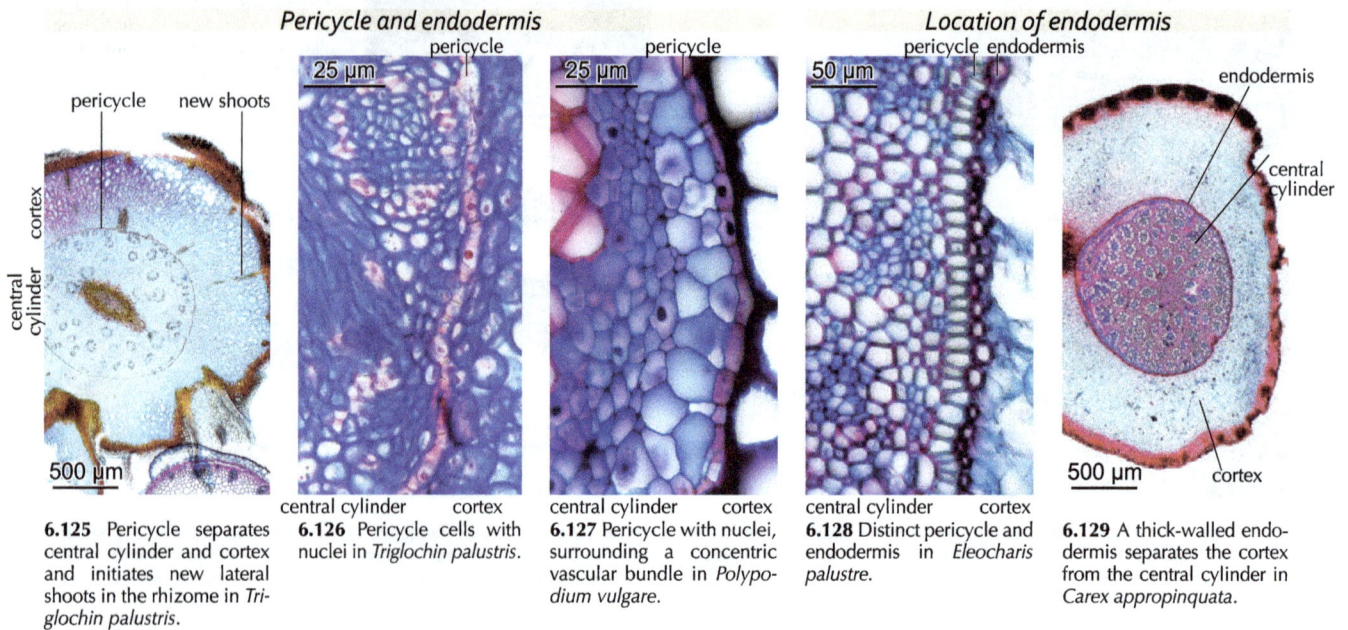

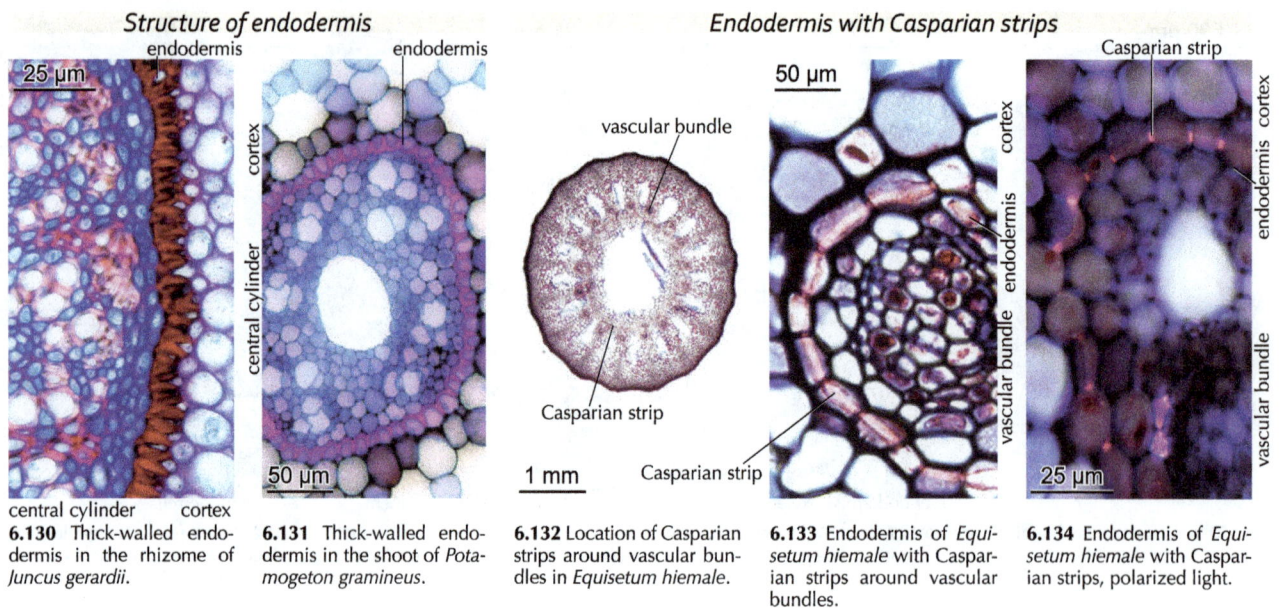

Pericycle and endodermis

6.125 Pericycle separates central cylinder and cortex and initiates new lateral shoots in the rhizome in *Triglochin palustris*.

6.126 Pericycle cells with nuclei in *Triglochin palustris*.

6.127 Pericycle with nuclei, surrounding a concentric vascular bundle in *Polypodium vulgare*.

Location of endodermis

6.128 Distinct pericycle and endodermis in *Eleocharis palustre*.

6.129 A thick-walled endodermis separates the cortex from the central cylinder in *Carex appropinquata*.

Structure of endodermis

6.130 Thick-walled endodermis in the rhizome of *Juncus gerardii*.

6.131 Thick-walled endodermis in the shoot of *Potamogeton gramineus*.

Endodermis with Casparian strips

6.132 Location of Casparian strips around vascular bundles in *Equisetum hiemale*.

6.133 Endodermis of *Equisetum hiemale* with Casparian strips around vascular bundles.

6.134 Endodermis of *Equisetum hiemale* with Casparian strips, polarized light.

6.2 Secondary and tertiary meristems and radial growth
– Cambium and cork cambium

6.2.1 Macroscopic aspect of radial growth and xylem coloration
– Stems get thicker

Stem thickening occurs through the lateral secondary meristem. This is the cambium, which is located between the xylem and phloem. In most conifers and dicotyledonous plants the cambium forms a mantle around the xylem. Plants with successive cambia (several active cambia) are a special case.

Years after wood formation, the inner part of the stem loses its conducting capacity. This is the moment when the stem differentiates into sapwood and heartwood. The peripheral sapwood conducts water and contains living parenchyma cells. In contrast, the heartwood does not conduct water and all cells are dead. Here, parenchyma cells often contain phenolic substances which play an important part in biological defense mechanisms. The width of the sapwood is generally proportional to the transpiring leaf area: the more leaves in the tree crown, the larger the sapwood.

A few groups of stem cross sections can be differentiated macroscopically.

Species with colored heartwood
◦ Species with high water content in the sapwood and low water content in the heartwood; e.g. in the genera *Pinus*, *Larix* and *Taxus*.
◦ Species without notable water content differences between sapwood and heartwood; e.g. in the genera *Quercus*, *Castanea*, *Robinia*, *Prunus* and *Juglans*.

Species without colored heartwood
◦ Species with high water content in the sapwood and low water content in the heartwood; e.g. in the genera *Picea* and *Abies*.
◦ Species with high water content in the whole stem; e.g. in the genera *Fagus*, *Carpinus* and *Alnus*.
◦ Species with higher water content in the sapwood than in the heartwood; e.g. in the genera *Acer* and *Citrus*.

Irregularly shaped discolorations are related to biological attacks. Different colors, textures and brilliance of heartwood, as well as color differences between heart- and sapwood are basic features for macroscopic wood identification. This is perfectly presented in the old *Woodbook* by R.B. Hough, republished in 2002.

The outline of stems varies from round (most trees), to eccentric (leaning trees), to fluted (buttressed stem basis) and square. Multiple stems occur mainly in perennial herbs. The bark thickness (phloem, cortex, cork) in relation to the xylem is very variable. The texture in transverse and longitudinal sections is species-specific, or mainly related to the structure of annual rings and rays.

Location of the cambia

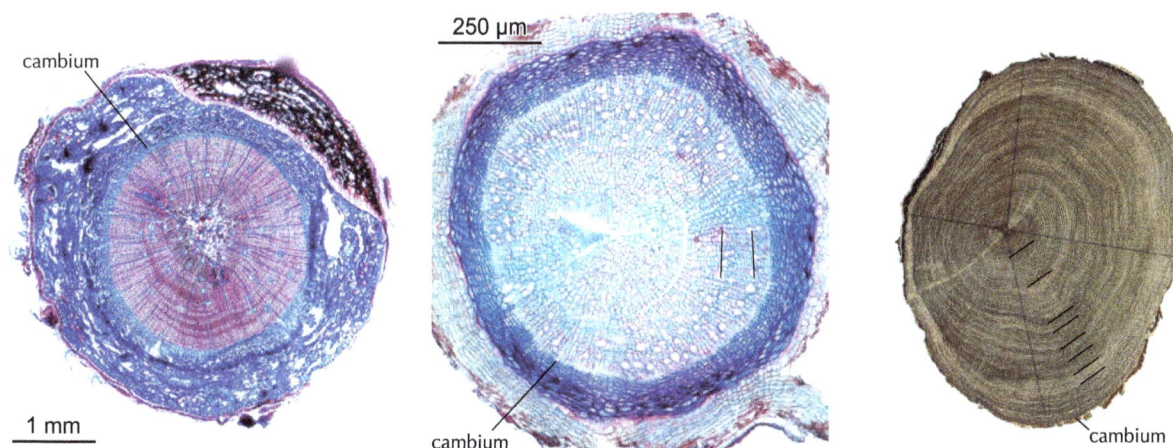

6.135 One cambium is located between the central xylem and the peripheral bark in an 11-year-old conifer twig of *Pinus sylvestris*.

6.136 One cambium is located between the central xylem and the peripheral bark of a four-year-old arctic herb, *Cerastium arcticum*.

6.137 Several peripheral cambia form several fiber and parenchyma bands during one year. This plant of *Haloxylon persicum* is approximately 10–12 years old.

Sapwood and heartwood

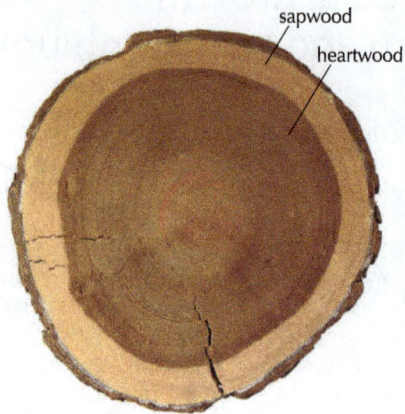

6.138 A belt of light sapwood surrounds the brown heartwood in the center of the conifer *Pinus sylvestris*.

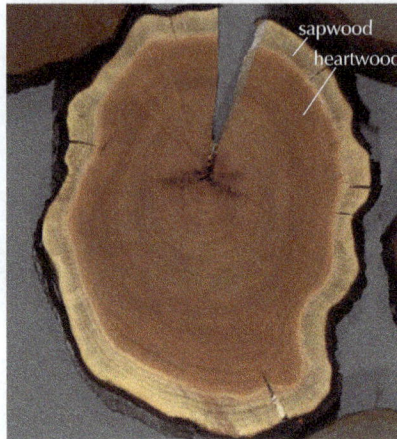

6.139 A belt of light sapwood surrounds the dark brown heartwood in the center of the deciduous tree *Rhamnus cathartica*.

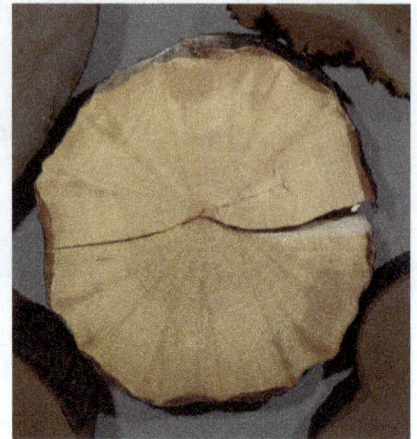

6.140 Heartwood and sapwood are not differentiated by color differences in the deciduous tree *Carpinus betulus*.

Discolorations are defense reactions

Biological resistance

6.141 Living parts of stems react to injuries with the formation of dark-stained phenolic substances. Compartmentalized overgrown injury in *Acer pseudoplatanus*.

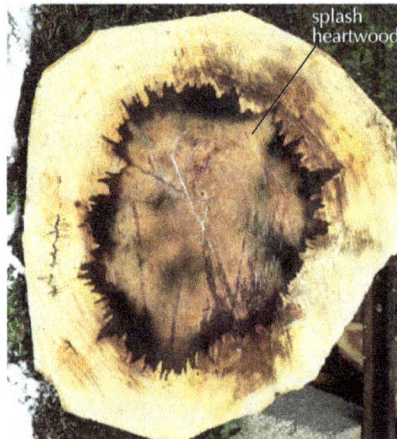

6.142 "Splash heartwood" (German "Spritzkern") in *Fagus sylvatica* is a sign of bacterial infections.

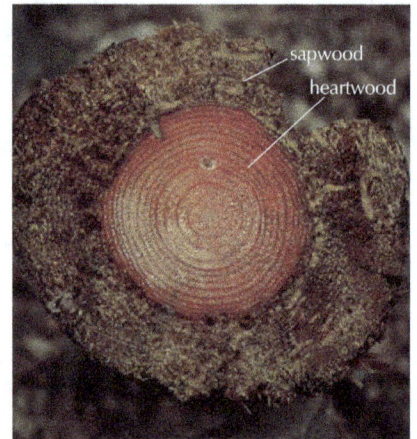

6.143 Heartwood of *Pinus sylvestris* is more resistant against fungal infestations than sapwood.

Outline of stems

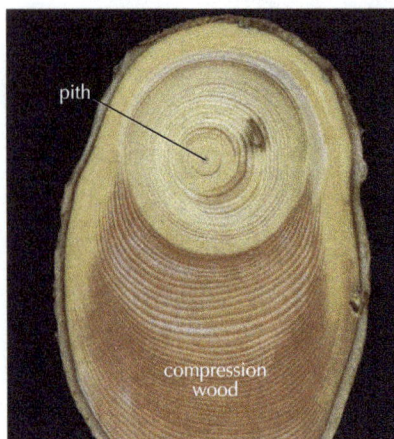

6.144 Eccentric stem due to compression wood formation in *Picea abies*.

6.145 Fluted stem of the shrub *Crataegus* sp.

6.146 Square stem of the tree-like succulent *Euphorbia ingens*.

Bark thickness

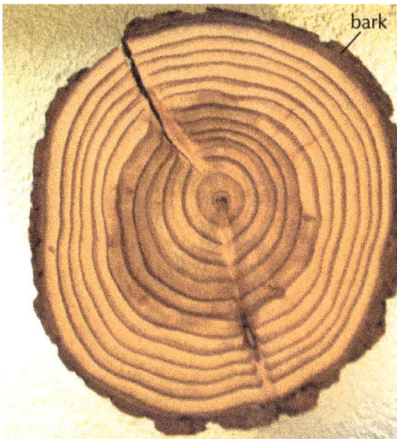

6.147 Thin bark in relation to the xylem in *Pinus mugo*.

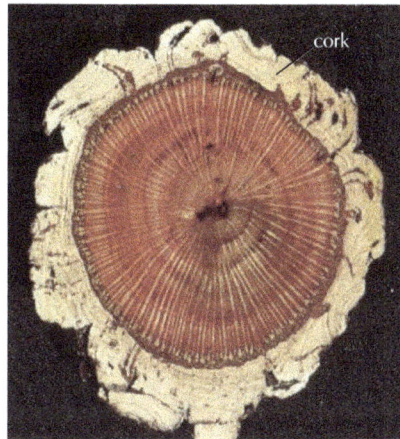

6.148 A large cork belt and a small phloem surround the xylem in *Quercus suber*.

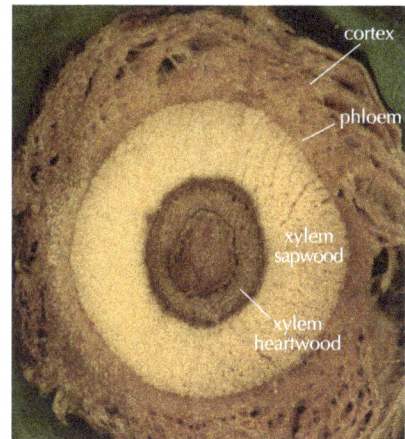

6.149 A large cortex surrounds the xylem of the herb *Heracleum pinnatum*.

Bark thickness

6.150 An extremely large cortex surrounds a very small xylem in the giant cactus *Carnegia gigantea* (dry cross section).

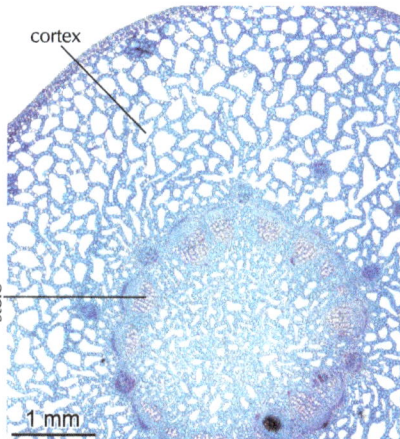

6.151 An extremely large aerenchymatic cortex surrounds a very small stele in the water plant *Menyanthes trifoliata*.

Multiple stems

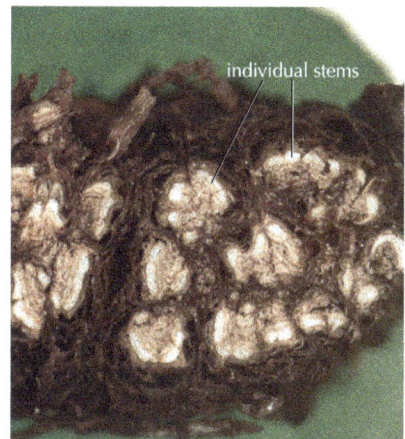

6.152 Stems of small cushion plants like *Saussurea glanduligera* in alpine zones are composed of many small individual stems.

Wood texture

6.153 The cutting direction of stems in parquet flooring highlights the annual ring structure of the wood of *Quercus* sp.

6.154 The radial cutting direction shows the structure of the rays in parquet flooring of *Quercus* sp.

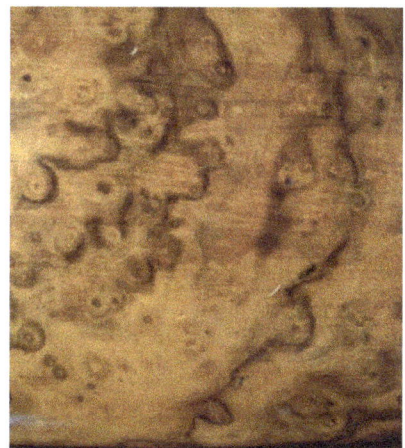

6.155 The section through burls with sleeping buds shows the unusual wood structure in an antique chest, made of a deciduous tree species.

6.2.2 Microscopic aspect of radial growth (conifers, dicotyledonous plants and palm ferns) – An overview

Radial growth of conifers and dicotyledonous plants with one cambium

As soon as the cambium is active it forms secondary tissue: the secondary xylem and the secondary phloem. The xylem is different from the one that was formed by the primary meristem: Tracheids and vessels do not have any annular or thick annular or spiral thickenings. The cambium transfers the single vascular bundles into a continuous ring of xylem and phloem.

Radial growth of some dicotyledonous plants with several (successive) cambia

Numerous species, especially those in the families of Amaranthaceae and Caryophyllaceae, form and maintain several cambia. As soon as the first cambium is formed it produces a xylem and a phloem like in all other dicotyledonous plants. However,

this stage lasts only for a short time. For growing in thickness, parenchyma cells outside of the phloem get reactivated and form a new cambium, which again produces a xylem and a phloem. This process repeats itself over many years. The lifetime of successive active cambia is limited but their effect is preserved in the anatomical structure of the stem.

Radial growth of a few monocotyledonous plants

Secondary radial growth occurs in a few families of monocotyledonous plants, e.g. in *Dracaena* sp. and *Yucca* sp. As in the group with successive cambia, parenchyma cells in the primary bark (cortex) get reactivated and form—towards the center—a continuous belt of parenchyma cells around the stem. A few of them remain active and form vascular bundles.

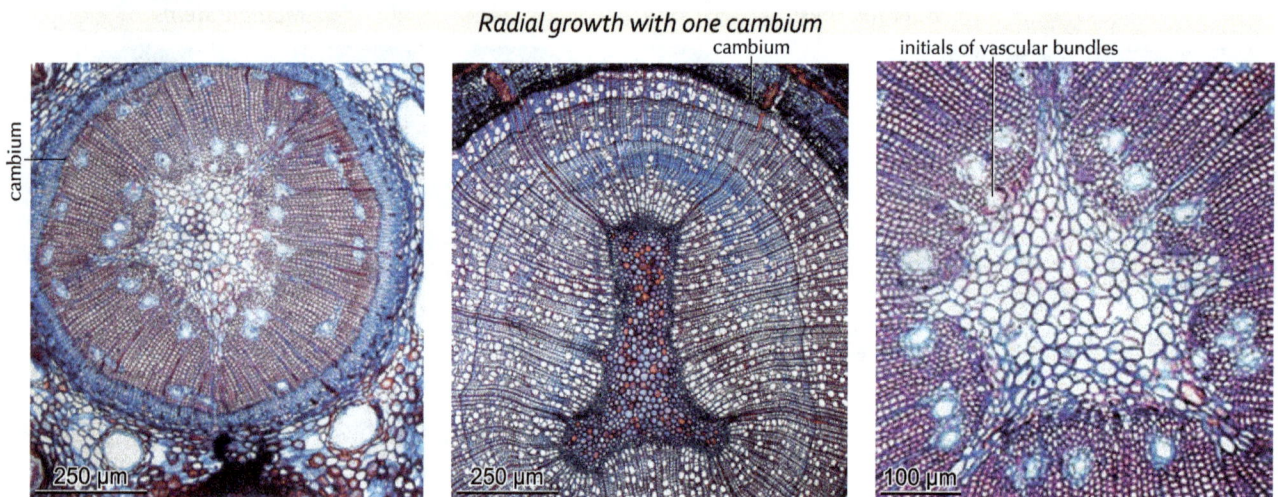

Radial growth with one cambium

6.156 One cambium produces the xylem and phloem. One-year-old shoot of the conifer *Pinus sylvestris*.

6.157 One cambium produces the xylem and phloem. Three-year-old shoot of the dicotyledonous tree *Alnus glutinosa*.

6.158 The secondary cambium merges the primary vascular bundles into a continuous belt in a twig of *Pinus sylvestris*.

Radial growth with several cambia (successive cambia)

6.159 The annual dicotyledonous herb *Chenopodium botrys*, Amaranthaceae.

6.160 Cross section of the basal stem of the annual dicotyledonous herb *Chenopodium botrys*. Several cambia (blue rings) produce xylem and phloem simultaneously.

6.161 The monocotyledonous tree *Dracaena draco*, Asparagaceae.

6.162 Cross section of the peripheral part of the stem of *Dracaena draco*. The cambium produces vascular bundles and rays towards the inside and parenchyma cells towards the outside.

6.2.3 Production and enlargement of new cells in the xylem of a thickening stem – The need for more and larger cells

Stem thickening is related to an increase and enlargement of axial elements, like parenchyma cells, tracheids and fibers, and an increase and enlargement of rays. The number of cells at the periphery is lower in smaller than it is in thicker stems. Due to increased leaf area and plant weight, larger plants need more water-conducting and stabilizing cells than smaller plants.

The process of stem thickening is anatomically expressed by axial cell initiations (tracheids, fibers, parenchyma cells and vessels), new ray initiations and dilating rays. This is underlined by Bailey 1923 who counted 794 tracheids in a one-year-old stem of *Pinus strobus* and 32,000 tracheids in a 60-year-old plant. The thickening process is also accompanied by cell death; cross sections of conifers show the disappearance of cell rows.

With the insertion of new ray cells and the enlargement of primary rays (ray dilatation), radial strength as well as storage capacity increases. The initial point for new cells is located in the cambial zone. New tracheids divide longitudinally. New rays are initiated in living tracheids, which change their mode; instead of longitudinal separation into tracheids, a small ray cell splits off laterally.

More cells and larger cells

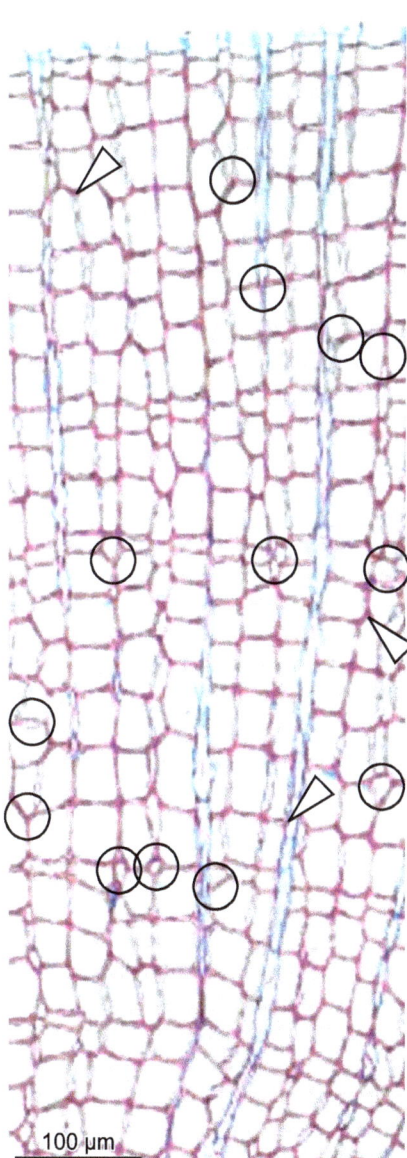

6.163 Initiation (circles) and disappearance (arrows) of tracheids in the young root of a 20 m-tall *Pinus nigra* tree.

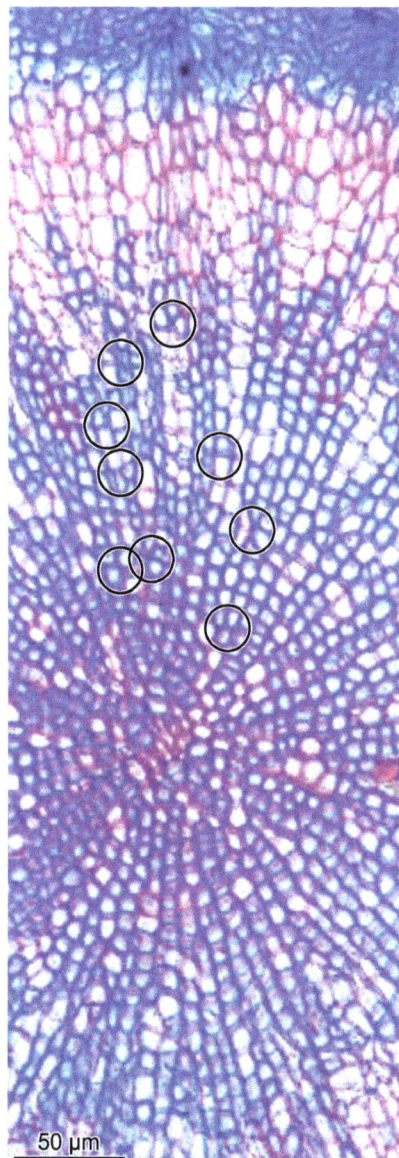

6.164 Initiation and enlargement of fibers in the stem of a 7 cm-tall annual herb *Erophila verna*.

15 cells wide

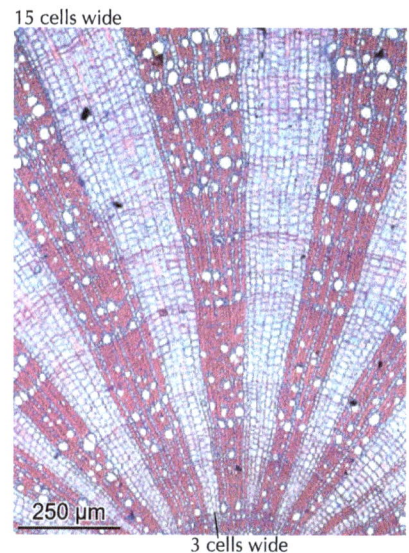

3 cells wide

6.165 Dilatation of large rays by insertion of new ray cells and enlargement of cells in *Rosa pendulina*.

Birth of cells

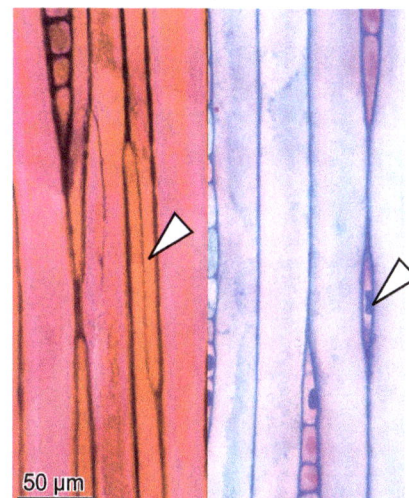

6.166 Dividing of tracheids in longitudinal direction (left) and separation of a single ray cell from a tracheid (right) in *Pinus sylvestris*.

6.2.4 Cell formation and differentiation in the *xylem* – The multifunctional stem center

Genetic information, physiological needs and ecological triggers form the background for all anatomical structures. The anatomical expression of many biological and biochemical processes are presented in the following. Basic wood and bark formation processes already existed in late Devonian times (370 million years ago) in conifer-like stem structures. These ancient principles have been transferred to the phylogenetically young angiosperms (140 million years ago).

The following basic processes can be observed in conifers and angiosperms: cell-type differentiation, cell-wall differentiation, nuclei differentiation, cell-wall enlargement, cell-wall thickening and lignification. Genetic information determines the general arrangement and distribution of cell types.

Cambium mother cells form anatomically undifferentiated phloem and xylem mother cells. These three cell types are anatomically combined in the cambial zone.

The first anatomical expression of **cell differentiation** appears within the cambial zone. Initial stages of conifers show tracheids, rays and resin ducts in the xylem and sieve cells and parenchyma cells in the phloem. In addition, angiosperms form vessels. In relation to space and physiological needs, some cell types have priority; resin ducts push aside tracheids and rays and, in angiosperms, vessels displace fibers and rays. The differentiation of the **nucleus form** takes place along with the cell-type differentiation.

Cambial zone

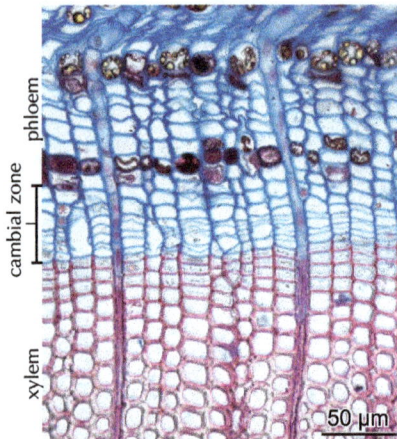

6.167 Cambial zone of the conifer *Picea abies* in the dormant state. Cambium initials, xylem mother cells and phloem mother cells are not anatomically differentiated.

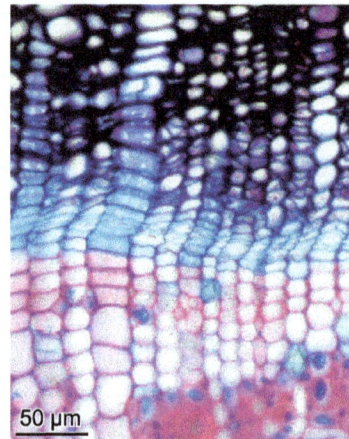

6.168 Cambial zone of the angiosperm *Ficus carica* in a dormant state. Cambium initials, xylem mother cells and phloem mother cells are not anatomically differentiated.

First-formed xylem cells

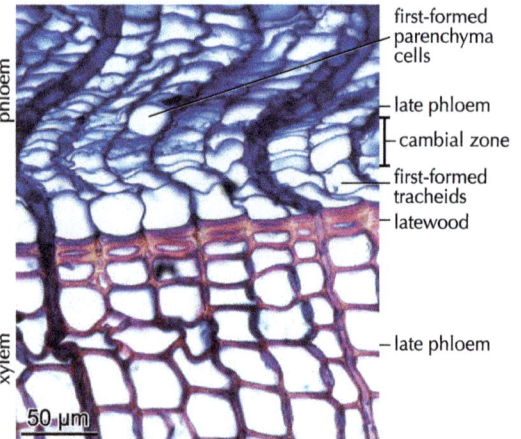

6.169 First-formed cells in the earlywood of a conifer. Tracheids and ray cells can be recognized on the xylem side and parenchyma cells on the phloem side.

Different priorites

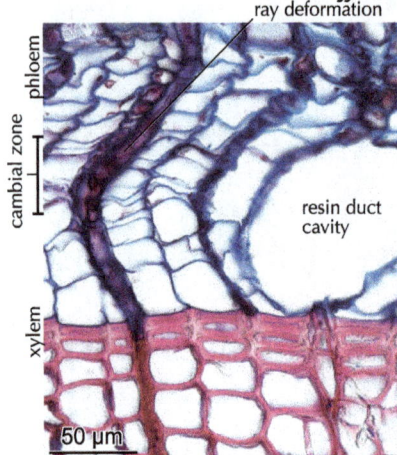

6.170 First-formed cells in the earlywood of the conifer *Larix decidua*. Formed are 2–3 rows of tracheids and a large cavity for a resin duct. The duct has the spatial priority.

6.171 First-formed cells in the earlywood of the angiosperm *Ficus carica*. Formed are 20 cells of fibers and a large vessel. The vessel has the spatial priority.

Nucleus differentiation

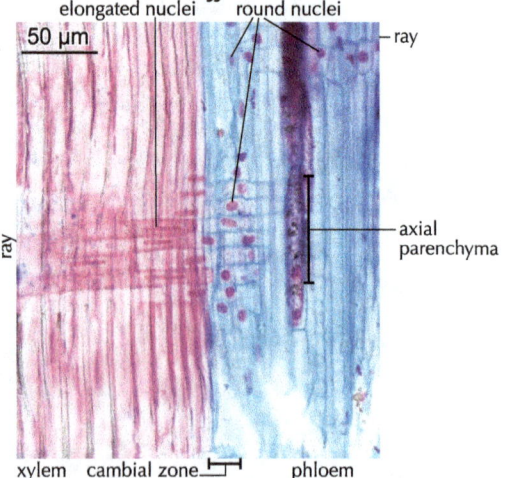

6.172 Round nuclei in cambial initials, phloem ray cells and axial phloem parenchyma cells; axially elongated nuclei in tracheids and phloem initials; radially elongated nuclei in xylem ray parenchyma cells in the conifer *Picea abies*.

The xylem and phloem mother cells already contain the information about their **cellular pathway** before their anatomical expression. The differentiation capacity of anatomically undifferentiated mother cells is very dynamic and changes within short time periods. This is very obvious in angiosperms. In one moment xylem mother cells divide into fibers and in the next into vessels, however, the change from fiber or vessel to ray cells is inexistent or rare.

Changing formation mode

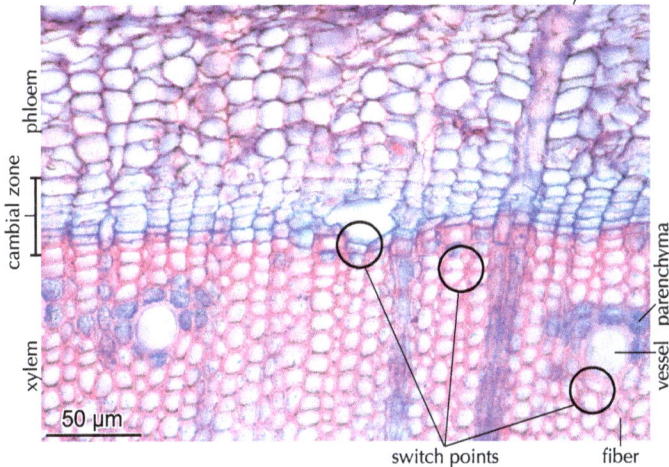

6.173 Cambial initials periodically determine which cell type has to be formed. Sometimes the initial differentiates into a fiber and sometimes into a vessel. *Fraxinus excelsior.*

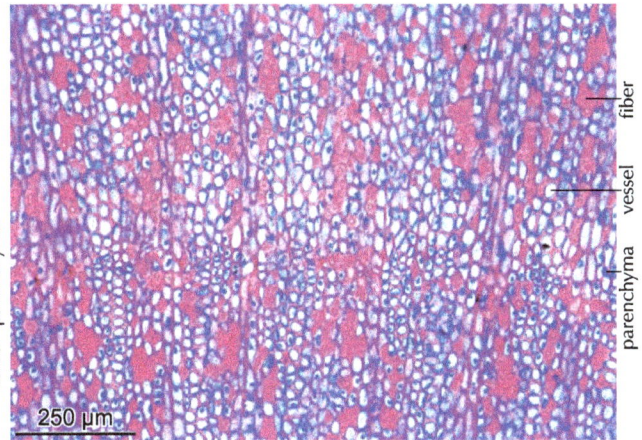

6.174 Cambial initials periodically determine which cell type has to be formed. The radial continuation of formed cell types (fibers, vessels and parenchyma cells) seems to be chaotic, however, the general pattern is typical for *Viscum album.*

The second phase of radial growth is radial and axial **cell enlargement**. The process takes place in the stage of primary wall formation, however, each cell type has its own expanding characteristic. Radially, fibers expand poorly, but longitudinally, they do so extensively (up to nine times), at which the axial ends become wedged. Axial parenchyma cells radially also expand poorly, and longitudinally only slightly; they generally remain in the state of the initials. Ray cells radially expand extensively, while longitudinally hardly at all. Vessels expand in radial, tangential and longitudinal direction.

Cell enlargement

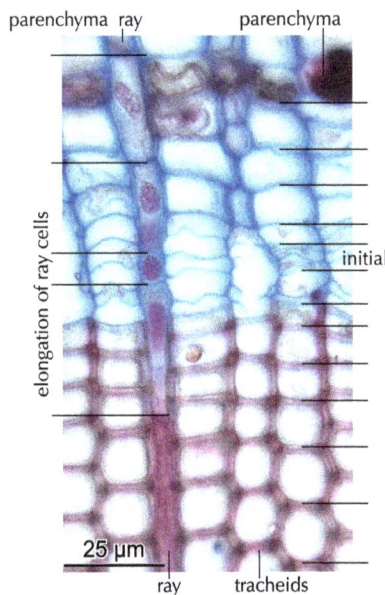

6.175 Derivates of xylem mother cells axially enlarge at different rates. Initial fiber and ray cells have the same radial dimension. Tracheids expand slightly; ray cells expand extensively. *Picea abies.*

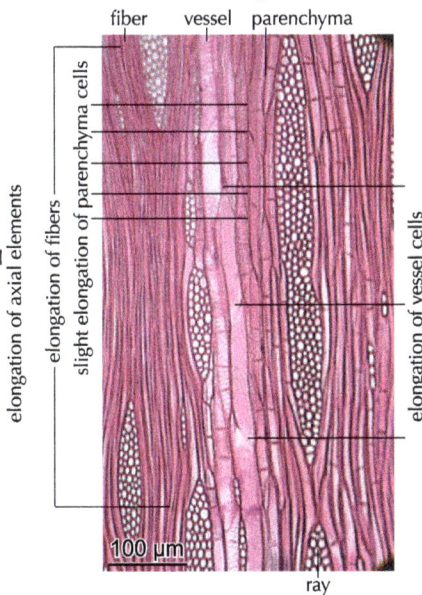

6.176 Derivates of xylem mother cells axially enlarge at different rates. Parenchyma cells stay more or less their initial length, however, fibers axially expand extensively. *Ulmus laevis.*

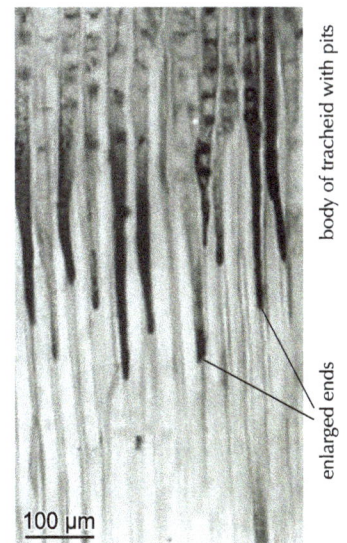

6.177 After elongation, the axially elongated tracheids in *Picea abies* become wedged.

Simultaneously with the wall expansion, *cell-wall differentiation* takes place. This is demonstrated here on bordered pits in conifers and dicotyledonous angiosperms. First, micro-fibrils form a submicroscopic comb-like pattern. Next, the outer border of the bordered pits in tracheids can be observed under a microscope.

It indicates the final size of the tracheid wall. Differentiation of the pits takes place over the course of a few weeks into pits with lignified borders and unlignified tori. Tori of conifers lignify when they are no longer involved in the water-conducting process. This normally occurs at the sapwood-heartwood boundary.

Cell wall differentiation and pit structure

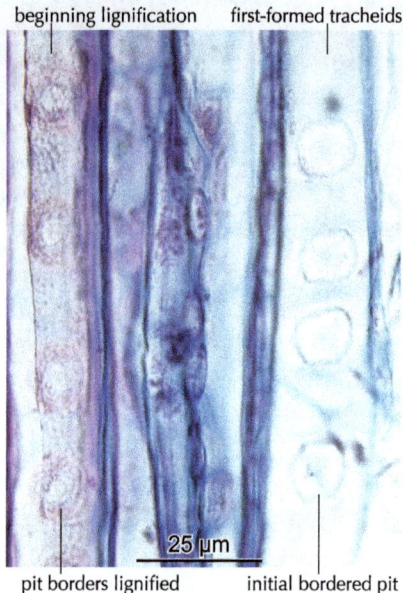

6.178 Formation of bordered pits in tracheids of the conifer *Picea abies*. First, unlignified, round, small pit borders appear. Development of the pit borders occurs simultaneously with lignification.

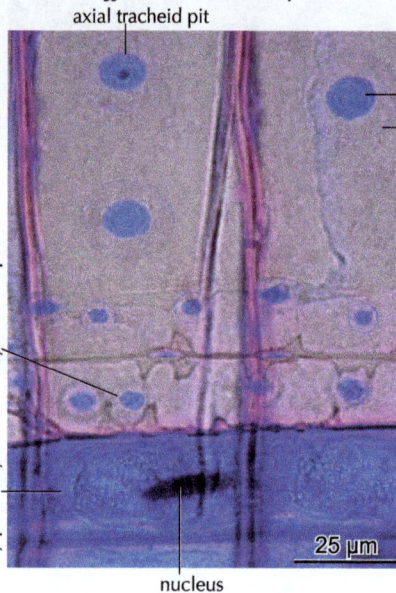

6.179 The final stage of the development near the cambium shows pits in axial tracheids and ray tracheids with large borders and unlignified tori in a radial section of *Pinus sylvestris*.

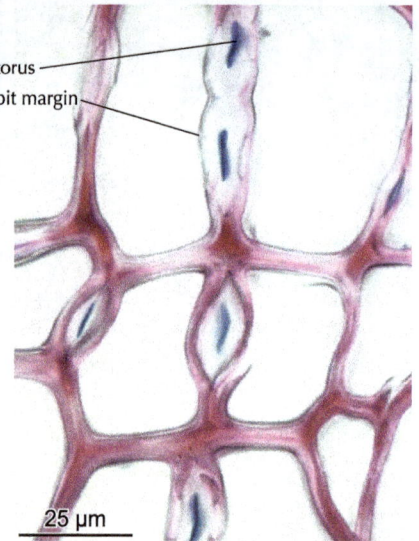

6.180 Final stage of the development of bordered pits in a transverse section of *Picea abies*. The borders are lignified, the tori are unlignified.

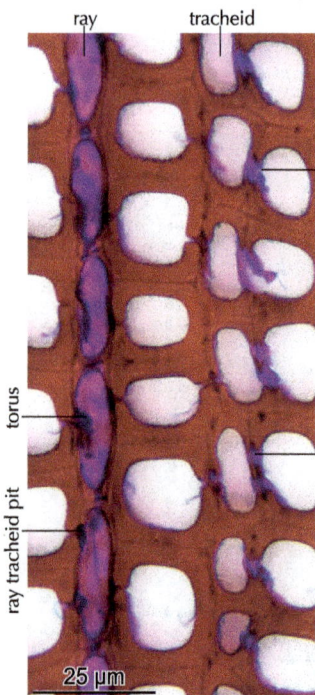

6.181 Tracheid pits with indistinct unlignified tori (blue) and ray tracheid pits with distinct, unlignified tori in a cross section (left) and radial section (right) of *Drimys piperita*.

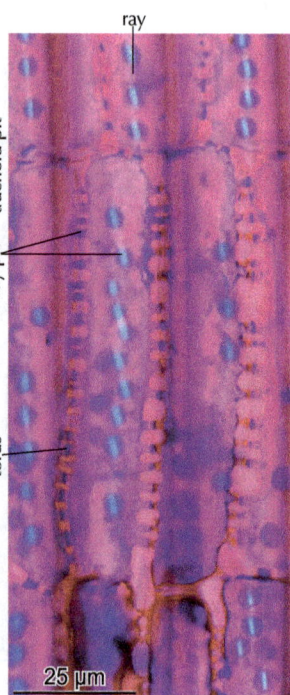

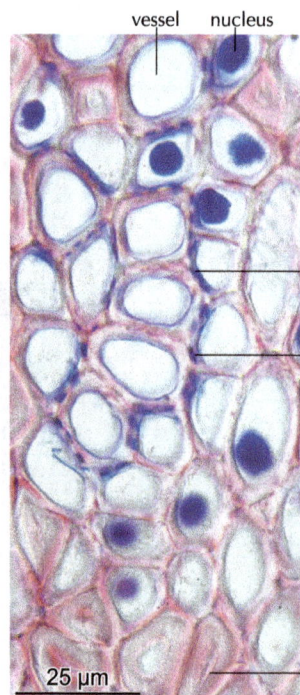

6.182 Intervessel pits with distinct tori and vessel-ray pits with distinct, unlignified tori in a cross section (left) and tangential section (right) of *Viscum album*.

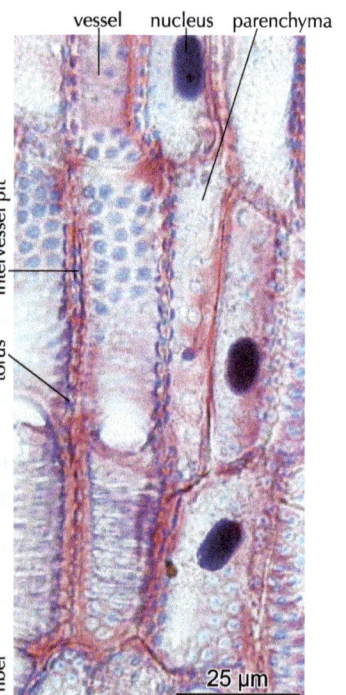

Cell-wall thickening and lignification are the last processes to take place. Lignification occurs immediately after or during the formation of the cellulose matrix. With the formation of the cellulosic matrix of the secondary wall, cell walls become thicker. This process is accompanied by the incrustation of lignin. It starts in primary walls in the corner of cells and expands towards the lumen of the cell. In conifers, all diffuse-porous species and in the latewood of ring-porous species, lignification occurs front-like behind the cambium. The lignification process is different in the earlywood of ring-porous species because the formation of earlywood vessels is asynchronous. Lignified vessels and the surrounding fibers of the first-formed cells stay side by side with newly formed, unlignified vessels.

Formation of the cellulose matrix and lignification of cells in angiosperms

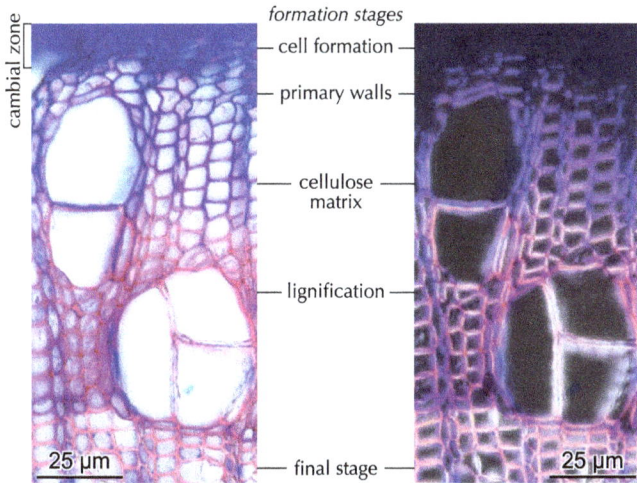

6.183 Lignification of cell walls of vessels, fibers and rays in *Salix fragilis*. Ontogenetically young cells near the cambium are thin-walled and unlignified.

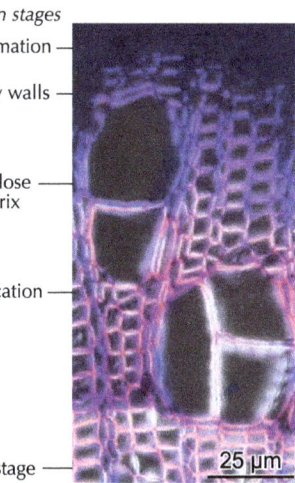

6.184 Cellulose-matrix formation is expressed by reflection in polarized light. Cellulose formation starts immediately after cell expansion is completed.

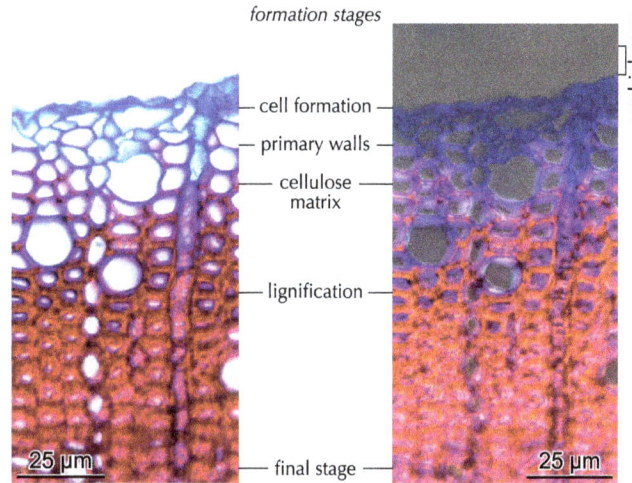

6.185 Cell-wall thickening, lignification of cell walls of vessels, fibers and rays in *Buxus sempervirens*. All cells near the cambium are thin-walled and unlignified. The thickening and lignification process occurs within 8–10 cell rows.

6.186 Cellulose-matrix formation is expressed by reflection in polarized light. Cellulose formation starts immediately after cell expansion is completed.

Conifers   Continuous processes   Discontinuous processes

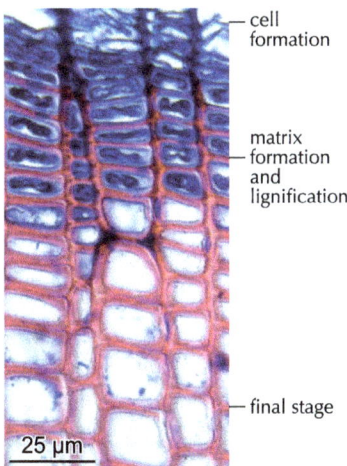

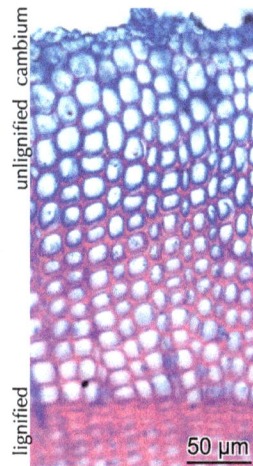

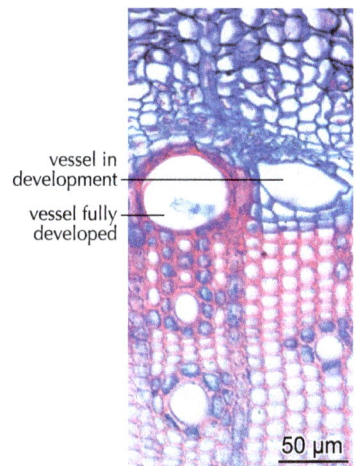

6.187 Cells with protoplasts continuously produce cellulosic matrix and lignin. This process occurs during several months in *Picea abies*.

6.188 Continuous cell-wall thickening and lignification in the conifer *Abies alba*.

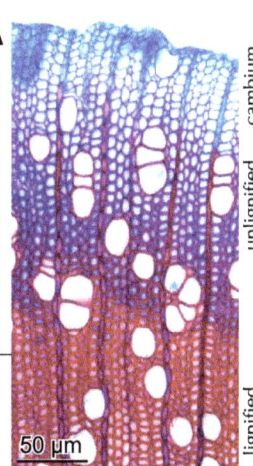

6.189 Continuous cell-wall thickening and lignification in the diffuse-porous angiosperm *Prunus padus*.

6.190 Cell formation and lignification occurs at different times in *Fraxinus excelsior*.

6.2.5 Timing of xylem formation

Ring formation in plants of seasonal climates is principally divided into a dormant and an active phase. Cell division by the cambium and cell growth (enlargement, lignification) are part of the active phase. The beginning of cambial activity is indicated by a large, anatomically undifferentiated cambial zone, while this zone is much smaller during dormancy. Genetic factors dictate the rhythm of cell-type formation (e.g. into fibers or vessels), and environmental factors modify the general principle and regulate the quantity and the size of cells.

The duration and occurrence of a ring-formation period varies*. It depends on:

○ **Taxonomy**. For example, in 2001, the cambial activity of *Prunus padus* trees in the lowland of temperate zones began in week 13 (late March), while that of *Juglans regia* trees began in week 23 (early June; Schweingruber & Poschlod 2005). Species-specific differences in timing in herbs can be much larger, e.g. *Erophila verna*, an annual small herb, fulfills its stem-formation cycle within three weeks in early March, and the small *Euphrasia cuspidata* within three weeks in August.

○ **Climate conditions**. Xylem formation in the arctic lasts for one to two months, in the temperate lowland four to five months, and in the tropical rain forests there often is no dormant period.

○ **Site conditions**. For example, the cambial activity of tall trees in the lowland of temperate zones begins mid-April and that of suppressed small individuals in June. Ring formation on south-facing slopes in the arctic starts in early July and in snow beds in early August.

Cambium width

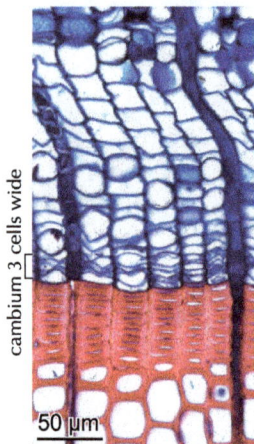

cambium 3 cells wide

50 µm

6.191 Small cambial zone during the dormant period in *Larix decidua*.

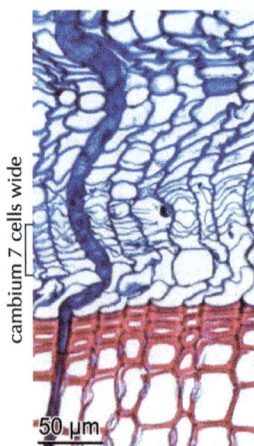

cambium 7 cells wide

50 µm

6.196 Large cambial zone in the active period in the earlywood of *Larix decidua*.

Modification of ring formation due to

taxonomy

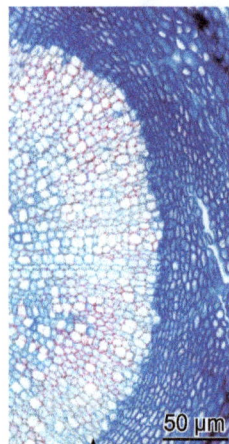

50 µm

↑ 1st March

6.192 Ring formation is completed within three weeks in February in the annual herb *Erophila verna*.

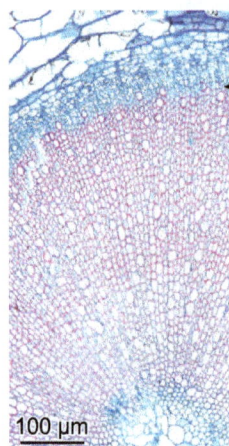

1st September

100 µm

6.197 Ring formation is completed within four weeks in August in the annual herb *Euphrasia cuspidata*.

climate conditions

1st October

100 µm   1st March

6.193 Ring formation occurs in temperate climates during four to five months from April to September. *Acer campestre*.

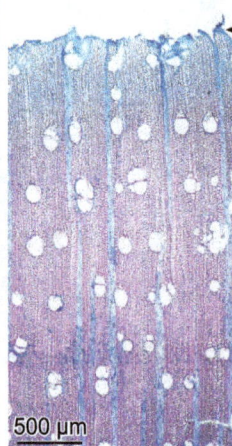

500 µm

6.198 Ring in formation in the tree *Schefflera abyssinica* in a tropical rain forest. Ring formation lasts 11–12 months.

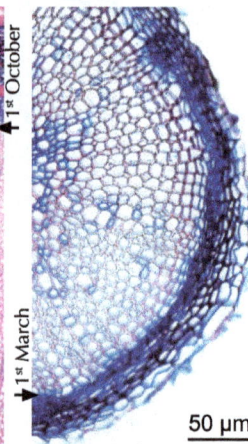

50 µm

6.194 Almost completed ring in the herb *Thlaspi perfoliatum* at the end of February in the Mediterranean zone of Cyprus.

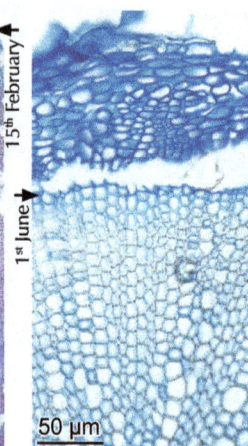

15th February

1st June

50 µm

6.199 Incomplete ring in the herb *Thlaspi perfoliatum* at the end of April in the temperate zone of Switzerland.

site conditions

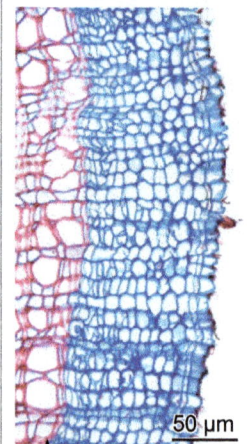

50 µm

↑ 1st August

6.195 Completed last ring in *Arctostaphylos alpina* at the beginning of August on a sunny slope in Greenland.

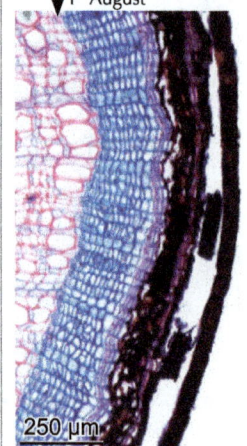

↓ 1st August

250 µm

6.200 Incomplete last ring in *Dryas octopetala* at the beginning of August in a snow bed in Greenland.

6.2.6 Cell differentiation in the *phloem* – The multifunctional stem periphery

Most cell formation processes described for the xylem also occur in the phloem: cell type differentiation, cell-wall differentiation, nucleus differentiation, cell-wall enlargement, cell-wall thickening, lignification and formation of crystals are the basic steps.

The following aspects are different than in the xylem (Huber 1961):
◦ Phloem mother cells normally produce less new cells than the xylem mother cells.
◦ Annual rings are mostly less distinct and smaller in the phloem than in the xylem.
◦ Vessels are replaced by sieve elements and companion cells.
◦ Spatial restrictions often lead to more intensive cell-wall enlargement and lateral cell divisions in rays (dilatation).
◦ Bordered pits in tracheids or vessels are replaced in the phloem by lateral sieve areas; perforation plates in vessels are replaced by axial sieve plates.
◦ The xylem normally forms a dense block of tissue onto which the phloem gets pushed, which results in collapsed sieve elements. As soon as sieve tubes die they collapse due to the higher turgor of neighboring parenchyma cells, the pressure from newly formed cells and/or the strength of the phellem belt. The processes primarily take place in the juvenile stage between the cortex and the xylem and in adult stages between the xylem and the rhytidome (isolated dead tissues formed by the phellogen). (Holdheide 1951)

Shown below are the principal changes during the thickening and aging process for a conifer, and a diffuse-porous and a ring-porous angiosperm.

Proportion of xylem and phloem

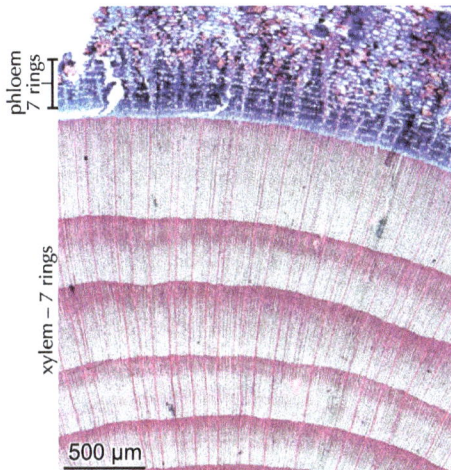

6.201 Comparison of xylem and phloem rings in *Abies alba*. Xylem/phloem = 7:1, xylem rings distinct, phloem rings only distinct in first few years.

Differentiation during stem thickening and aging in conifers

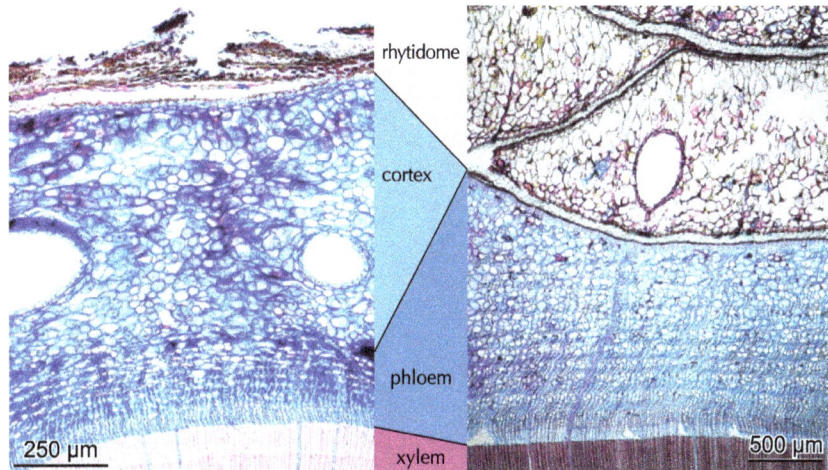

6.202 Juvenile bark of the conifer *Pinus sylvestris*. Characteristic are the small phloem, a large cortex and a small rhytidome.

6.203 Adult bark of the conifer *Pinus sylvestris*. Characteristic are the large phloem, an absent cortex and a large rhytidome.

Differentiation during stem thickening and aging in angiosperms

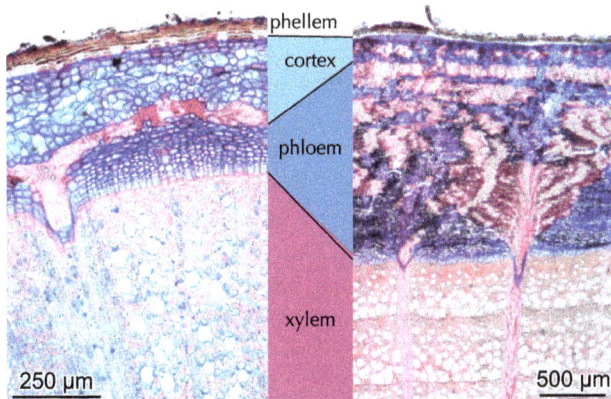

6.204 Juvenile bark of the deciduous angiosperm *Fagus sylvatica*. Characteristic are a small phloem, a large cortex containing a continuous fiber/sclereid belt and a small periderm.

6.205 Adult bark of *Fagus sylvatica*. Characteristic are a large phloem with sclereid groups, a very small cortex with remnants of the juvenile fiber/sclereid belt and a small periderm; rhytidome is absent.

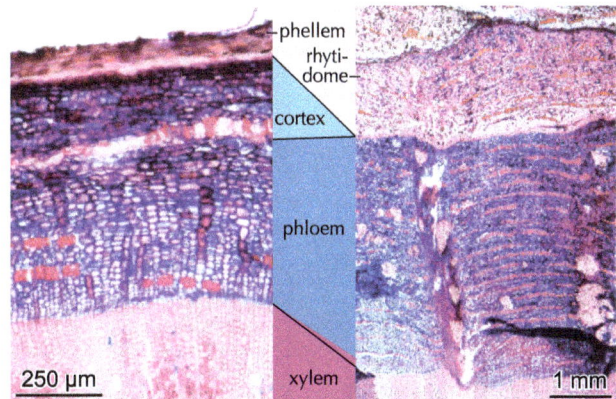

6.206 Juvenile bark of the deciduous angiosperm *Quercus robur*. Characteristic are a large phloem containing groups of fibers, a large cortex with a continuous fiber/sclereid belt and a small periderm; rhytidome is absent.

6.207 Adult bark of *Quercus robur*. Characteristic are a large phloem, consisting of many bands of groups fibers and a few groups of sclereids, and a rhytidome; cortex is absent.

64

Plant Stem Anatomy: An Illustrated Atlas

Changing formation mode

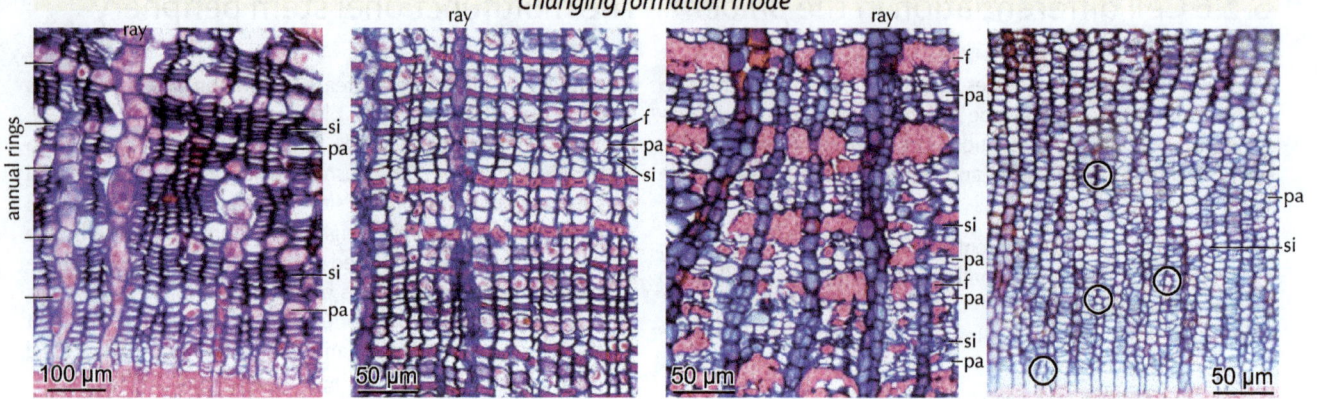

6.208 Annual rhythms in *Abies alba* are indicated by a tangential row of early-bark parenchyma cells and several rows of late-bark sieve cells.

6.209 Regular rhythms in *Juniperus communis* are indicated by thin-walled tangential rows of sieve cells, parenchyma cells (with nuclei) and thick-walled fibers. Annual growth rates are indistinct.

6.210 Rhythms are indicated in *Sorbus chamaemespilus* by poorly differentiated zones of sieve tubes and parenchyma cells and distinct groups of fibers. Fibers probably develop in the second year.

6.211 Arrhythmic formation of sieve tubes, companion cells and parenchyma cells in *Buxus sempervirens*. Radial rows are not permanent due to aperiodic lateral cell divisions and cell death (circles).

Different priorities

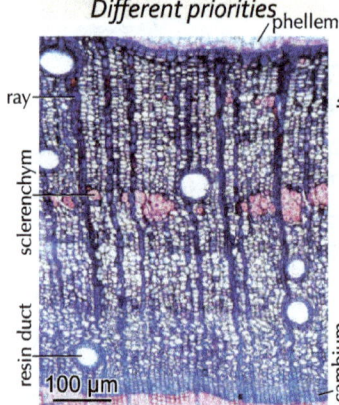

Cell enlargement, tissue and ray dilatation

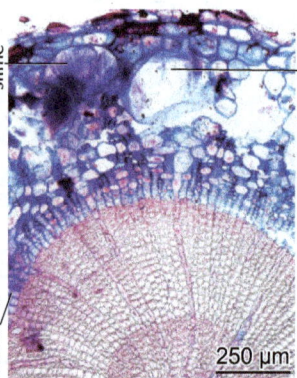

6.212 The general pattern in *Cotinus coggygria* changes periodically when the cambial mode changes to the production of resin ducts. In later stages, living parenchyma cells produce thick secondary walls (sclereids).

6.213 Extensive cell enlargements of parenchyma cells in the cortex of *Abies alba*. The enlarged parenchyma cells produce slime.

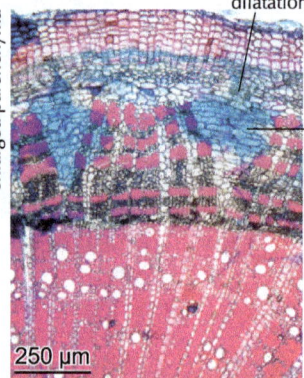

6.214 Lateral cell divisions and cell-wall expansion increase the circumference of the stem wedge-like in *Lavatera acerifolia*.

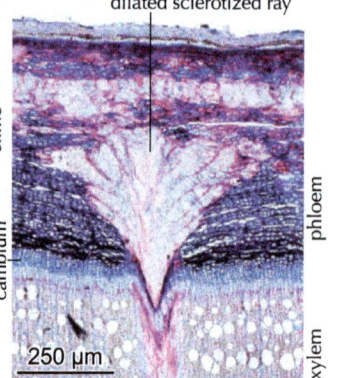

6.215 Ray dilatation and sclerotization in *Fagus sylvatica*.

Cell-wall differentiation and pit structure

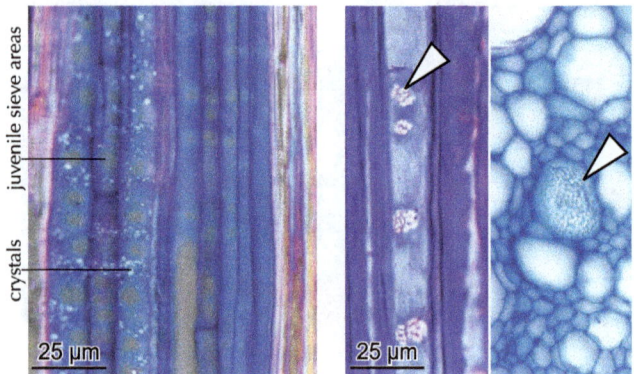

Collapse of sieve tubes

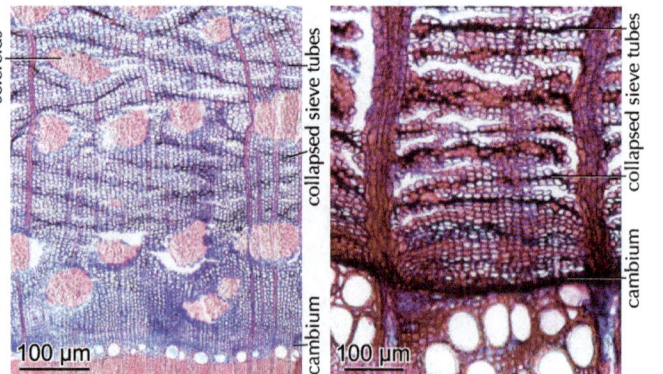

6.216 Sieve tubes in the cambial zone of *Metasequoia glyptostroboides*. Calcium oxalate crystals seem to play a physiological role in the formation of the primary wall.

6.217 Left: Adult sieve plates on radial walls in *Larix decidua*.
Right: Sieve plate on the axial end of a sieve tube in *Nelumbo nucifera*.

6.218 Irregular pattern of collapsed sieve tubes in *Hippophae rhamnoides*. Sieve tubes in the cambial zone are not collapsed.

6.219 Tangential lines of collapsed sieve tubes in *Laburnum anagyroides*.

6.2.7 Formation of tertiary meristems, the cork cambium – A new skin

Tertiary meristems determine the face of tree stems because the formation mechanisms are species-specific.

Periderms in the bark

Most plants with secondary growth form a tertiary meristem, which is located somewhere in the bark: the **phellogen**. Toward the inside, the phellogen produces a few long-lived parenchyma cells, the **phelloderm**, and towards the outside, it produces various amounts of short-lived cork cells, the **phellem**. Their walls consist of cutin or suberin. Their origin are living parenchymatic cells. In young shoots, parenchyma cells of the cortex, and in older shoots, parts of the phloem get reactivated to meristems. The number of formed cells is normally much bigger towards the outside than towards the inside. The zone of phellogen, phelloderm and phellem is called **periderm**. All dead phloem and cortex parts outside of the phellogen are called **rhytidome**. This formation mode occurs in all growth forms of conifers and dicotyledons.

With continuous stem thickening and the associated tension, the external phellogen and adjacent phloem and cortex parts die and normally flake off. Rhytidomes are species-specific. Therefore tree species can be identified macroscopically by their bark: the face of the tree. Godet 2011 presents the bark of central European tree species.

Cork formation is essential for most perennial terrestrial plants because cork layers build a continuous mantle around the plant. It protects the plant lifelong against mechanical and biological damages.

Morphology of the bark

6.220 Juvenile shoots and adult bark in *Prosopis* sp.

6.221 The phellem in *Acer griseum* thinly flakes off.

Size of the cork mantle

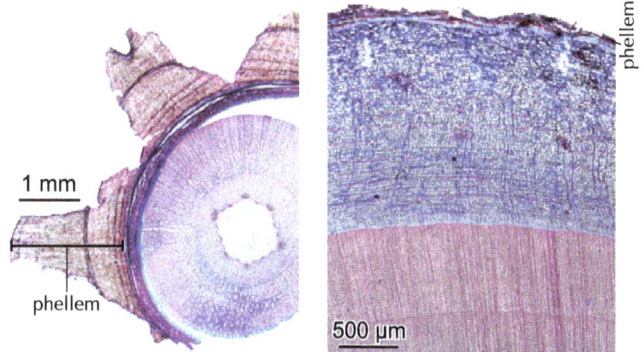

6.222 Large phellem in *Acer campestre*.

6.223 Small phellem in *Taxus baccata*.

Definition of the formation zone

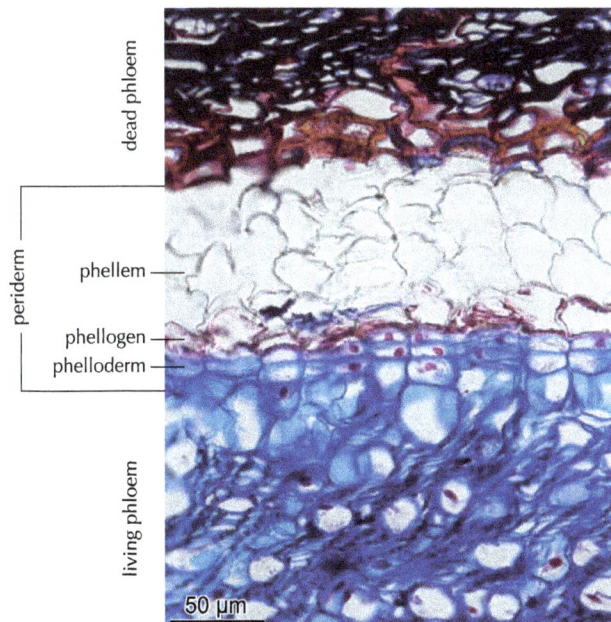

6.224 Bark of *Pinus mugo*.

Periderms formed in the cortex

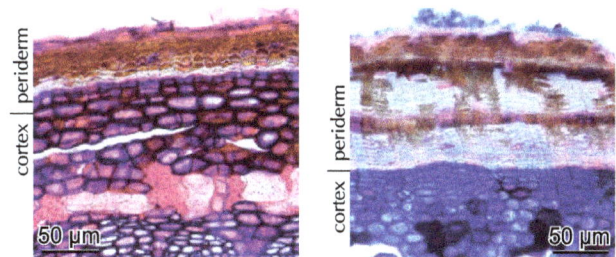

6.225 Bark of *Carpinus betulus*.

6.226 Bark of *Betula pendula*.

Periderms formed in the phloem

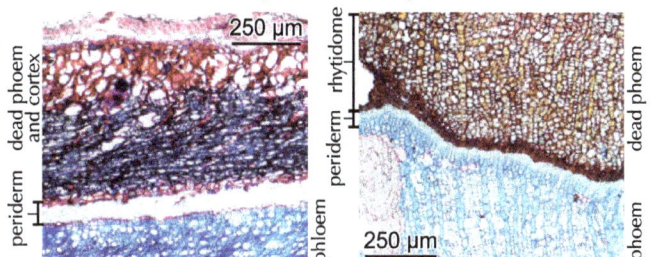

6.227 Bark of *Pinus mugo*.

6.228 Bark of *Alnus glutinosa*.

Periderms form lenticels

Phellem layers seal the stem. The phellogen locally creates perforations in the young twigs by accelerated cork-cell production: the lenticels. Lenticels occur on young twigs and especially on roots in wet environments. The phellogen locally forms an external tissue with numerous intercellulars, which permit the entrance of air to the cortex.

Lenticels as air portals

6.229 Twig with flowers of *Forsythia suspensa*.

6.230 Annual twig of *Acer pseudoplatanus* with lenticels.

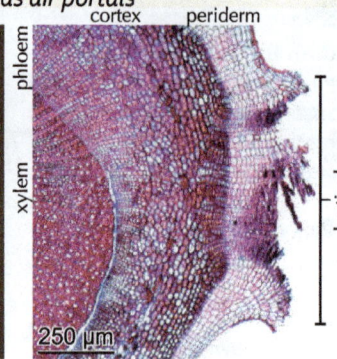

6.231 Lenticel in a twig of *Forsythia suspensa*.

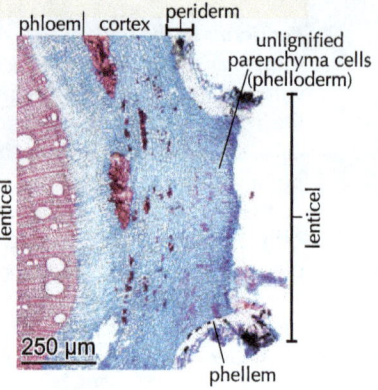

6.232 Lenticel in a root of *Alnus glutinosa*.

Periderms in stem centers

A special form of cork formation can occur, mainly in small, long-lived plants of certain families (e.g. Lamiaceae, Rosaceae, Fabaceae or Aceraceae) at high altitudes and northern latitudes. As soon as a plant is unable to maintain the metabolism of the stem as a whole, living parenchyma cells in the xylem get reactivated to form a cork cambium, the products of which separate part of the living tissue towards the inside. This process occurs repeatedly and forms an internal rhytidome inside the stem.

Periderms in stem centers as protection layers

6.233 *Potentilla nitida*, Rosaceae, a 5 cm-tall alpine plant with a long-lived rhizome.

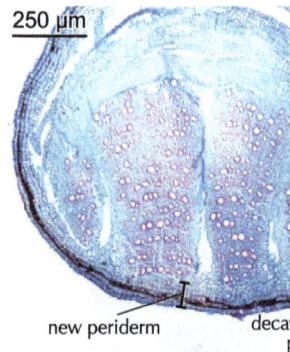

6.234 Rhizome of *Potentilla nitida* with a re-shaped, round stem. The original central part disappeared and the wood was sealed by a periderm.

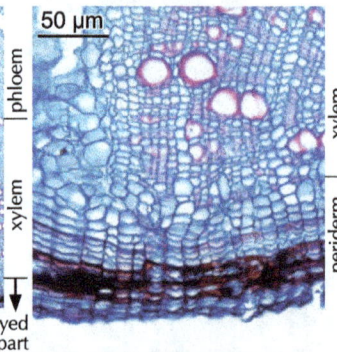

6.235 The new periderm in *Potentilla nitida* bridged vessel/parenchyma parts and enlarged rays.

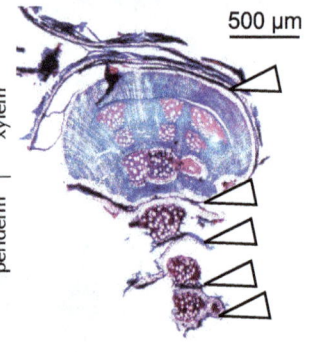

6.236 Rhizome of the alpine herb *Nepeta discolor*, Lamiaceae, with one active and four inactive central periderms.

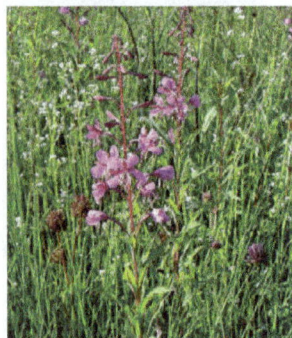

6.237 *Epilobium angustifolium*, Onagraceae, with a long-lived rhizome.

6.238 Rhizome of *Epilobium angustifolium* with a small living part and many central dead periderms.

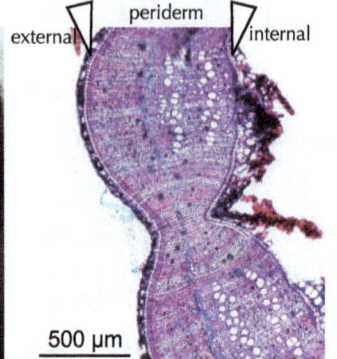

6.239 Living part of the rhizome of *Epilobium angustifolium*, with a central periderm.

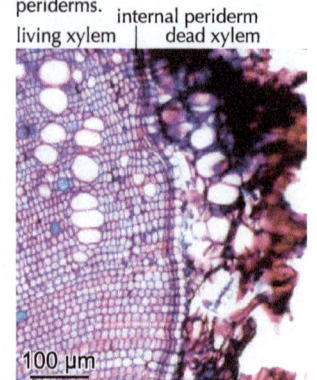

6.240 The central periderm of *Epilobium angustifolium*.

Periderms as protection layers

6.241 *Saussurea gnaphalodes*, Asteraceae, a 5 cm-tall alpine plant with a long-lived rhizome.

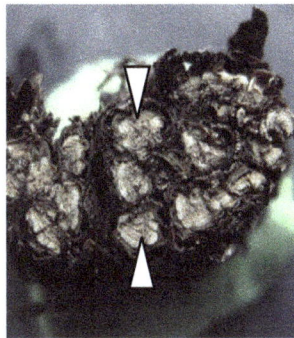

6.242 Rhizome of *Saussurea glanduligera*, composed of many separated partial rhizomes.

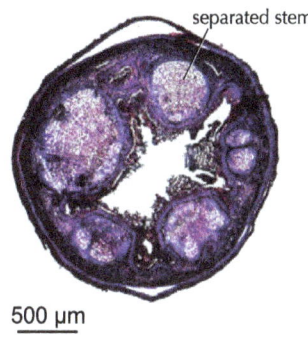

6.243 Stem separation by central periderms in *Potentilla crantzii*, Rosaceae.

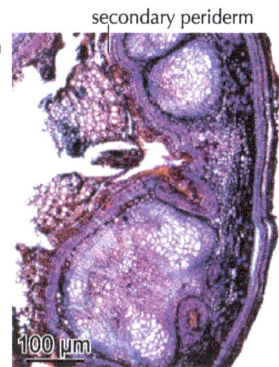

6.244 Stem compartments separated by secondary periderms in *Potentilla crantzii*.

Periderms as breaking zones for leaves

Just as important are the accelerated cork-formations at the break-off zones of leaves. Long before the leaves drop, the phellogen becomes active and forms a layer of phellem cells. As soon as the leaves drop, the potential wound is already sealed.

Breaking zones for leaves

6.245 Leaf scar below a bud of *Acer pseudoplatanus*.

6.246 Leaf scars in a cabbage stem, *Brassica oleracea*.

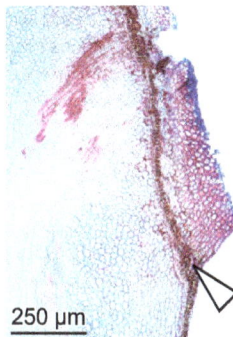

6.247 Leaf of *Castanea sativa*, separated by a periderm.

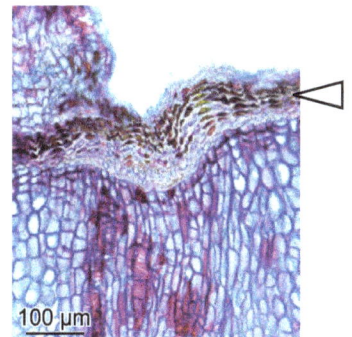

6.248 Periderm on a leaf scar in *Castanea sativa*.

Periderms as breaking zones for spines

Spines are products of local periderms. In contrast to all other periderms they differentiate into special forms. Spines occur on stems, e.g. of *Bombax ceiba* (cotton tree), roses and others, and on fruits, e.g. of *Aesculus hippocastanum* (horse chestnut). The spine itself is a product of the hypodermis (cells just below the epidermis) and the breaking zone is the outermost part of the periderm, the phellem.

Breaking zones for spines

6.249 Spines on the stem of *Bombax ceiba*, Malvaceae.

6.250 Spines on a twig of *Rosa arvensis*, Rosaceae.

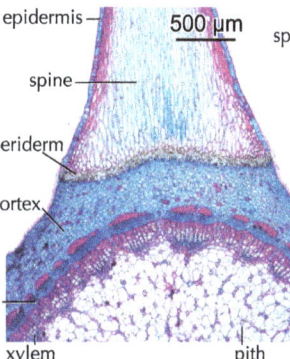

6.251 Spine on a twig of *Rosa arvensis*.

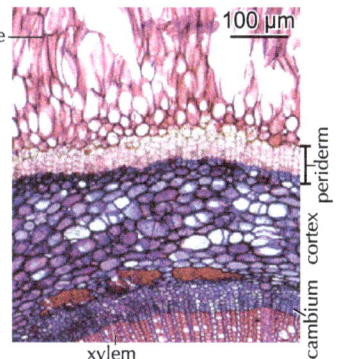

6.252 Breaking zone of a spine of *Rosa arvensis*.

6.2.8 Life span and death of cells – Cells must die

In the following, the **programmed cell death** or **apoptosis** is anatomically described. **Genetically predetermined cell death** is behind all phenomena in which plants shed leafs, twigs or fruits. These processes have partially been described in Chapter 6.2.7 "Tertiary meristems". Death is part of any living organism. The physiologically driven dying process is called apoptosis or programmed cell death. The live span of the entire plant body is also genetically predetermined.

Aging processes, called **senescence**, lead to the death of central parts of the stem (heartwood formation) or of entire plant bodies. Annual plants sometimes survive for only a few weeks, while perennials live for up to 5,000 years.

Externally induced cell death is behind all phenomena in which pathological factors or extreme ecological conditions determine cell death. This dying process is called necrosis or necrobiosis and is described in Chapter 10.6.

Programmed cell death within living parts of plants

A healthy, functioning plant body is based on a perfectly designed balance of living and dead cells. Genetically induced programs activate enzymes (caspases), which determine the longevity of cells. Meristematic cells of clonal plants theoretically can live forever. The life span of their derivates varies within a time range of a few days up to more than 100 years; in reality only parenchymatic cells have such a long life span. Xylem and phloem mother cells, conducting tissues (tracheids, vessels, cork cells and sieve elements) and sclereids have a short live span. Illustrated in the following is the age mosaic of juvenile and adult tissues in conifers and deciduous angiosperms.

Programmed cell death separates living and dead parts – Heartwood formation

The macroscopic characteristics of heartwood are described in Chapter 6.2.1, and of heartwood substances in Chapter 5.6.5 (Fig. 5.99–5.118). For deeper insight into heartwood formation processes see Fromm 2013.

Longevity (days, months and years) of different cells in the xylem, phloem, cortex and periderm of conifers

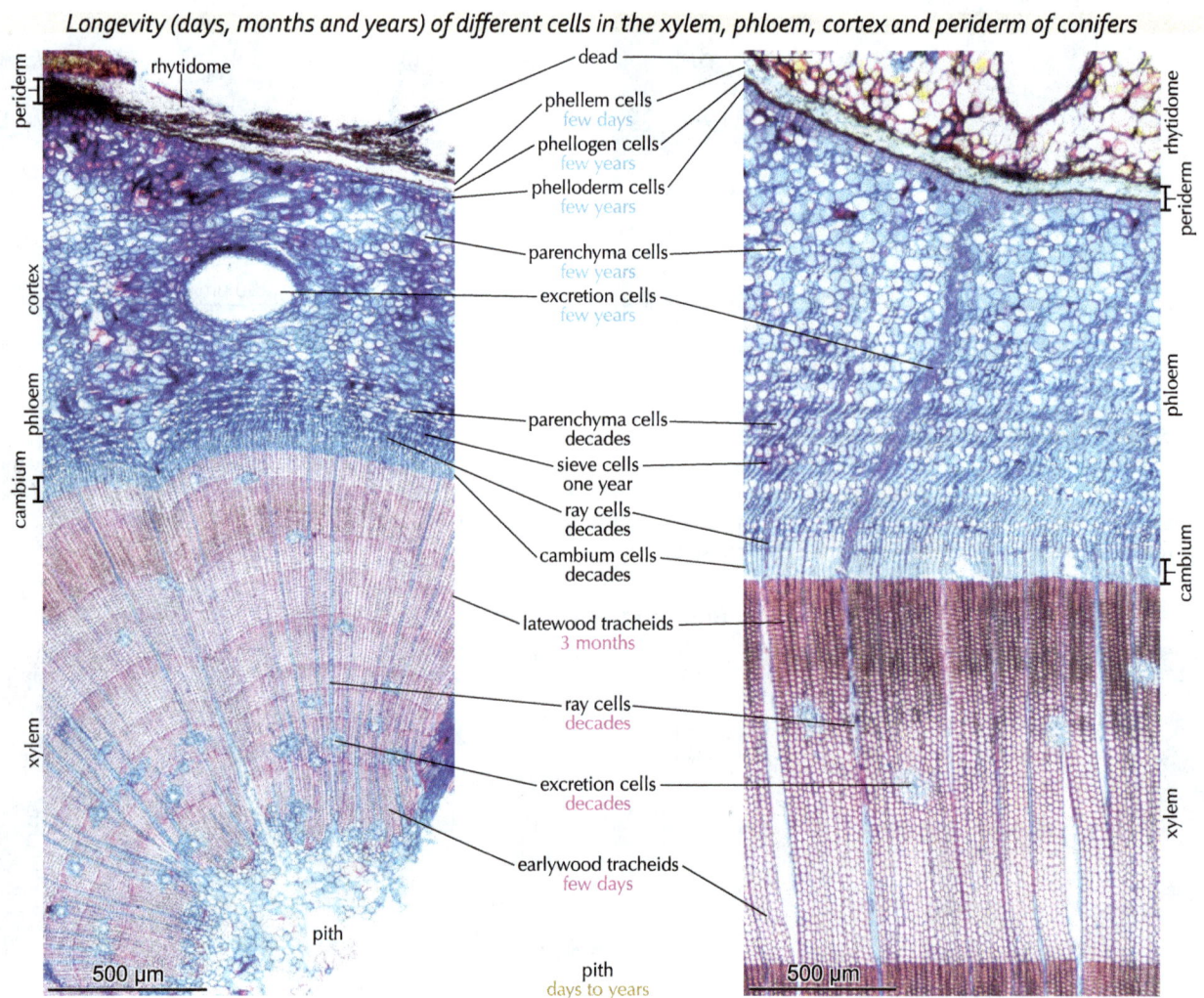

6.253 Juvenile tissues in a nine-year-old twig of *Pinus mugo*.

6.254 Adult tissue in a 60-year-old stem of *Pinus sylvestris*.

Longevity (days, months and years) of different cells in the xylem, phloem, cortex and periderm of dicotyledons

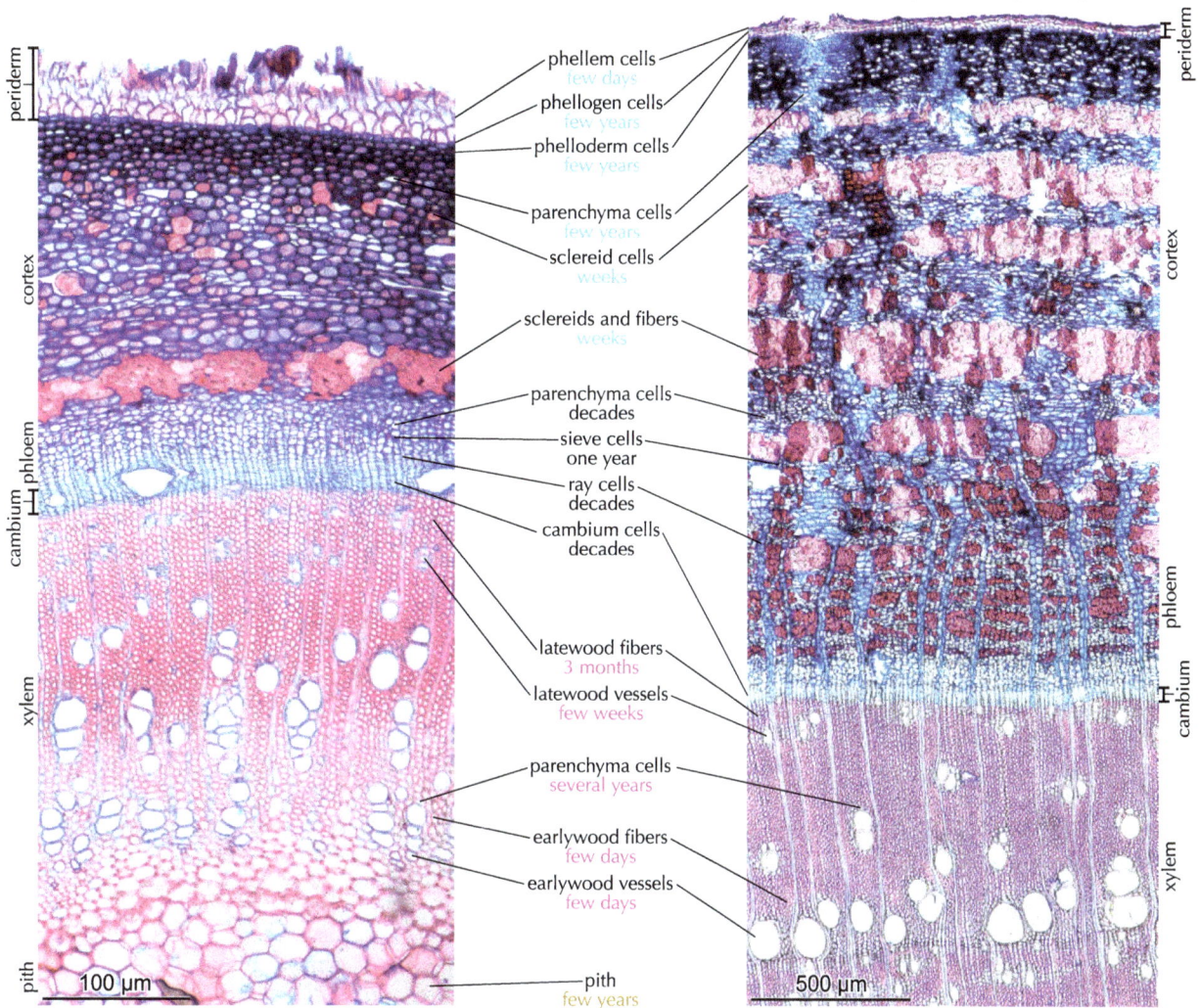

periderm

cortex

phloem

cambium

xylem

pith

phellem cells
few days

phellogen cells
few years

phelloderm cells
few years

parenchyma cells
few years

sclereid cells
weeks

sclereids and fibers
weeks

parenchyma cells
decades

sieve cells
one year

ray cells
decades

cambium cells
decades

latewood fibers
3 months

latewood vessels
few weeks

parenchyma cells
several years

earlywood fibers
few days

earlywood vessels
few days

pith
few years

periderm

cortex

phloem

cambium

xylem

100 µm

500 µm

6.255 Juvenile tissue in a one-year-old twig of *Fraxinus excelsior*.

6.256 Adult tissue in a 30-year-old stem of *Fraxinus ornus*.

Longevity of meristematic cells in the cambial zone

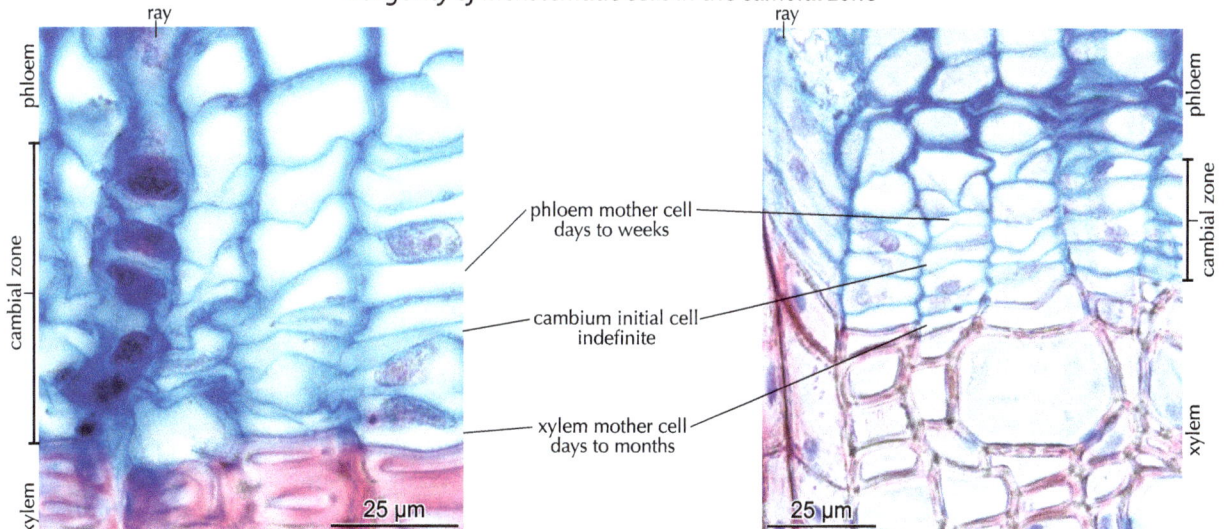

ray

phloem

cambial zone

xylem

ray

phloem mother cell
days to weeks

cambium initial cell
indefinite

xylem mother cell
days to months

phloem

cambial zone

xylem

25 µm

25 µm

6.257 Conifer *Pinus sylvestris*.

6.258 Dicotyledonous *Sambucus nigra*.

6.3 Cambial variants – Phloem elements within the xylem

Within the groups of palm ferns (Cycadopsidae) and dicotyle-donous angiosperms species exist in which the cambium does not constantly produce a centripetal xylem and a centrifugal phloem. This group principally contains two formation modes which each include many different subtypes.

One cambium periodically produces centripetal bark elements
A phloem containing sieve cells, companion cells and paren-chyma cells, or cork cells. The normal formation mode (vessels, fibers, parenchyma) is expanded to elements of the phloem or the periderm.

Several circular arranged cambia simultaneously produce xylem and phloem
This group comes under the term plants with **successive cambia**. Within this unit there are principally two groups:
- The cambia periodically produce single collateral vascular bundles, which are located within a parenchymatic tissue. However, specific taxonomic groups anatomically modify these modes. Of special interest are monocotyledonous representatives (Agavaceae), which continuously produce concentric vascular bundles.
- The cambia produce tangential bands of xylem and phloem within a parenchymatic tissue.

One cambium produces fibers, vessels and, periodically, groups of sieve tubes and parenchyma cells or bands of cork

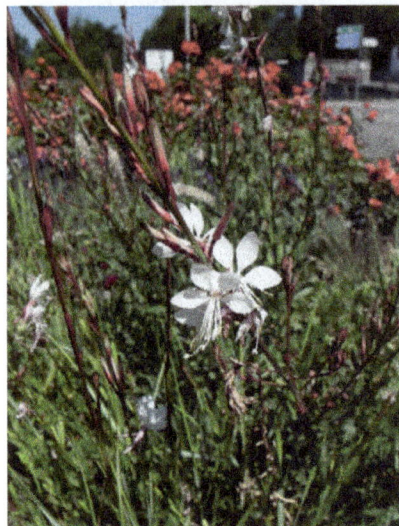

6.259 The herb *Gaura lindheimeri*, Onagraceae.

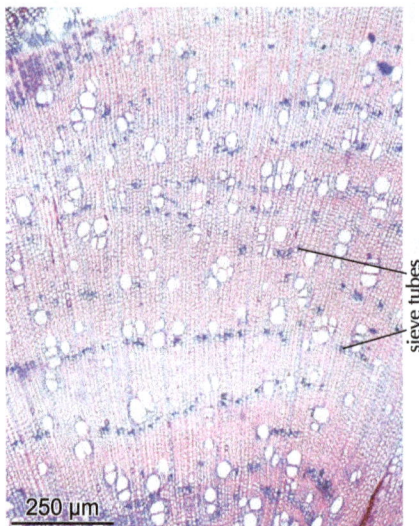

250 µm

6.260 Tangential rows of groups of sieve tubes in *Gaura lindheimeri*.

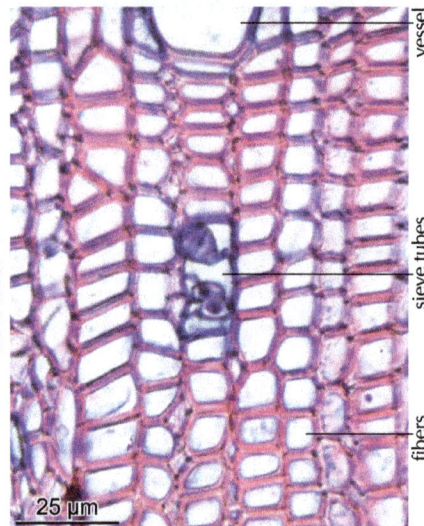

vessel
sieve tubes
fibers
25 µm

6.261 Group of sieve tubes with companion cells containing nuclei in *Gaura lindheimeri*.

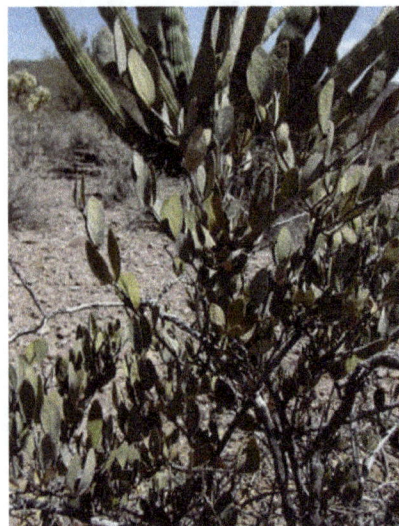

6.262 The shrub *Simmondsia chinensis*, Simmondsiaceae.

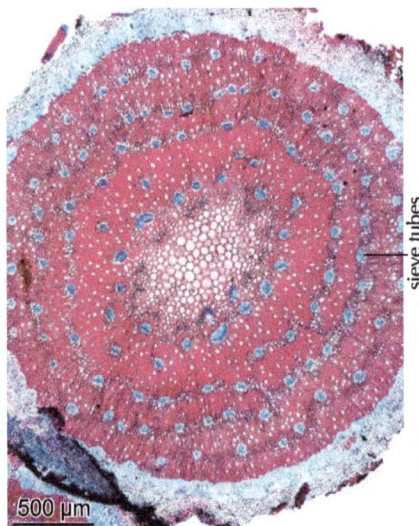

sieve tubes
500 µm

6.263 Circular rows of groups of sieve tubes in *Simmondsia chinensis*.

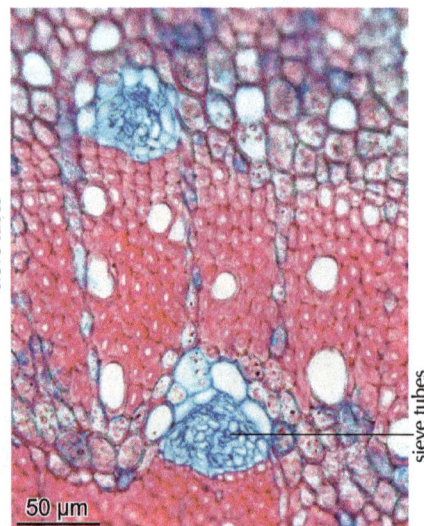

sieve tubes
50 µm

6.264 Groups of sieve tubes and parenchyma cells within a dense fiber/vessel tissue in *Simmondsia chinensis*.

One cambium periodically produces xylem and bands of cork

6.265 The dwarf shrub *Artemisia tridentata*, Asteraceae.

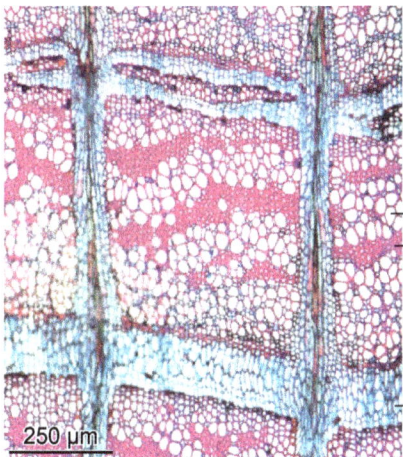

6.266 Tangential bands of cork cells between a vessel/fiber tissue in *Artemisia tridentata*.

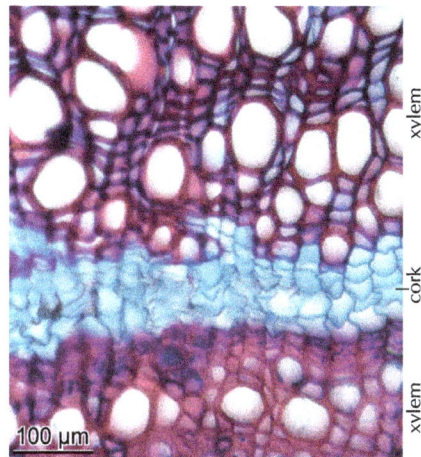

6.267 Thin-walled cork cells in the xylem of *Tanacetum millefolium*.

One cambium produces concentric vascular bundles

6.268 The monocotyledonous tree *Dracaena serrulata*, Asparagaceae.

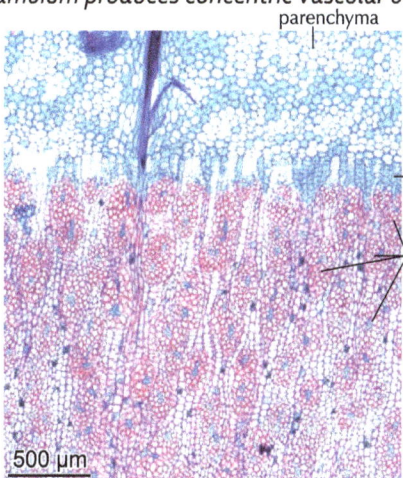

6.269 Xylem and cortex of *Dracaena serrulata*.

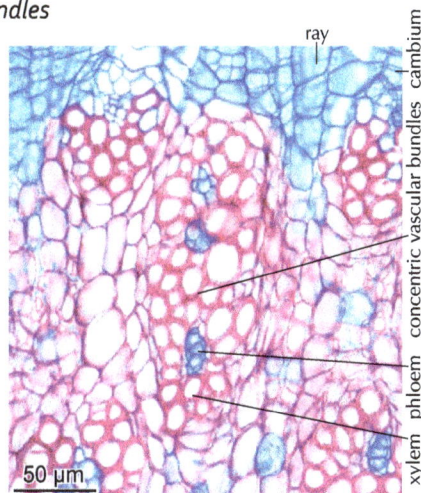

6.270 Single concentric vascular bundles between rays in *Dracaena serrulata*.

Several cambia periodically produce single collateral vascular bundles, fibers and parenchyma cells

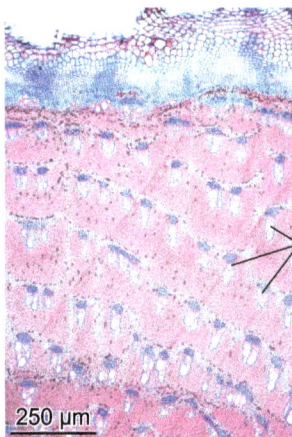

6.271 Isolated vascular bundles within a dense fiber tissue in *Bassia prostrata*, Amaranthaceae.

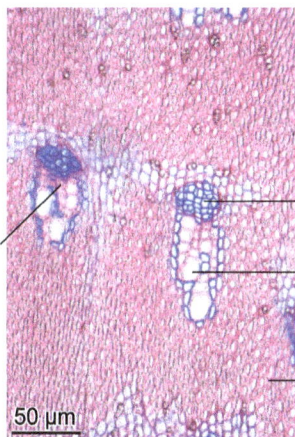

6.272 Vascular bundles within a dense fiber tissue in *Bassia prostrata*.

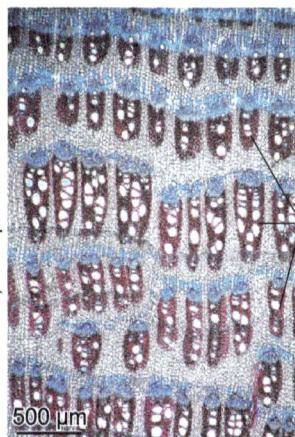

6.273 Tangentially arranged vascular bundles in *Bosea cypria*, Amaranthaceae.

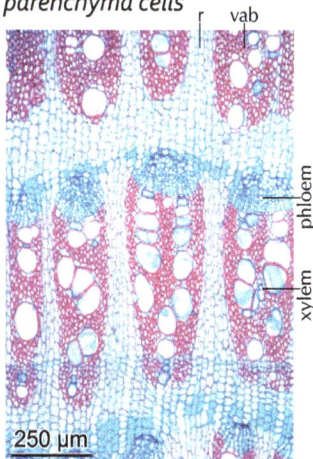

6.274 Vascular bundles between rays in *Bosea cypria*.

Several cambia periodically produce bands of xylem and phloem

6.275 *Welwitschia mirabilis*, Welwitschiaceae. Photo: P. Poschlod.

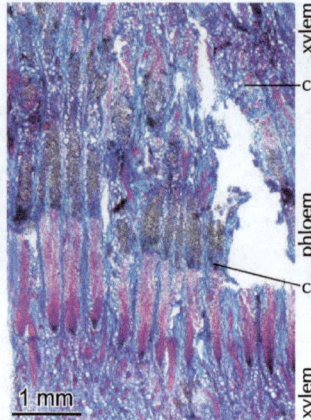

6.276 Two tangential rows of vascular bundles in *Welwitschia mirabilis*.

6.277 *Macrozamia moorei*, Cycadaceae.

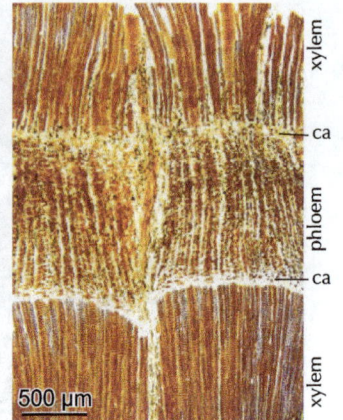

6.278 A band of phloem between two bands of xylems in *Macrozamia moorei*.

Several cambia periodically produce bands of xylem and phloem

Several cambia produce irregular bands of xylem and phloem

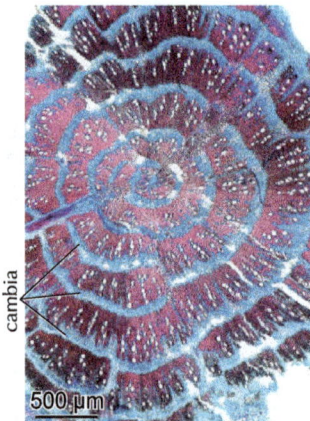

6.279 Several concentric rows of cambia produce bands of xylem and phloem in the herb *Atriplex prostrata*, Amaranthaceae.

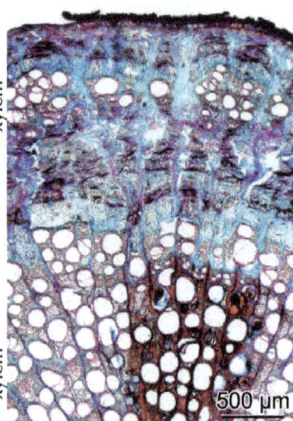

6.280 Two cambia produce xylem/phloem belts in the subtropical liana *Pueraria hirsuta*, Fabaceae.

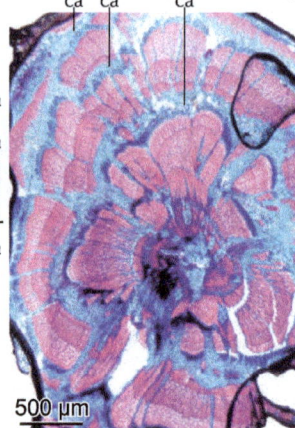

6.281 Irregular bands of internal cambia in the herb *Polycarpaea divaricata*, Caryophyllaceae.

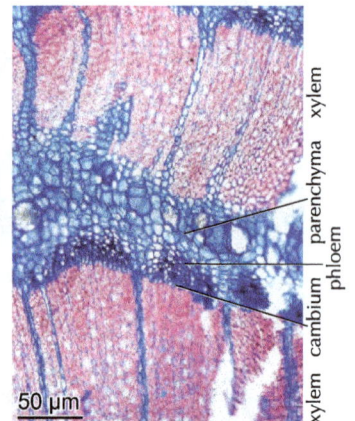

6.282 The zone between xylem belts in *Polycarpaea nivea* consists of a cambium, a phloem and an unlignified parenchymatic belt.

Several cambia periodically produce irregular bands of xylem and phloem

6.283 Alpine cushion plant *Silene acaulis*, Caryophyllaceae.

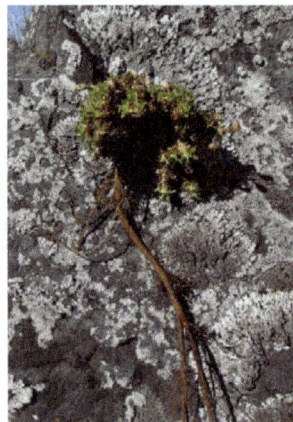

6.284 Cushion of *Silene acaulis* with taproot.

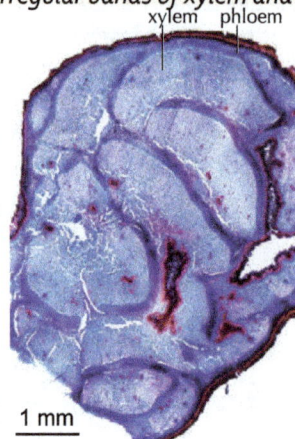

6.285 Irregular bands of internal cambia in *Silene acaulis*.

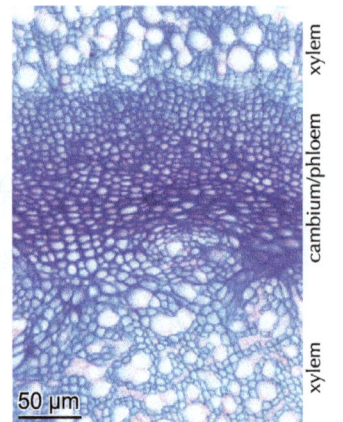

6.286 The zone between the two xylems in *Silene acaulis* consists of an anatomically undifferentiated cambium-phloem belt.

6.4 Intercalary meristems – Longitudinal growth far behind the tips in shoots and roots

Intercalary meristems are a special form of meristems. Intercalary meristems occur in grasses above nodes between leaf initials, in nodes on horsetails, and in root collars of mistletoes. Theses meristems originally are a product of apical meristems, which retain their meristematic activity far behind the inactivated apical meristems. Their activity is obvious in the elongation phase of the culm of grasses and horsetails. A long time after the formation of the flowers and the inactivation of apical meristems the culms are getting longer and longer due to the activity of intercalary meristems. In grass species with several nodes, multiple intercalary meristems are active until the culm reaches its final length.

Mistletoes can only survive if the elongation of root collars follows the thickening of radial growth of the host. As soon as a haustorium touches the cambium of the host xylem, cells incorporate the foreign body. The mistletoe's strategy is to avoid isolation by forming new tissues in between its shoot and root. The original place of haustoria attachment remains, but the root elongates and enlarges in the cambial zone of the host.

Macroscopic aspect of intercalary meristems

Horsetails

6.287 Initial phase of stem elongation in a fertile shoot of *Equisetum telmateia*.

Monocotyledons

6.288 Node of a Poaceae culm. The intercalary meristem is located above the node.

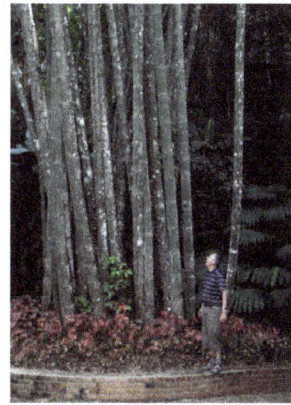

6.289 Adult phase of stem elongation in shoots of a giant bamboo.

Mistletoe

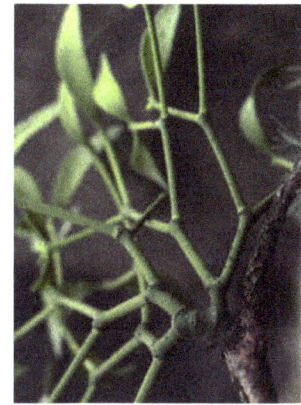

6.290 Mistletoe *Viscum album* on a branch of *Pinus sylvestris*.

Microscopic aspect of intercalary meristems

Horsetails

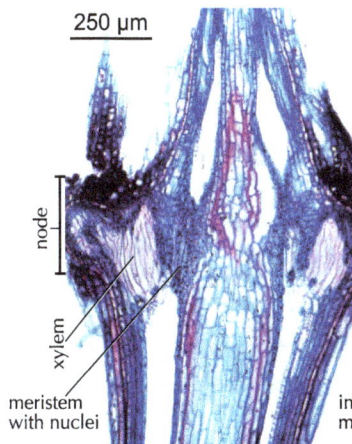

250 µm

node

xylem

meristem with nuclei

6.291 Internal structure of a node in *Equisetum arvense*.

Monocotyledons

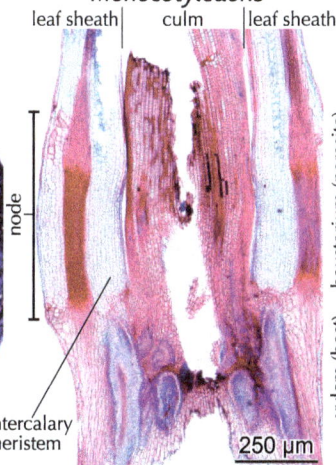

leaf sheath | culm | leaf sheath

node

intercalary meristem

250 µm

6.292 Elongation zone with unlignified meristematic cells in leaf sheaths of the grass *Milium effusum*.

Mistletoes

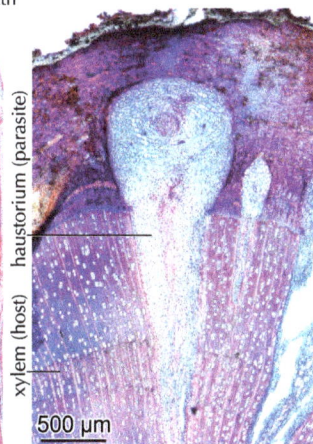

haustorium (parasite)

xylem (host)

500 µm

6.293 Mistletoe haustoria of *Viscum album* in the xylem and phloem of an apple tree (*Malus domestica*).

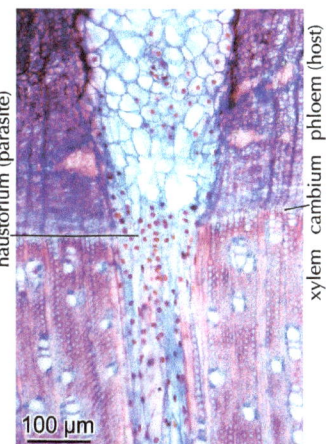

haustorium (parasite)

xylem cambium phloem (host)

100 µm

6.294 Concentration of nuclei in the thin-walled haustorium of the parasite in the cambial zone of the host.

7

Anatomy of Various Species

This chapter describes the anatomy of major stem-forming taxa within the taxonomic hierarchic system. This system has seven principal categories with specific endings on Latin names:

Domain		Eukarya
Kingdom		e.g. Plantae, Fungi
Division	–phyta (plants), –phycota (algae), –mycota (fungi)	e.g. Basidiomycota
Subdivision	–phytina (plants), –phycotina (algae), –mycotina (fungi)	e.g. Lycopodiophytina
Class	–opsida (plants), –phyceae (algae), –mycetes (fungi)	e.g. Bryopsida, Coniferopsida
Order	–ales	e.g. Magnoliales
Family	–aceae	e.g. Rosaceae
Genus		e.g. *Prunus*
Species		e.g. *avium*

Described in this chapter are species of the following groups:

- Basidiomycota (fungi)
- Phaeophyceae (brown algae)
- Lichenes (lichens)
- Bryopsida (mosses) and Sphagnopsida (*Sphagnum* moss)
- Psilophytina (whisk ferns)
- Lycopodiophytina (several species of lycopods)
- Equisetophytina (several species of horsetails)
- Filicophytina (several species of ferns)

- Spermatophyta (many species of seed plants):
  Cycadopsida (some species of palm ferns)
  Ginkgopsida (*Ginkgo biloba*)
  Coniferopsida including Gnetales (some species of conifers, *Ephedra*, *Gnetum*, *Welwitschia*)
  Angiosperms (many species of flowering plants) – monocotyledons and dicotyledons (old term)

Fungi	*Brown algae*	*Lichens*	*Mosses*

   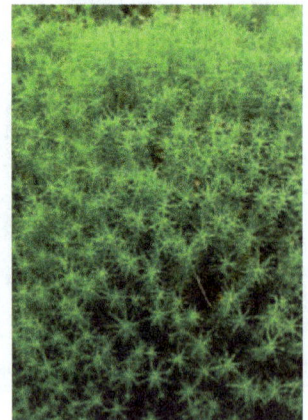

7.1 Division: Basidiomycota (*Boletus edulis*)

7.2 Class: Phaeophyceae

7.3 Lichenes

7.4 Class: Bryopsida

Whisk ferns

7.5 Subdivision: Psilophytina

Lycopods

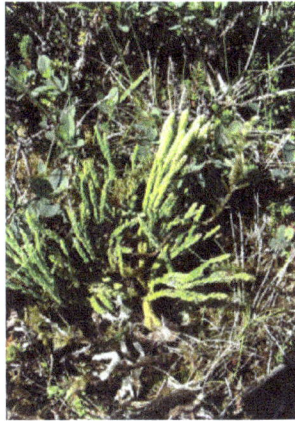

7.6 Subdivision: Lycopodiophytina

Horsetails

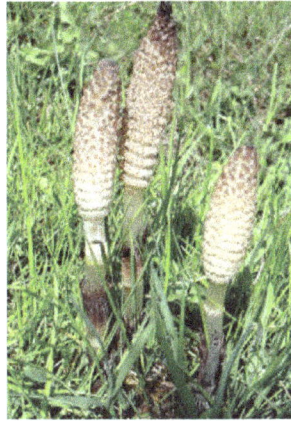

7.7 Subdivision: Equisetophytina

Ferns

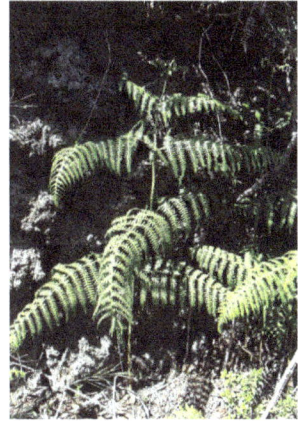

7.8 Subdivision: Filicophytina

Seed plants

Palm ferns

7.9 Class: Cycadopsida

Ginkgo

7.10 Class: Ginkgopsida

Conifers

7.11 Class: Coniferopsida

Ephedra

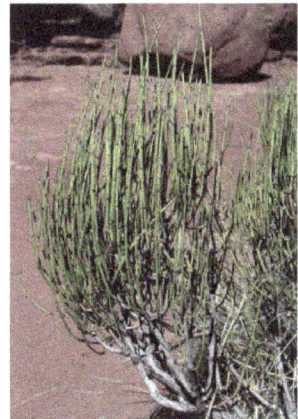

7.12 Order: Gnetales

Monocotyledons

7.13 Angiosperms

7.14 Angiosperms

Dicotyledons

7.15 Angiosperms

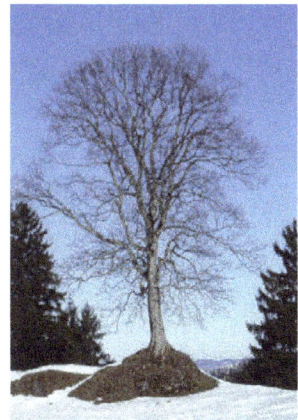

7.16 Angiosperms

7.1 Stem-forming fungi and algae

7.1.1 Sporophytes of fungi

Described are microscopic structures of a few larger stems of fruiting bodies of Eubasidiomycetes. One cell type, the hyphae, form mushroom stems. The images demonstrate that the anatomical characteristics of hyphae create a great morphological diversity. Distribution, variation of density, orientation, diameter, wall thickness, chemical composition and different cell types determine the aspect and the construction of mushroom stems.

Macroscopic aspect of fungal fruiting bodies with stems

7.17 A 5 cm-tall *Polyporus brumalis* mushroom on a branch.

7.18 A 12 cm-tall *Ganoderma carnosum* mushroom, which grows on dead wood of *Abies alba*.

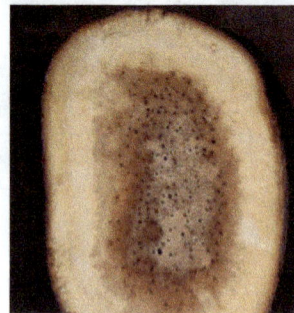

7.19 Cross section of a stem of *Ganoderma carnosum* with a dark peripheral layer, lighter outer part and brown center with longitudinal tubes.

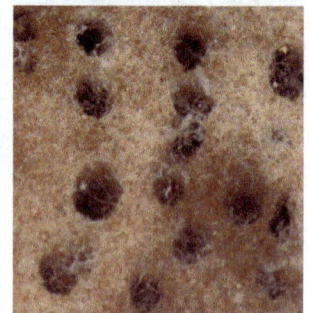

7.20 Irregular distribution of hyphae around the holes in the central stem part of *Ganoderma carnosum*.

Microscopic aspect overview

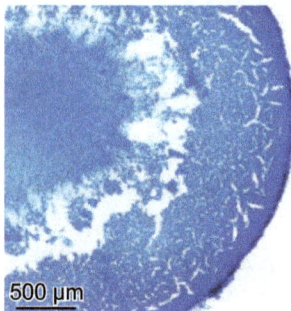

500 µm

7.21 Cross section of a stem of *Polyporus brumalis* with a dense peripheral layer, middle layer with few hyphae, and central strand of hyphae.

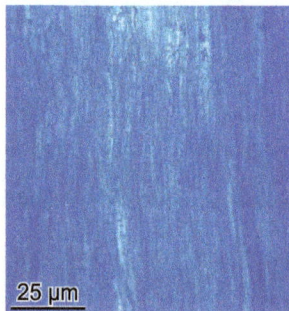

25 µm

7.22 Axially and parallel orientated hyphae in the peripheral zone of *Polyporus brumalis*.

Orientation of hyphae

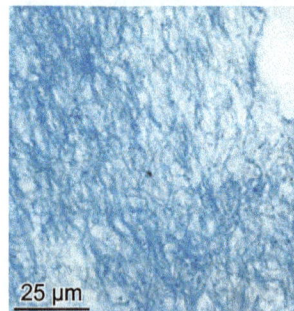

25 µm

7.23 Radially oriented hyphae in the peripheral zone of *Ganoderma carnosum*.

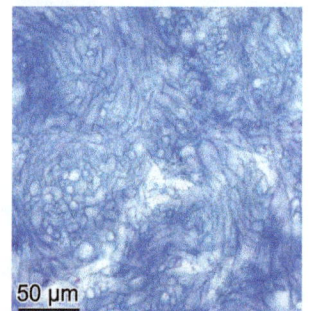

50 µm

7.24 Irregularly oriented hyphae in the peripheral zone of *Boletus edulis*.

Diameter and wall density of hyphae

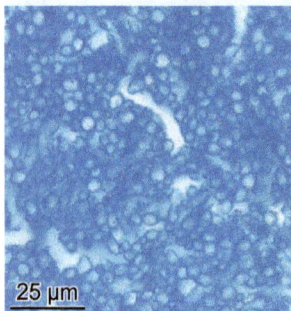

25 µm

7.25 Cross section of large and thick-walled hyphae with 2–3 µm diameter in the peripheral layer of *Polyporus brumalis*.

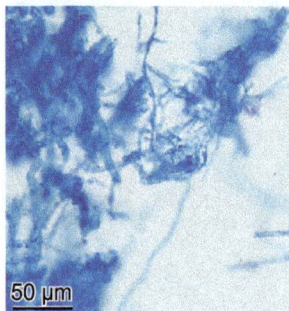

50 µm

7.26 Occurrence of hyphae with different diameters in the middle layer of a cross section of *Polyporus brumalis*.

Composition of stems

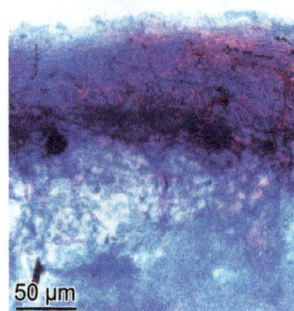

50 µm

7.27 Red-stained peripheral layer and blue-stained central parts of the stem of *Boletus edulis* (Astrablue/Safranin-stained).

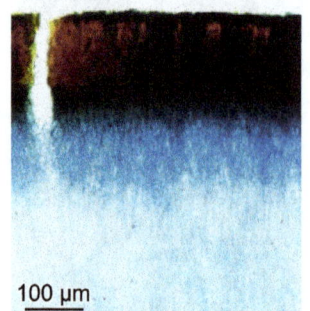

100 µm

7.28 Brown-stained peripheral and blue-stained central layers of *Ganoderma carnosum* (Astrablue/Safranin-stained).

7.1.2 Thalli and stems of brown algae

Described are a thallus and stems of brown algae. Large thallus-forming brown algae primarily occupy rocky, permanently or temporally submerse coastal sites (benthos) of temperate and cold oceans.

Coastal brown algae are anchored with rhizoids, and form stems (cauloids) and leaves (phylloids) of various lengths. Small plants remain a few centimeter tall, very large ones, e.g. the giant kelp, can reach a length of up to 40 meters. The principal stem construction of all types is similar. The cortex consists of cells filled with chloroplasts and brown substances (fucoxanthin) and a large center with less stained cells. The cellular structure is homogeneous in stems of small algae. Stems of *Laminaria* species in cold oceans are special, in that central, living cells form annual rings in a seasonal rhythm. It is a type of primitive secondary growth. Cell walls consist of a dense layer of cellulose fibrils and a mucilaginous layer of alginate (a polysaccharide). Stability is provided by the cellulosic layer, while flexibility is provided by the alginate. Peripheral cells contain photosynthetically active chloroplasts.

Macroscopic aspect of large brown algae

7.29 *Fucus serratus* at a sea shore in Iceland.

7.30 Washed-up, 6 m-long *Macrocystis pyrifera* (giant kelp) on the coast of South Africa.

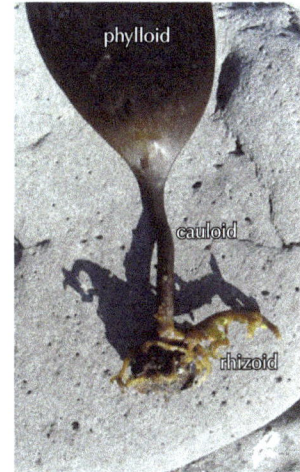

7.31 Phylloid (leaf), cauloid (stem) and rhizoids (roots) of *Laminaria* sp. on a rocky site in Iceland.

Annual rings

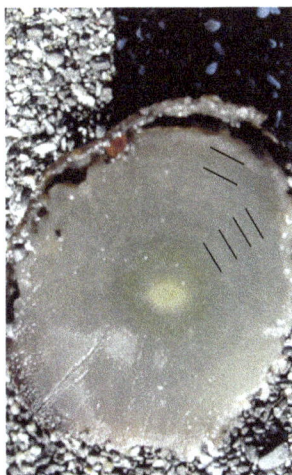

7.32 Cross section of a *Laminaria* sp. stem with 6 annual rings.

Microscopic aspect of stems and cells

7.33 Cross section of a 5 cm-tall stem forming tree-like brown algae.

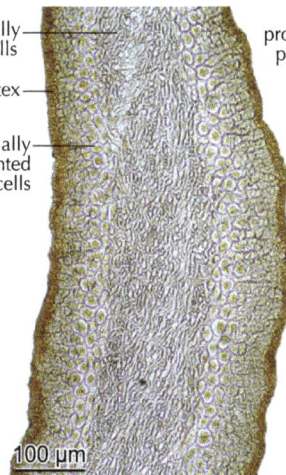

7.34 Cross section of a 15 cm-tall tallus of *Fucus serratus*. Peripheral cells are axially, central cells longitudinally oriented.

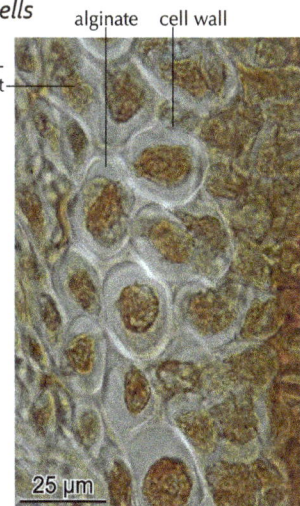

7.35 *Fucus serratus* cells with chloroplasts in the protoplast and double-layered walls.

7.2 Mosses – The oldest living plants

Described are stems of mosses (Bryopsida), a peat moss (Sphagnopsida) and a liverwort (Marchantiopsida). They include approximately 20,000 species and occupy all sites from the tropics to the arctic, and from dry to submersed sites.

The principal stem structure of mosses and peat mosses is similar. An epidermis and a mantle of fiber-like cells with various wall thickness surrounds a parenchymatic center. Parenchyma cells can be perforated by simple pits. This is the structure of the simplest types. Some species have developed a central strand consisting of leptoids, and others of leptoids and hydroids. Leptoids conduct photosynthetic products, the hydroids conduct water. The central parenchyma cells of most species are not perforated. Simple pits with large apertures could only be observed in a few walls. Leptoids are unlignifed, longitudinally enlarged

cells with horizontal walls. Hydroids in *Polytrichum* sp. are longitudinally enlarged, lignified, fiber-like cells without pits.

The peat mosses belong to the simplest anatomical type, however, the mantle is divided into a very thin-walled peripheral and a thick-walled inner part.

The thallus of *Marchantia* sp. resembles a leaf rather than a stem. It consists of a layer of large parenchyma cells with large, simple pits and few oil cells. The top layer is composed of small cells with chloroplasts.

Mono- and multicellular extrusions of the epidermis cells, the rhizoids, occur on all observed species.

Principal structure of moss stems

Types with an undifferentiated center

7.36 *Neckera crispa*. Photo: A. Bergamini.

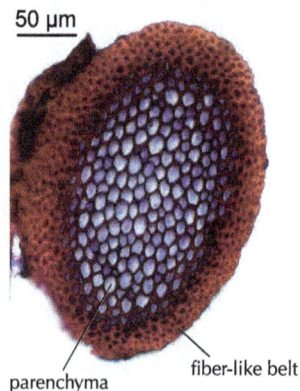

7.37 Stem of *Neckera crispa* with a thick-walled mantle.

7.38 *Sphagnum compactum*.

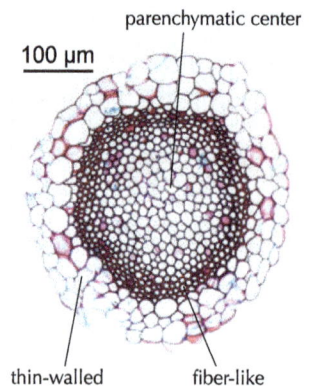

7.39 Stem of *Sphagnum subnitens* with a mantle of thin- and thick-walled cells.

Types with leptoids in the center (sieve-tube-like cells)

7.40 *Funaria hygrometrica*.

7.41 Stem of *Funaria hygrometrica* with a mantle of thick-walled cells and leptoids in the center.

Types with leptoids and hydroids in the center

7.42 *Polytrichum commune*.

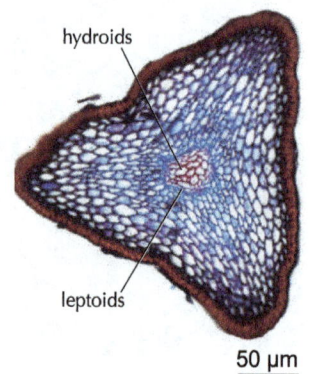

7.43 *Polytrichum commune* with a mantle of thick-walled cells and a center of hydroids and leptoids.

Types with a thallus

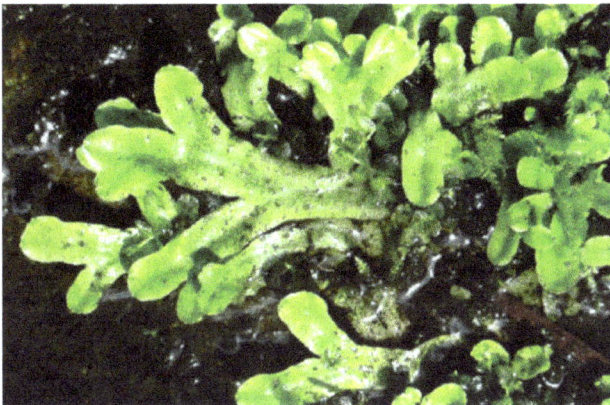

7.44 *Marchantia polymorpha*.

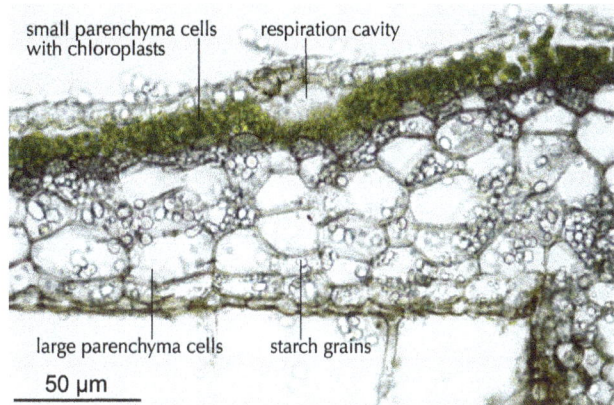

small parenchyma cells
with chloroplasts   respiration cavity

large parenchyma cells   starch grains

50 µm

7.45 Thallus of *Marchantia polymorpha* with chloroplasts and a respiration cavern in the top layer. Starch grains in large parenchyma cells and rhizoids a the base of the thallus. Unstained slide.

Principal cell structure of moss stems

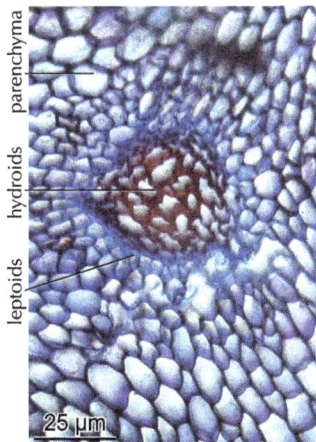

parenchyma   parenchyma   leptoids   leptoids

simple pit   transversal wall

25 µm   25 µm   25 µm   25 µm

7.46 *Thamnobryum alopecurum*. Photo: A. Bergamini.

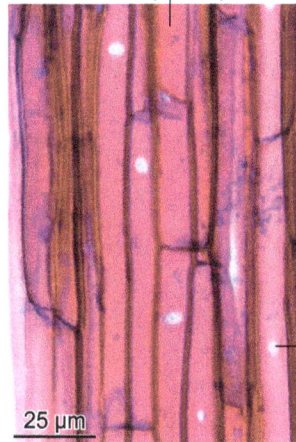

7.47 Parenchyma cells with large simple pits in *Thamnobryum alopecurum*.

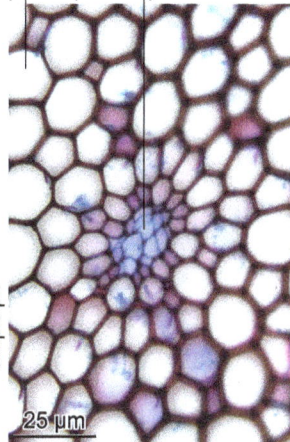

7.48 Unlignified leptoids in *Thamnobryum alopecurum*.

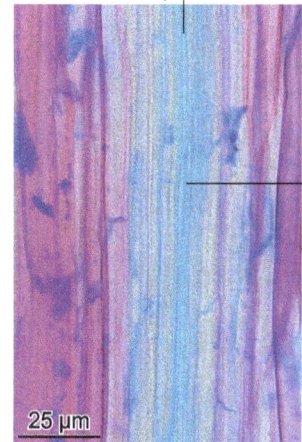

7.49 Unlignified, 300 µm long leptoid with transverse walls in *Thamnobryum alopecurum*.

Principal cell structure of moss stems

Rhizoids

parenchyma

hydroids

leptoids

hydroids   leptoids

parenchyma

25 µm   25 µm   100 µm

7.50 Unlignified leptoids and red-stained (lignified?) hydroids are surrounded by unlignified parenchyma cells in *Polytrichum commune*.

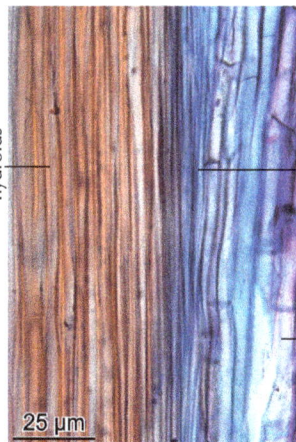

7.51 Lignified fiber-like hydroids, unlignified leptoids and parenchyma cells with horizontal walls in *Polytrichum commune*.

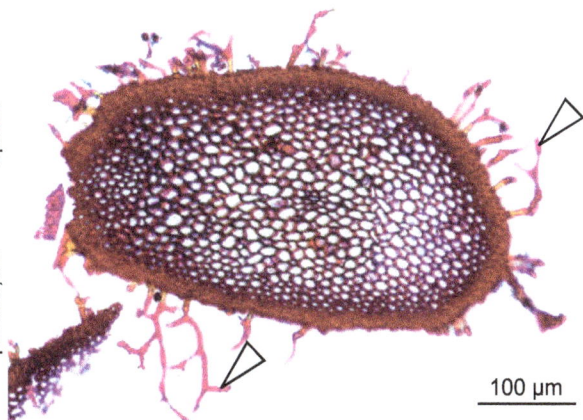

7.52 Rhizoids are specialized epidermis cells. *Thuidium tamariscinum*.

7.3 Fern-like plants

7.3.1 Spikemosses, quillworts and clubmosses

Described are a few stems of the terrestrial genera *Selaginella* (spikemoss), *Lycopodium* (clubmoss) and the bulb of the swamp plant *Isoetes* (quillwort). They all have in common the absence of secondary growth, as well as the presence of concentric vascular bundles with a central xylem and a peripheral phloem. The xylem consists of tracheids with sclariform pits.

The families Selaginellaceae, Lycopodiaceae and Isoetaceae are distinguishable by the distribution of vascular bundles in the central strand (stele), and the individual species by the composition of the cortex and the form of vascular bundles.

Selaginellaceae (spikemosses)

This family contains approximately 40 species. All of them belong to the genus *Selaginella*. Their stems are characterized by the presence of single, laterally extended vascular bundles (polystele), which are surrounded by an endodermis.

7.53 *Selaginella denticulata.*

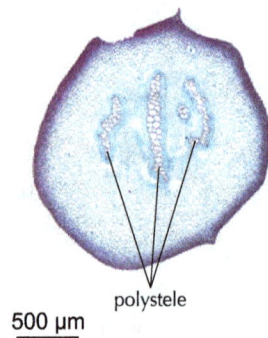

7.54 Stem of *Selaginella* sp. with isolated concentric vascular bundles.

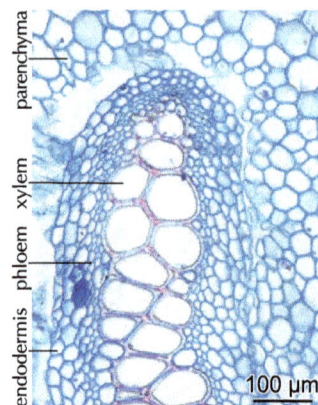

7.55 Part of a vascular bundle in *Selaginella* sp.

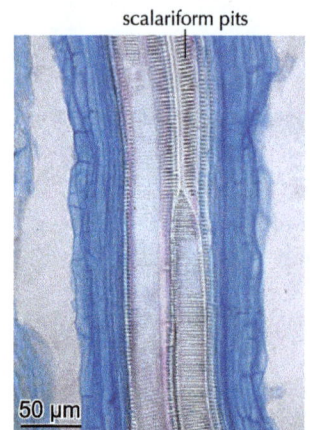

7.56 Tracheids with scalariform pits in a longitudinal section of *Selaginella* sp.

Isoetaceae (quillworts)

This family contains approximately 50 species. All of them belong to the genus *Isoetes*. They grow on wet sites. The bulb of *Isoetes lacustris* contains a central circle of tracheids where many laterally oblique emerging shoots are initiated. The vascular bundles are embedded in a very thin-walled parenchymatic tissue. The walls of tracheids are characterized by intensively lignified annular structures. An endodermis is absent.

7.57 *Isoetes lacustris.* Photo: M. Ctvrtlikova.

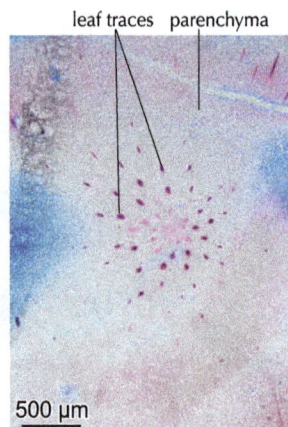

7.58 A central strand and many leaf traces are embedded in a thin-walled parenchymatic tissue in *Isoetes lacustris*.

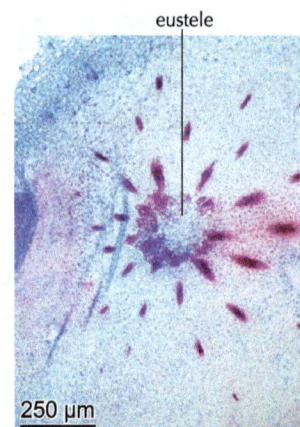

7.59 The central ring consists of circular arranged tracheids in *Isoetes lacustris.*

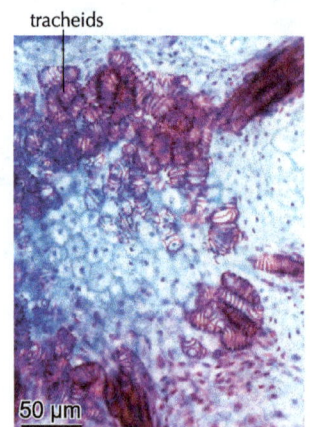

7.60 Tracheids with scalariform pits in the central strand in *Isoetes lacustris.*

Lycopodiaceae (clubmosses)

This family contains more than 10 genera and approximately 1,000 species. Stems of clubmosses are characterized by a large cortex and a central strand with vascular bundles. The strand is surrounded by an endodermis. Different compositions of the cortex and distribution patterns of the central vascular bundles allow species identification.

7.61 *Lycopodium alpinum*, an alpine cushion plant on dry sites.

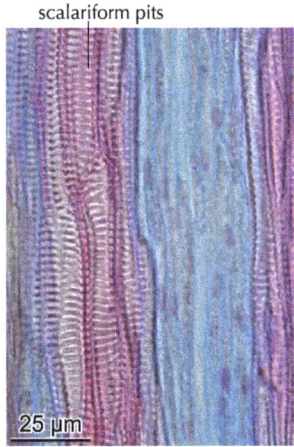

7.62 Scalariform pits on the walls of tracheids in *Lycopodium alpinum*.

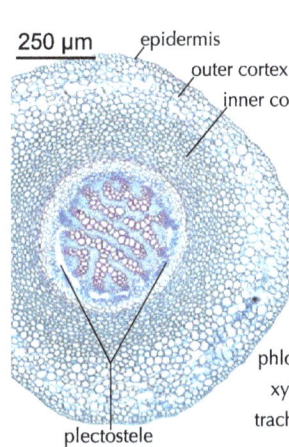

7.63 The central strand (stele) with a large surrounding cortex in *Lycopodium alpinum*.

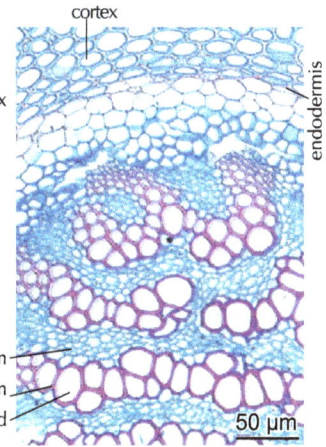

7.64 Vascular bundles within the central plectostele in *Lycopodium alpinum*.

7.65 *Lycopodium annotinum*, a subalpine plant with long rhizomes.

7.66 Scalariform pits on the walls of tracheids in *Lycopodium annotinum*.

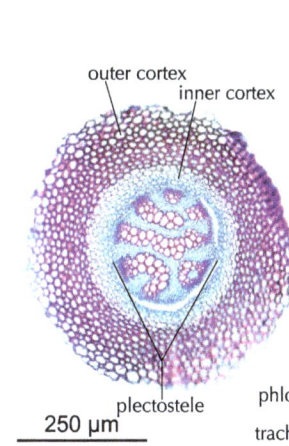

7.67 The central strand (stele) with a surrounding cortex in *Lycopodium annotinum*.

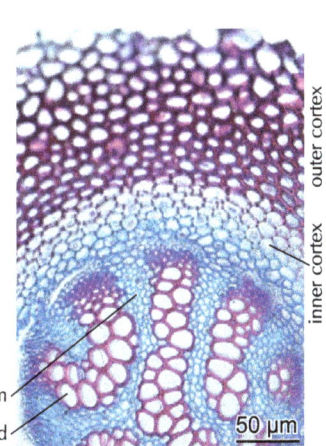

7.68 Vascular bundles with a central plectostele in *Lycopodium annotinum*.

7.69 *Lycopodiella cernua*, an upright plant on subalpine bogs (Azores).

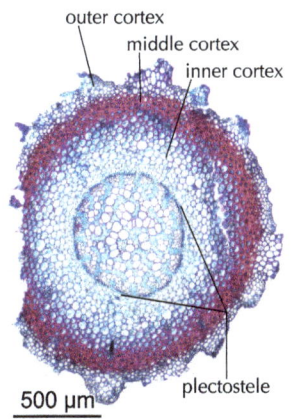

7.70 The central strand with a surrounding three-part cortex in *Lycopodiella cernua*.

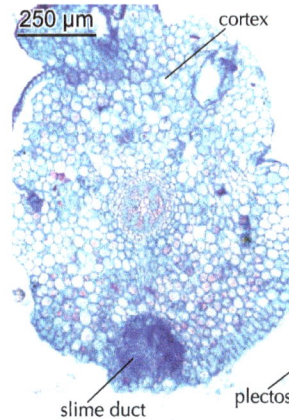

7.71 Large, thin-walled cortex and slime ducts in *Lycopodiella innundatum*, a bog plant.

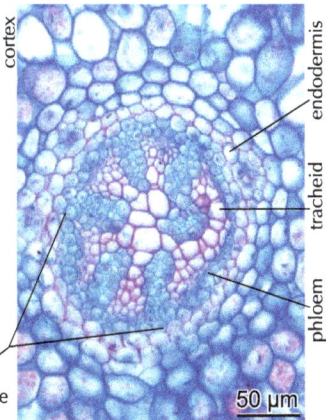

7.72 Vascular bundles with a central plectostele in *Lycopodiella innundatum*.

7.3.2 Whisk ferns and moonworts

Described are two terrestrial species: *Psilotum nudum* of the family Psilotaceae (whisk ferns) and *Botrychium lunaria* of the family Ophioglassaceae (moonworts). The family Psilotaceae contains two genera with three species, and the family Ophioglossacea contains four genera with approximately 80 species.

Representatives of the two families differ significantly from each other. *Psilotum nudum* is anatomically close to the lycopods, however, two features are different: the tracheids have scalariform and round pits, and the vessels are arranged star-like around a sclerified pith (actinostele). This is characteristic for

the family. Absent are endodermis, fibers and parenchyma cells within the xylem.

Stems of *Botrychium lunaria* are anatomically closer to dicotyledons rather than to lycopods. Characteristic is a central strand of xylem and phloem (siphonostele) with a cambium in the upper part of the plant. It centripetally produces a xylem, which consists of vessels with distinct perforation plates with large pits and rays. Fibers and axial parenchyma cells are absent. The central strand in the root is a closed concentric vascular bundle consisting of tracheids with bordered pits.

7.73 *Psilotum nudum.*

7.74 Scalariform vessel pits (left) and bordered pits (right) in *Psilotum nudum.*

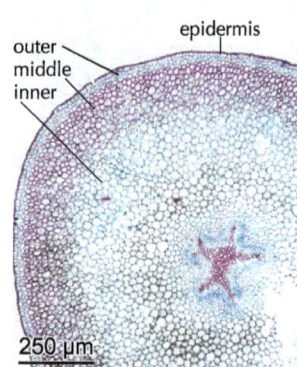

7.75 Central cylinder surrounded by a large cortex and an epidermis in *Psilotum nudum.*

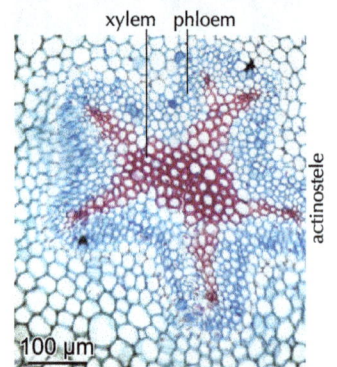

7.76 Star-like arrangement of vessels (actinostele) in *Psilotum nudum.*

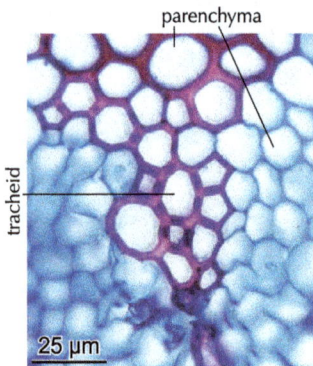

7.77 Groups of vessels at the tips of the star in *Psilotum nudum.*

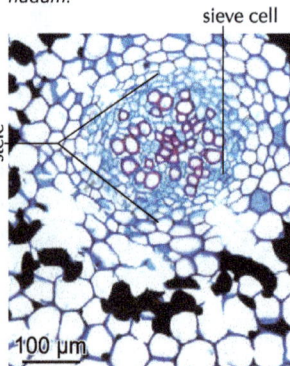

7.78 Irregular distribution of tracheids and sieve cells in the root of *Psilotum nudum.*

7.79 *Botrychium lunaria.*

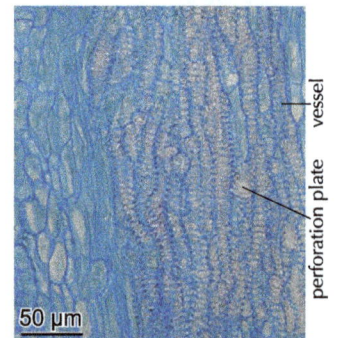

7.80 Vessels with perforation plates in *Botrychium lunaria.*

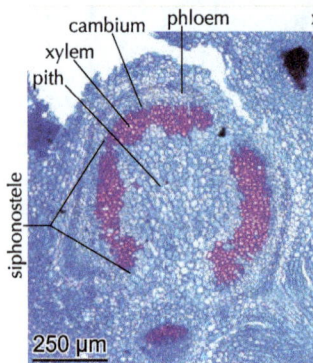

7.81 Tube-like arrangement of vessels (siphonostele) in *Botrychium lunaria.*

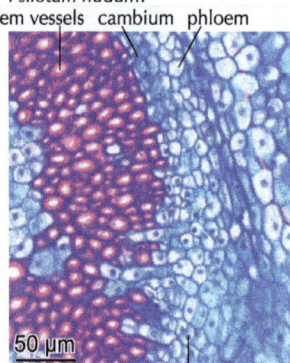

7.82 Cambium between xylem and phloem in *Botrychium lunaria.*

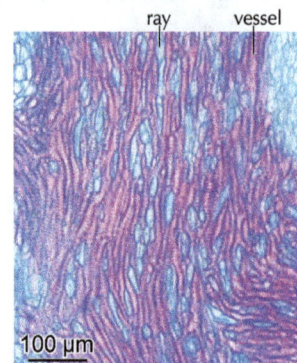

7.83 Lignified vessels and unlignified rays in *Botrychium lunaria.*

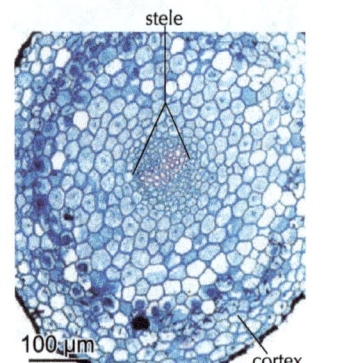

7.84 Root of *Botrychium lunaria* with a protostele.

7.3.3 Horsetails

The genus *Equisetum*, with approximately 20 species, is the only living genus within the Equisetophytina. Horsetails are anatomically chimeric: tracheids/vessels in vascular bundles within the internodes indicate a relationship to ferns, those in the nodes to dicotyledons (intercalary meristem) and the roots to the rhizomes of monocotyledons.

Vertical shoots with reduced leafs (sheaths) above nodes and long internodes are characteristic for all horsetails. Secondary growth is absent. Common for all species are intercalary meristems in the nodes, circular arranged vascular bundles (siphonostele) centripetal of the epidermis and a large cortex. The cortex consists of an epidermis, an outer part with thick-walled, often lignified fibers and inner, thin-walled part, with large, axially elongated schizogenous intercellulars (vallecular canals). The anatomy of the central strand is different in the internodes and nodes.

Structure of the internodes
In one group of species the central strand is bordered by an endodermis, e.g. in *Equisteum arvense* or *E. sylvaticum*. In the other group, the single vascular bundles are directly surrounded by the endodermis, e.g. in *E. limosum* or *E. hiemale*. The xylem of the vascular bundles of all species is reduced to a few isolated thick-walled tracheids or vessels. Their walls consist of lignified rings or of wide-spaced scalariform pits. Perforations plates and scalariform pits with very large apertures are hard to differentiate. The term tracheid/vessel is therefore used. Parenchyma cells surround the phloem. The sieve tubes have horizontal sieve plates. Lateral plates are absent. Within the area of the xylem is a large opening, the carinal canal, which occasionally contains tyloses.

Macroscopic aspect

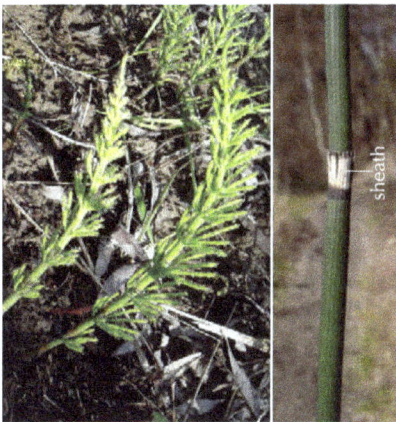

7.85 *Equisetum arvense* (left) and node of *Equisetum hiemale* (right).

Structure of internodes

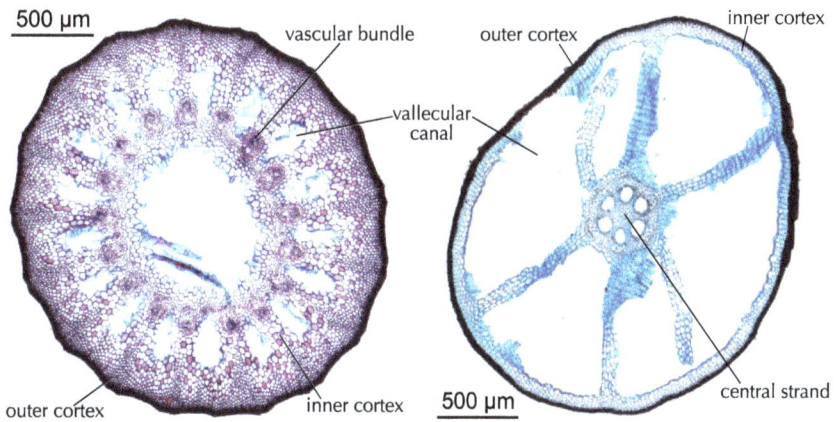

7.86 *Equisetum arvense* with dense outer cortex, small vallecular canals, circular arranged vascular bundles and a hollow center.

7.87 *Equisetum palustre* with very large vallecular canals (aerenchyma) in the cortex and a small central strand.

Vascular bundles in internodes

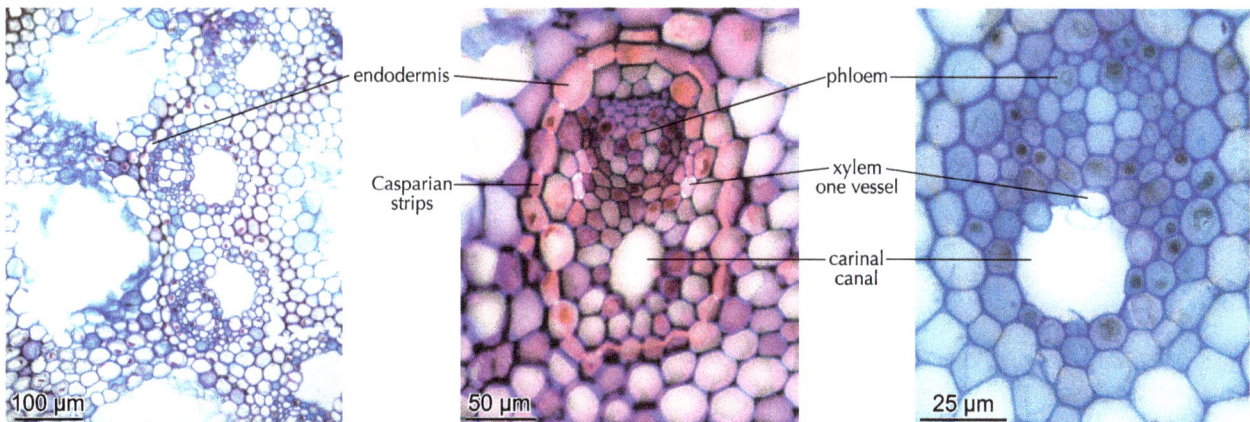

7.88 An endodermis separates the central strand and the cortex in *Equisetum sylvaticum*.

7.89 An endodermis separates a single closed collateral vascular bundle from surrounding parenchyma in *Equisetum hiemale*.

7.90 A single vascular bundle without distinct endodermis in *Equisetum limosum*.

Structure of the nodes

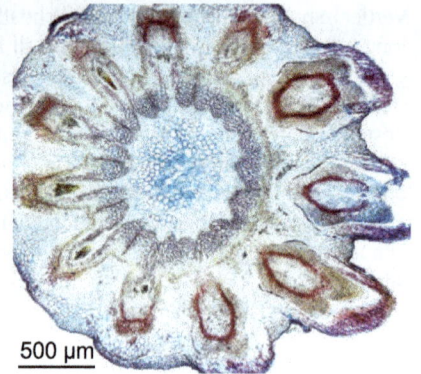

The origin of lateral shoots is in the nodes. Collateral, probably open vascular bundles are arranged in a compact circular belt of xylem and phloem. The xylem of the bundles consists of tracheids/vessels with bordered pits. A layer of small parenchyma cells in the pith and horizontally oriented vessels and an intercalary meristem divide the nodes axially. There is a continuum between bordered pits in the internodes and scalariform pits in the nodes.

Structure of the root

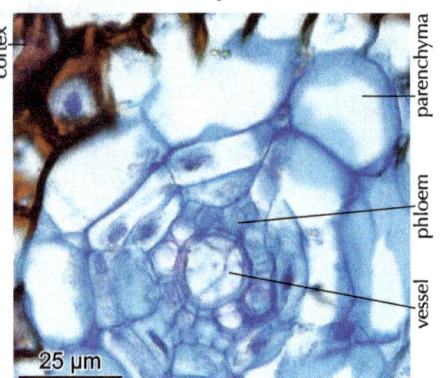

A single concentric vascular bundle with a single vessel is surrounded by a cortex (Carlquist 2011).

Cell walls in vessels of internodes

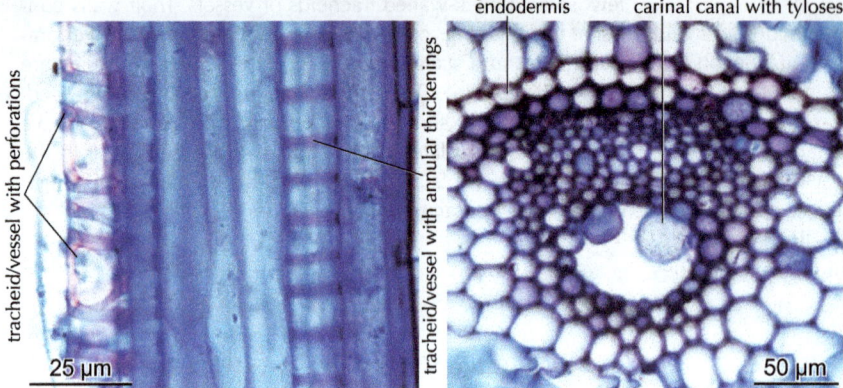

7.91 Vessel walls with widely spaced pit apertures or perforation plates and annular thickenings in *Equisetum arvense*.

7.92 Tyloses in a carinal canal of a vascular bundle in *Equisetum sylvaticum*.

Structure of nodes

7.93 Central strand and lateral shoots of *Equisetum arvense*.

Structure of vascular bundles in nodes

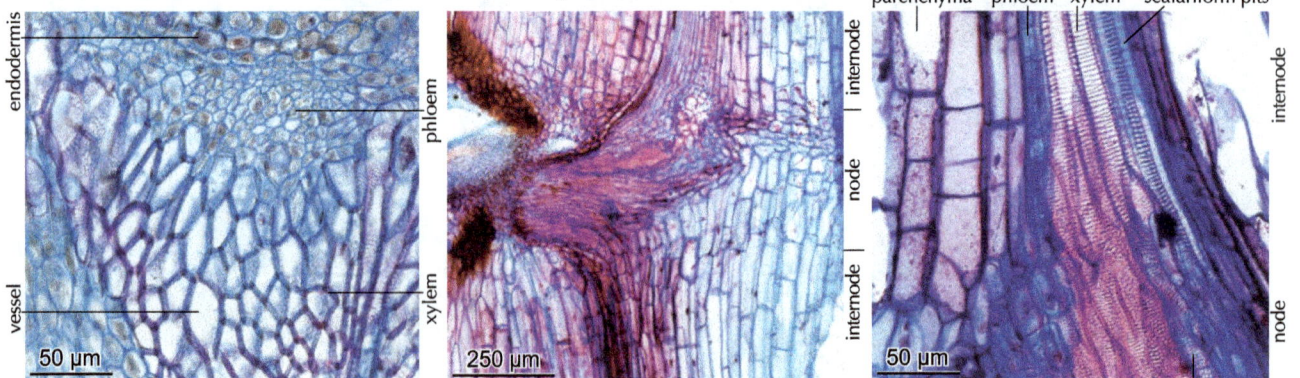

7.94 A single open? collateral vascular bundle with a xylem and phloem in *Equisetum arvense*.

7.95 Node with a lateral branch in *Equisetum arvense*.

7.96 Vessel transition from an internode to a node in *Equisetum arvense*.

Structure of vessel walls in nodes

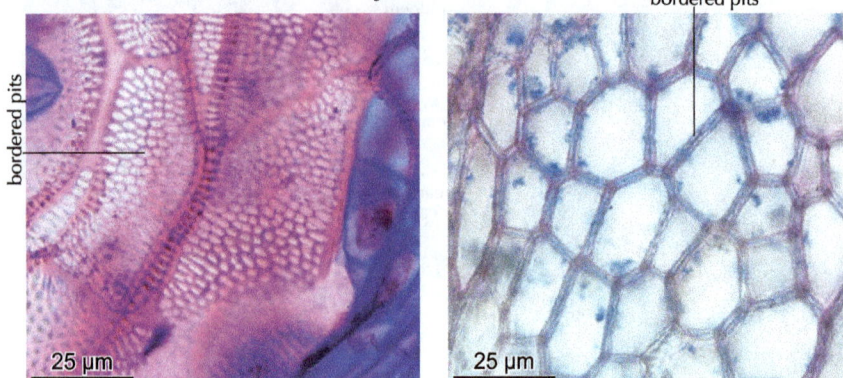

7.97 Bordered pits in vessels of a node of *Equisetum arvense*, longitudinal section.

7.98 Bordered pits in vessels of a node of *Equisetum arvense*, cross section.

Structure of the root

7.99 Concentric vascular bundle with a single vessel in *Equisetum arvense*.

7.3.4 Ferns

The Filicophytina include approximately 9,000 species. Presented here are common anatomical traits of ferns. Tree ferns, hemicryptophytes and water ferns occur from the tropics to the arctic zone and grow on very dry to very moist sites. Ferns form petioles, stems, rootstocks and rhizomes. Secondary growth is absent. Vascular bundles are arranged solitary, or in irregular groups lateral of petioles; they are circular arranged in root stocks and stems, or they form bands. The arrangement of vascular bundles varies within the plant; it is different in rhizomes and in petioles.

Most ferns have closed amphiversal vascular bundles with a lignified xylem and solitary parenchyma cells in the center. The unlignified surrounding belt consists of sieve tubes, companion cells, groups of parenchyma cells and an endodermis. The form and the arrangement of the cell types vary. Closed collateral vascular bundles occur in hydrophytes.

A very thin-walled endodermis, often with Casparian strips, surrounds vascular bundles. The parenchyma cells which separate the cortex from the vascular bundle are mostly centripetally very thick-walled and lignified. The majority of tracheids in all observed species have scalariform pits, however, transitions of scalariform pits to bordered pits occur frequently. Perforation plates on distal ends of tracheids occur in a few species. These types are real vessels. In a few species sieve fields occur, as far as could be observed, only on lateral walls.

The structure of the cortex of petioles, rhizomes and stems greatly varies. In all cases one or more layers of thick-walled parenchyma cells surround a layer of more or less thin-walled parenchyma. The thick-walled cells can be either parenchyma cells or fiber-like cells with slightly bordered pits. Thin-walled parenchyma cells contain starch. Specialized cells or ducts of some species produce slime. See also White 1963.

Growth forms of ferns

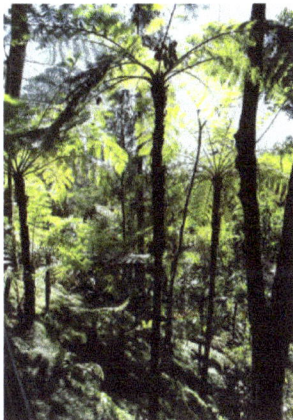

7.100 Tree fern *Cyathea cooperi*.

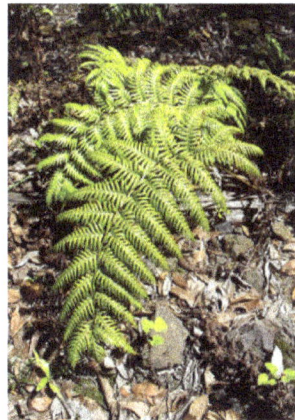

7.101 Large hemicryptopyte *Woodwardia radicans*.

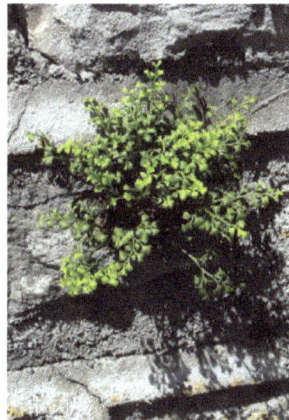

7.102 Small hemicryptophyte *Asplenium ruta-muraria*.

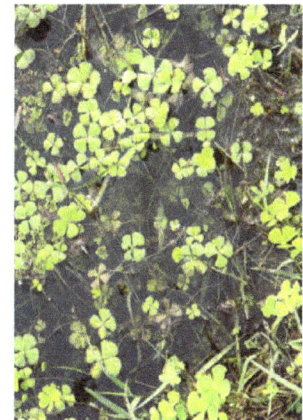

7.103 Hydrophyte *Marsilea quadrifolia*.

Stems, root stocks and rhizomes

vascuar bundle

7.104 Stem of the tree fern *Cyathea cooperi*.

7.105 Rootstock with leaf bases of the hemicryptophyte *Dryopteris filix-mas*.

leaf bases

7.106 Rootstock with leaf bases of the hemicryptophyte *Matteuccia struthiopteris*.

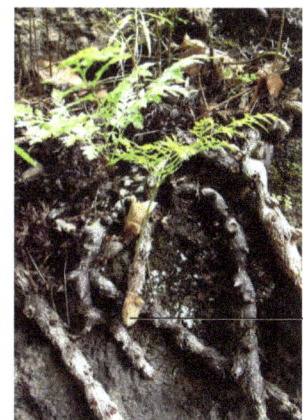

rhizomes

7.107 Rhizomes of *Davallia canariensis*.

Arrangement of vascular bundles

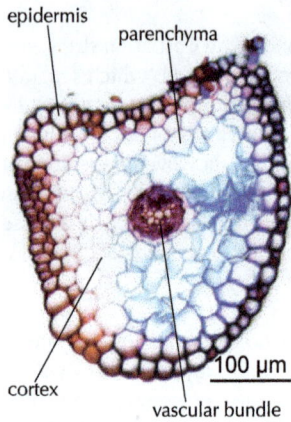

7.108 Solitary in the center of the petiole of an annual stem of *Hymenophyllum tunbrigense*. Protostele.

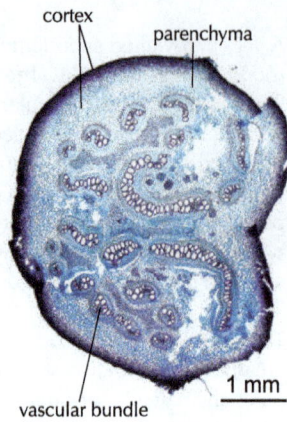

7.109 Irregularly distributed in the annual stem of *Pteridium aquilinum*. Polystele.

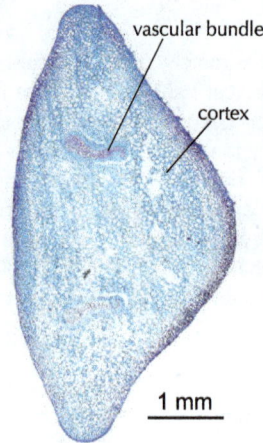

7.110 Lateral in a petiole of *Athyrium filix-femina*. Polystele.

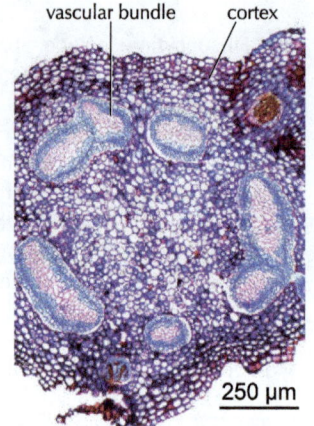

7.111 Circular in the rhizome of *Asplenium septentrionale*. Siphonostele.

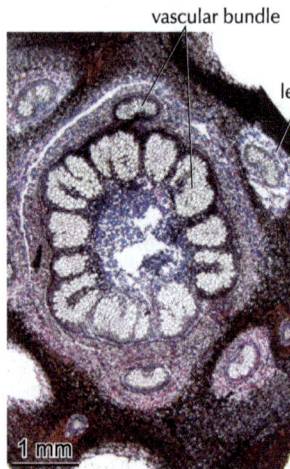

7.112 Circular in the root stock of *Osmunda regalis*. Siphonostele.

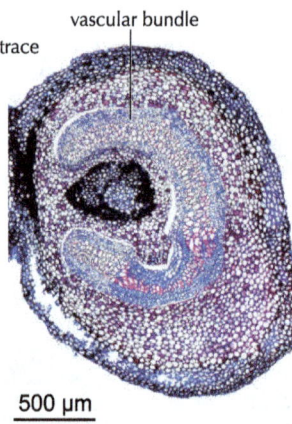

7.113 Arc-like in the rhizome of *Cryptogramma crispa*. Siphonostele.

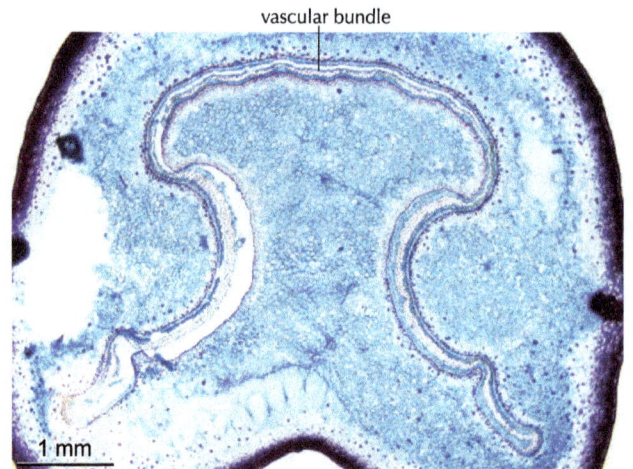

7.114 Arc-like in the petiole of *Culcita macrocarpa*. Siphonostele.

Internal variation of vascular bundle arrangement

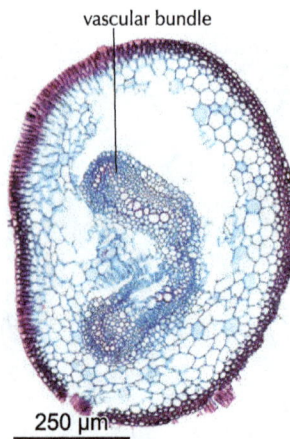

7.115 Two bundles in the petiole of *Gymnocarpium robertianum*. Polystele.

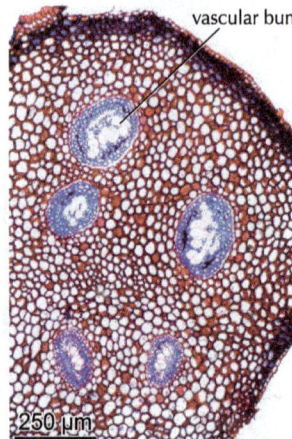

7.116 Round bundles, circular arranged in the root stock of *Gymnocarpium robertianum*. Siphonostele.

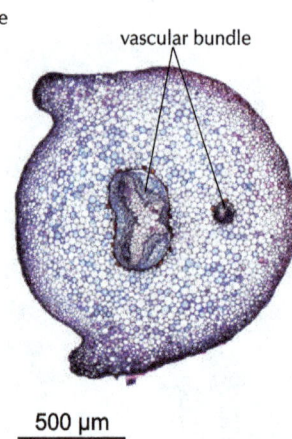

7.117 A large and a small bundle in a petiole of *Polypodium vulgare*.

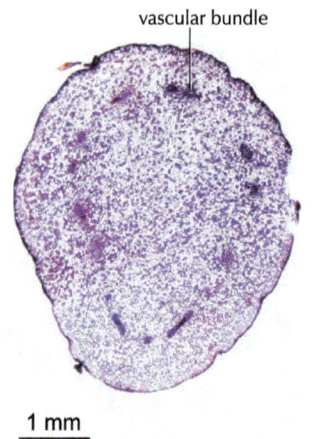

7.118 Different forms of bundles, circular arranged in the root stock of *Polypodium vulgare*. Siphonostele.

Structure of closed amphiversal vascular bundles

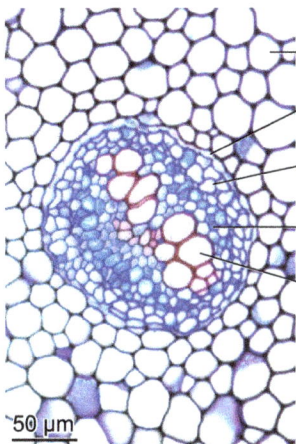

7.119 Laterally elongated xylem in a round vascular bundle of *Gymnocarpium robertianum*.

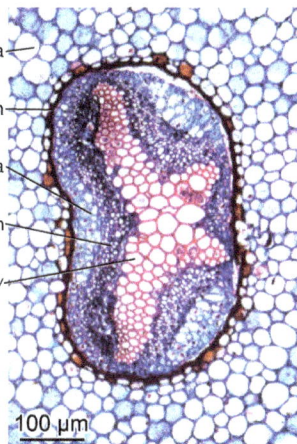

7.120 Eagle-shaped xylem in rhizome of *Polypodium vulgare*.

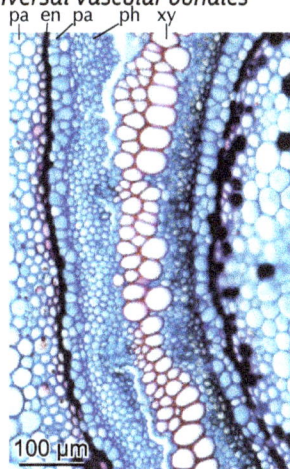

7.121 Band-like xylem in a vascular bundle of the petiole of *Culcita macrocarpa*.

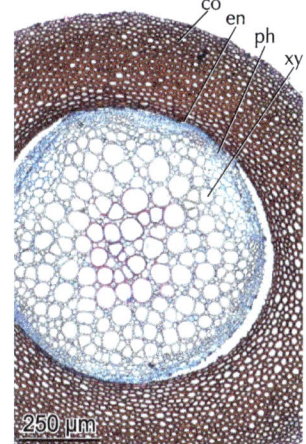

7.122 Round xylem/phloem strand in the stem of the liana *Lygodium* sp.

Endodermis of vascular bundles

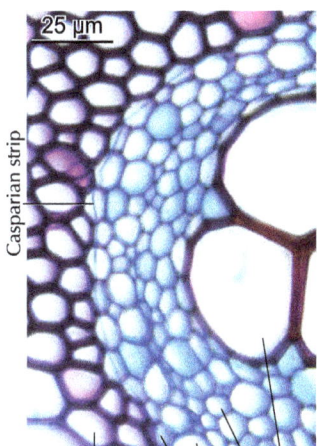

7.123 Endodermis with Casparian strips in *Cyathea cooperi*.

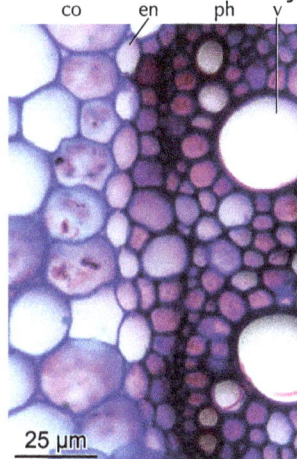

7.124 Endodermis without distinct Casparian strips in *Marsilea quadrifolia*.

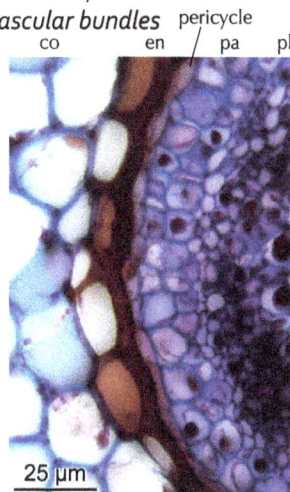

7.125 Thin-walled pericycle and thick-walled endodermis separate central cylinder and cortex in *Polypodium vulgare*.

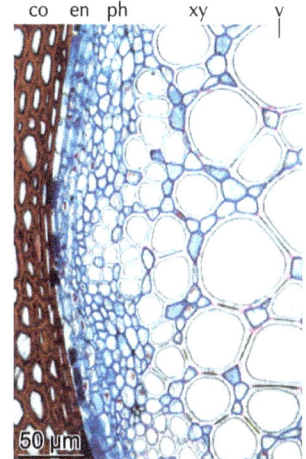

7.126 Thin-walled endodermis between a lignified cortex and an unlignified parenchyma tissue in *Lygodium* sp.

Wall structure of tracheids

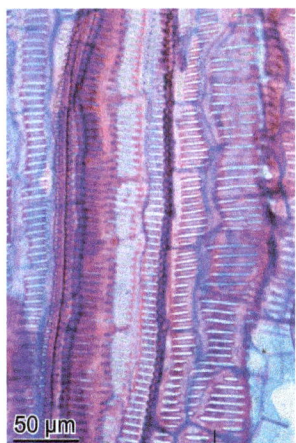

7.127 Scalariform pits in *Cyathea cooperi*.

7.128 Scalariform pits in *Blechnum spicant*.

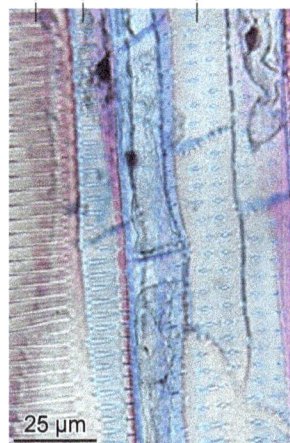

7.129 Scalariform and bordered round pits in *Lygodium* sp.

Wall structure of sieve elements

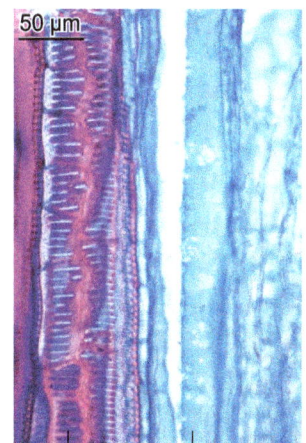

7.130 Sieve fields on lateral walls in *Cyathea cooperi*.

Structure of the cortex

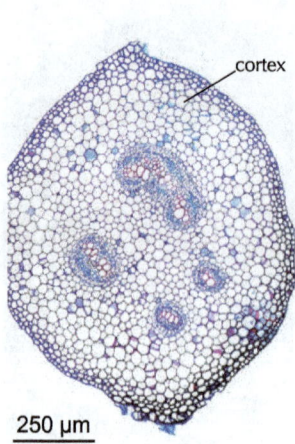

7.131 Thin-walled parenchyma cells are surrounded by an epidermis in *Gymnocarpium robertianum*.

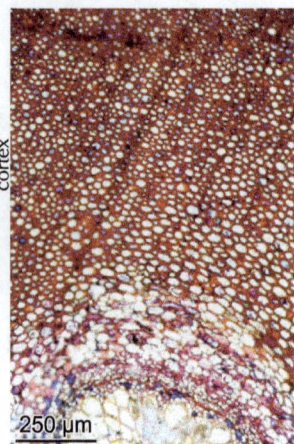

7.132 A layer of thick-walled parenchyma cells surrounds the vascular bundles in *Osmunda regalis*.

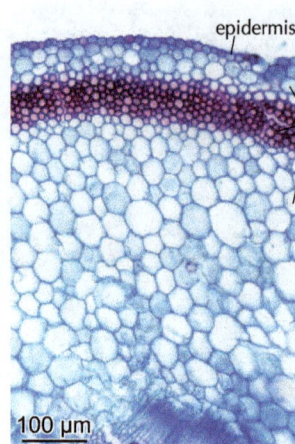

7.133 A layer of thick- and thin-walled cells occurs outside of a large, thin-walled parenchymatic zone in *Marattia fraxinea*.

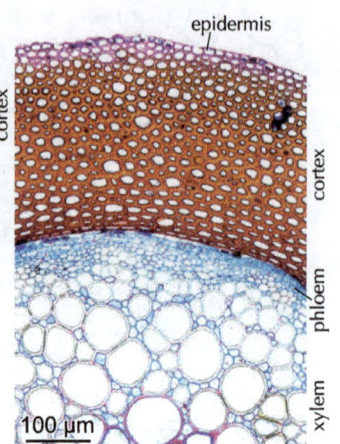

7.134 A dense belt of thick-walled cells surrounds the central xylem/phloem strand in *Lygodium* sp.

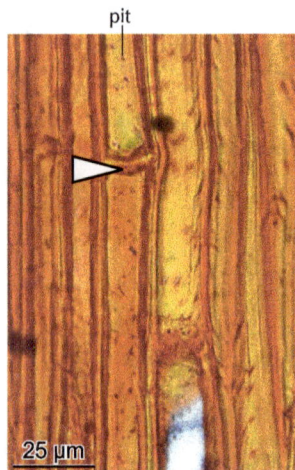

7.135 Parenchyma cells with horizontal walls and slightly bordered pits in *Lygodium* sp.

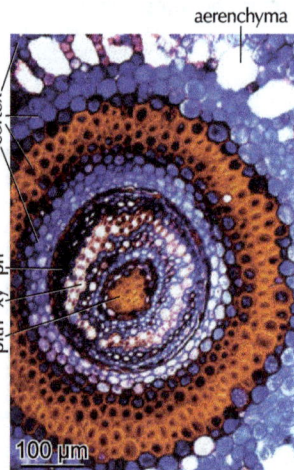

7.136 Multilayered cortex in the rhizome of *Marsilea strigosa*.

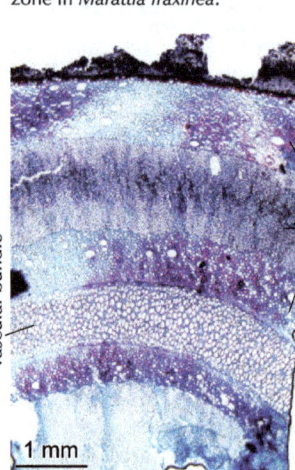

7.137 Multilayered cortex in the stem of *Cyathea cooperi*.

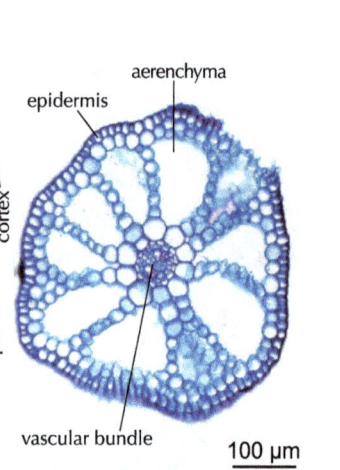

7.138 Cortex with large aerenchymatic spaces of *Pilularia globulifera*.

Content of cortex cells

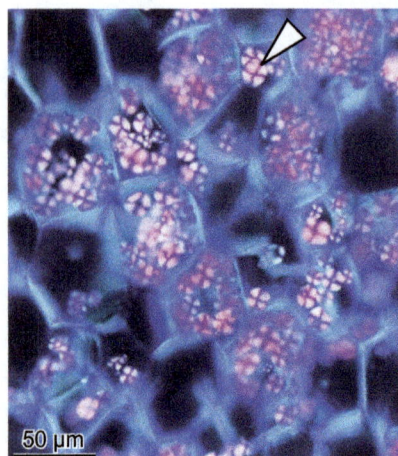

7.139 Starch in thin-walled parenchyma cells in *Culcita macrocarpa*, polarized light.

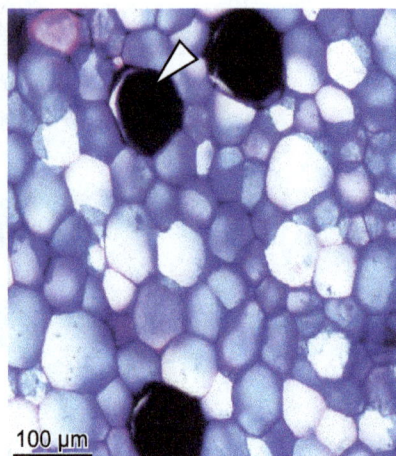

7.140 Dark-stained substances in parenchyma cells in *Marattia fraxinea*.

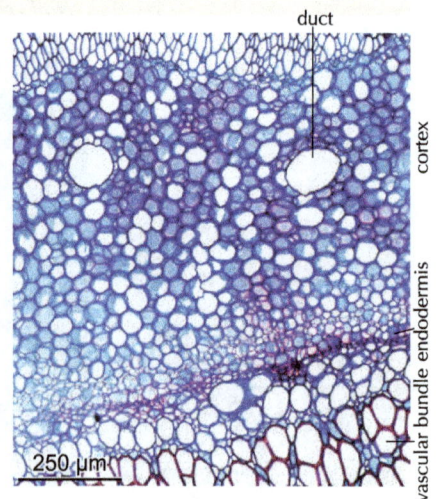

7.141 Slime-conducting ducts in parenchyma cells of *Cyathea cooperi*.

7.4 Seed plants

7.4.1 Palm ferns

Within the Cycadopsida, the familes Cycadaceae and Zamiaceae exist worldwide today, with ten genera and approximately 90 species. Most species occur in the tropics. The following presentation is primarily based on the collection of Greguss 1968.

Secondary growth is characteristic for all palm ferns. The presence of circular arranged tracheids is a common feature. Within the Cycadopsida, two radial growth types occur: one forms a simple siphonostele containing isolated vascular bundles or a closed xylem/phloem ring, the other has successive cambia, which form single vascular bundles or xylem/phloem rings. Collateral vascular bundles occur in petioles. The xylem is composed of radially arranged roundish tracheids. Annual rings have not been observed. Tracheids have scalariform or bordered pits with slit-like apertures. Pits are arranged in one or more axial rows. Rarely, tracheids with ephedroid perforation plates (per definition vessels) occur. Rays are homogenous and composed of thin-walled parenchyma cells, arranged in one to several rows. The phloem contains fibers and sieve elements with lateral sieve fields. Companion cells were not observed.

Slime ducts occur in the pith and the cortex of some species. They are surrounded by small, unlignified excretion cells or, in a few species, also with lignified tracheid-like cells. Short, lignified cells, often with bordered pits, occur in the pith of some species (transfusion cells, terminus Greguss). Crystals in the form of druses, prisms or sand are frequent in parenchyma cells.

Macroscopic aspect of palm ferns

Pál Greguss' slide collection

7.142 *Cycas revoluta*

7.143 *Cycas* sp.

7.144 *Encephalartos* sp.

7.145 Cycad slide collection at the Hungarian Natural History Museum Budapest, Department of Botany.

Types of radial growth

7.146 One cambium forms circular arranged vascular bundles in *Ceratozamia mexicana*. Eustele.

7.147 One cambium forms a closed belt of xylem and phloem in *Cycas* sp. Eustele.

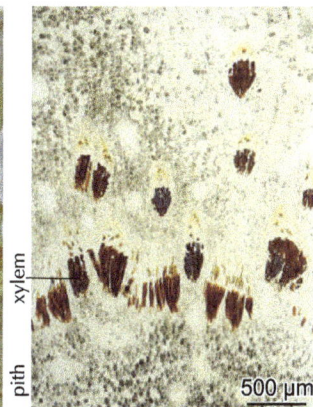

7.148 Successive cambia form single vascular bundles in *Cycas* sp.

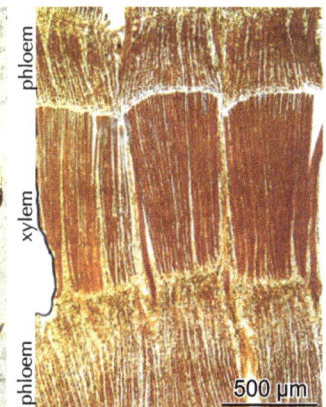

7.149 Successive cambia form several xylem/phloem rings in *Macrozamia moorei*.

Collateral vascular bundles

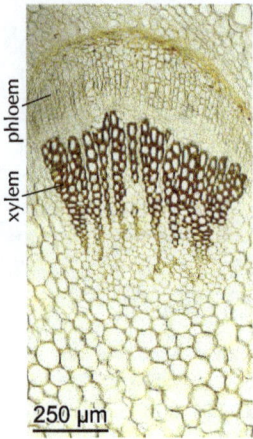

7.150 Stem of *Ceratozamia mexicana*.

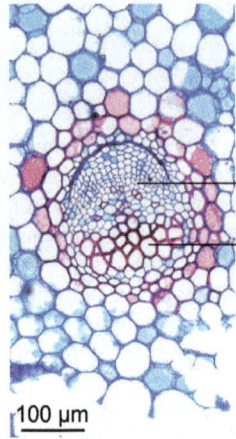

7.151 Petiole of *Cycas revoluta*.

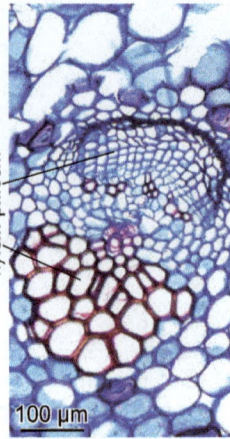

7.152 Petiole of *Zamia pygmaea*.

Xylem with tracheids

7.153 Roundish tracheids, strictly radially arranged, in *Dioon spinulosum*.

7.154 Roundish tracheids in *Zamia skinneri*.

Pitting of tracheids

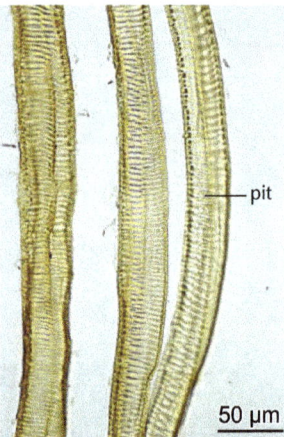

7.155 Scalariform pits in *Zamia furfuracea*.

7.156 Bordered pits with slit-like apertures in *Encephalartos hildebrandtii*.

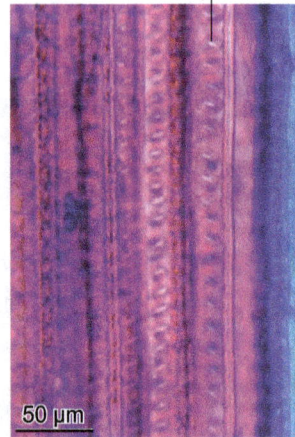

7.157 Bordered pits with slit-like apertures in *Cycas revoluta*.

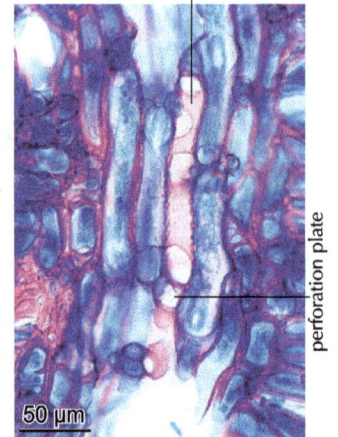

7.158 Vessels with ephedroid perforation plates in *Zamia* sp.

Width of rays

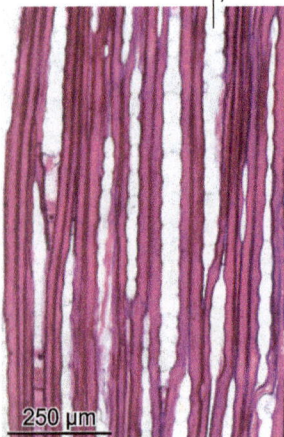

7.159 Uniseriate rays in *Cycas media*.

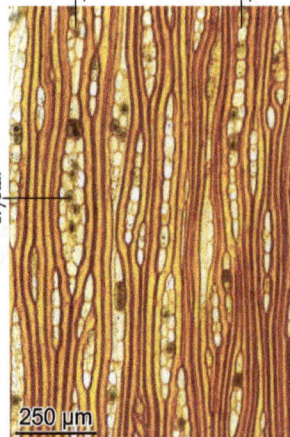

7.160 One- to triseriate, homogeneous rays in *Dioon spinulosum*.

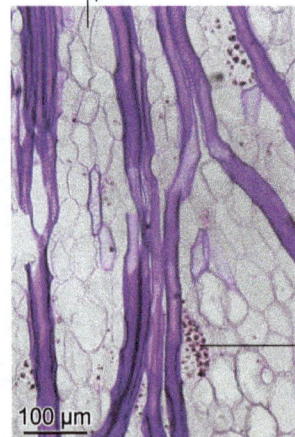

7.161 Multiseriate rays with thin-walled cells and a few transfusion cells in *Macrozamia pauli-guilielmi*.

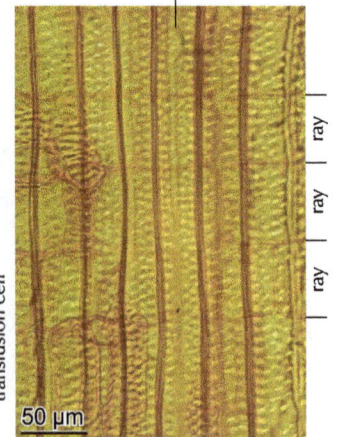

7.162 Ray with square, thin-walled cells, radial section of *Encephalartos septentrionalis*.

Phloem

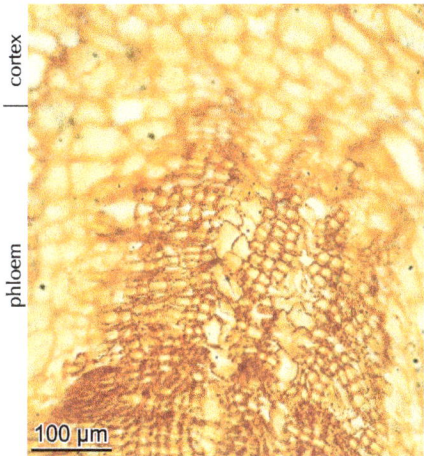

7.163 Collapsed sieve elements and square fibers in *Zamia skinneri*.

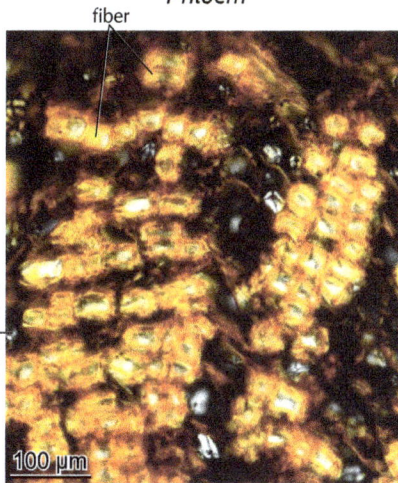

7.164 Distribution of fibers (yellow) and thin-walled sieve elements in *Zamia skinneri*.

7.165 Sieve element with lateral sieve plates in *Encephalartos hildebrandtii*.

Slime ducts

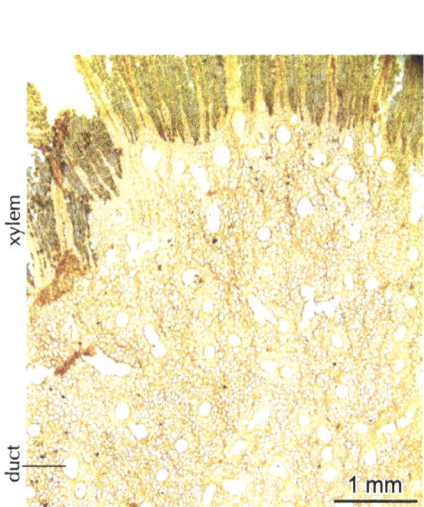

7.166 Ducts in the pith of *Zamia skinneri*.

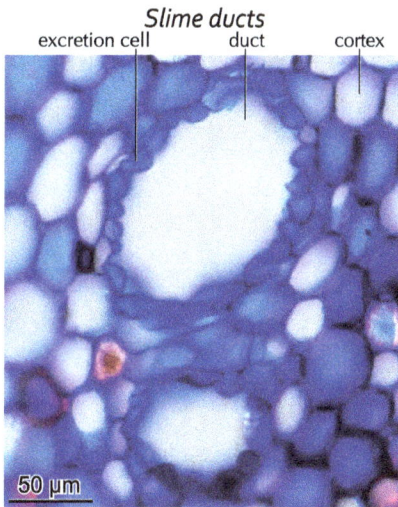

7.167 Ducts with small, unlignified excretion cells in *Encephalartos hildebrandtii*.

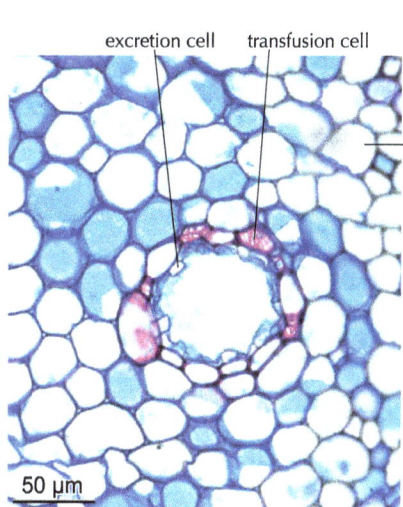

7.168 Ducts with small, unlignified excretion cells and lignified pitted cells in *Dioon edule*.

Isolated lignified cells in the pith (transfusion cells)

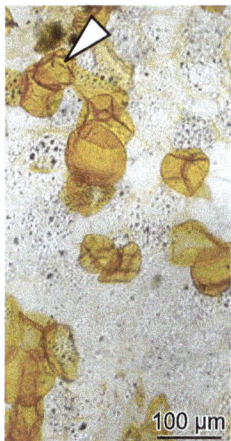

7.169 Cells with unstructured walls in *Macrozamia moorei*.

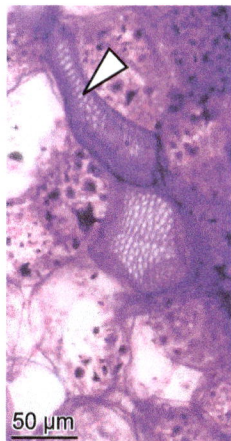

7.170 Cells with pitted walls in *Macrozamia pauliguilielmi*.

Crystals in parenchyma cells

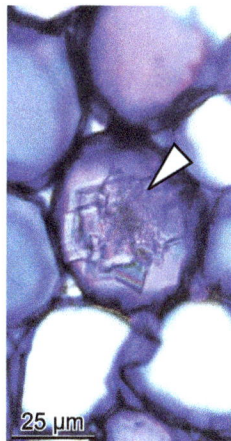

7.171 Druses in *Zamia pygmaea*.

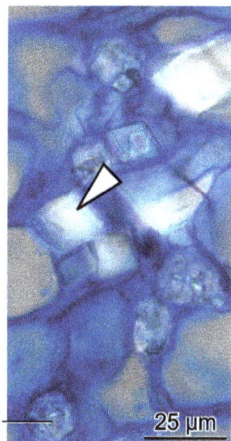

7.172 Prismatic crystals in *Cycas revoluta*.

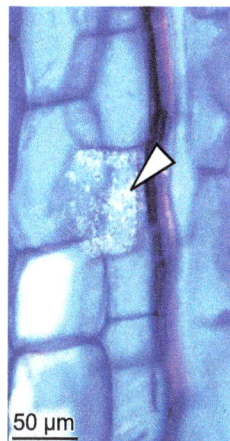

7.173 Crystal sand in *Encephalartos hildebrandtii*.

7.4.2 Ginkgoaceae

Ginkgo biloba is the only living species in the family of Ginkgoaceae. Fan-shaped leaves are characteristic for the deciduous tree. The species is native to southwestern China. This "living fossil" is frequently cultivated in temperate zones.

The stem/root anatomy of *Ginkgo biloba* has been described in detail by Greguss 1955.

The conifer-like xylem with annual rings is a product of secondary growth. Large earlywood and small latewood tracheids separate the square, radially arranged tracheids within the annual ring. Bordered tracheid pits are arranged in axial uni- to triseriate rows. Pit apertures are round in the earlywood and oval in the latewood. Axial parenchyma cells do not exist. Ray height varies between two and five cells. Ray cells are not lignified. Ray pits are intensively bordered and have slit-like apertures (taxodioid). Crystal druses occur in axial elongated chambers.

The phloem is characterized by alternating layers of sieve cells and lignified, thick-walled fibers. Companion cells are absent. Sieve fields on sieve cells occur on radial walls. A few resin ducts occur in the pith and the phloem. Sclereids occur, but are rare. Large crystal druses and large quantities of crystal sand are characteristic. The rhytidome of older bark contains many layers of phellem.

Morphological aspect

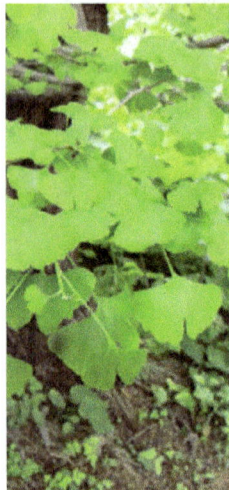

7.174 Leaves of *Ginkgo biloba*.

Xylem

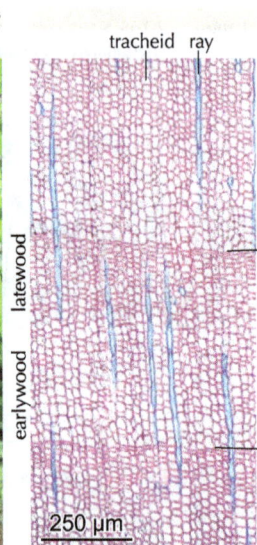

7.175 Tracheids in radial rows.

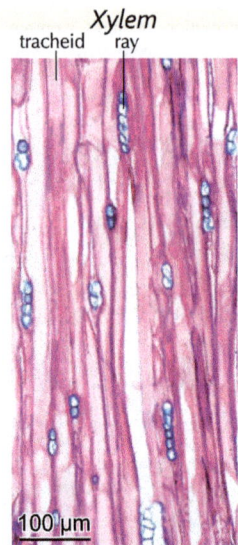

7.176 Short, unlignified, uniseriate rays.

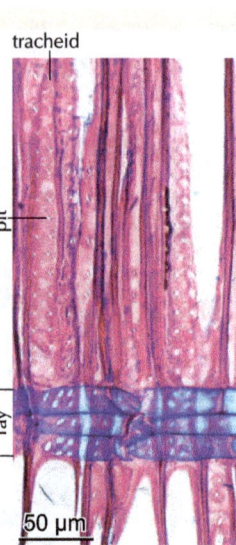

7.177 Tracheids and rays with pits.

Periderm

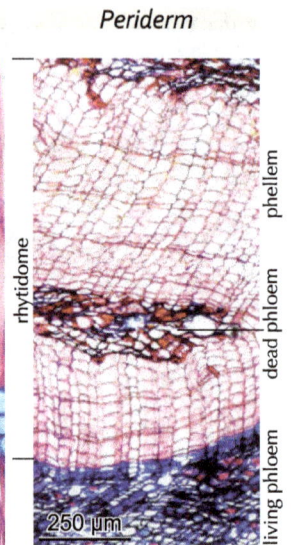

7.178 Phloem and rhytidome (phellogen and phellem).

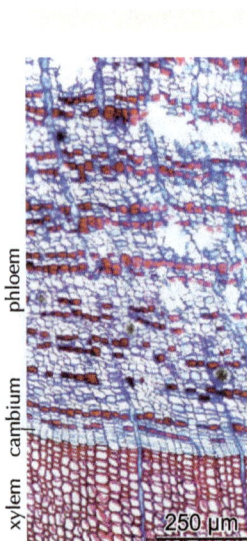

7.179 Alternating sieve cells and fibers.

Phloem

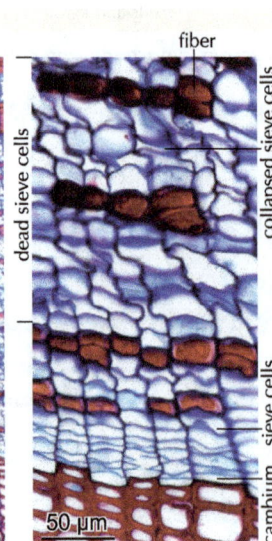

7.180 Active and dead phloem.

7.181 Sieve fields on radial walls of sieve cells.

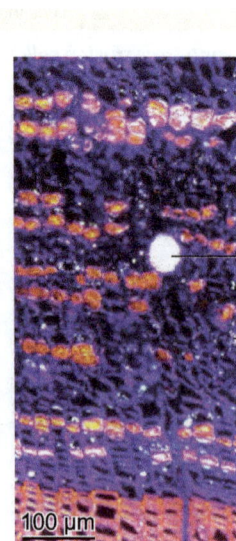

7.182 Crystal druses and crystal sand.

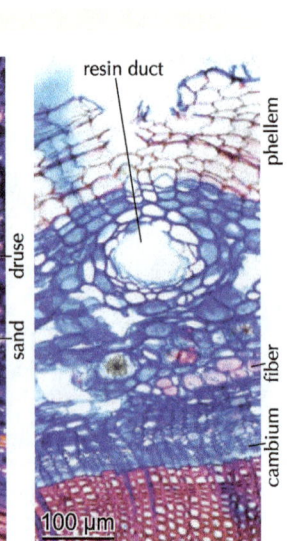

7.183 Resin duct in the phloem.

7.4.3 Conifers

Today, seven families (Pinaceae, Araucariaceae, Podocarpaceae, Cephalotaxaceae, Cupressaceae, Taxodiaceae, Sciadopityaceae) are recognized within the conifers worldwide, together containing approximately 630 species. Conifers of the Pinaceae dominate the boreal zone in the Northern Hemisphere. Araucariaceae and Podocarpaceae are families of the Southern Hemisphere.

Plant growth forms and the forms of reproduction organs greatly vary. The presence or absence of heartwood is characteristic for many species. Stems on extreme sites reduce the xylem to radial strips.

Secondary growth is characteristic for all conifers. In common is the presence of square, radially arranged tracheids, often separated into earlywood and latewood. Only species growing in seasonal climates form more or less distinct annual rings. The presence or absence of axial parenchyma and axial and radial resin ducts in the xylem is species-specific. Pits on axial tracheids occur in uniseriate (e.g. Pinaceae) or muiltiseriate rows (Araucariaceae). Pit apertures are mostly circular. The absence or presence of ray tracheids differentiates large groups. The form of the ray parenchyma cells is very variable.

The phloem is characterized by sieve cells with radial sieve fields, parenchyma cells, fibers and sclereids.

Conifers with one stem

7.184 *Picea abies*, Norway spruce, on a subalpine meadow.

7.185 *Pinus sylvestris*, Scots pine, on an Atlantic meadow.

Reproduction organs

7.186 Fruits of *Juniperus nana*.

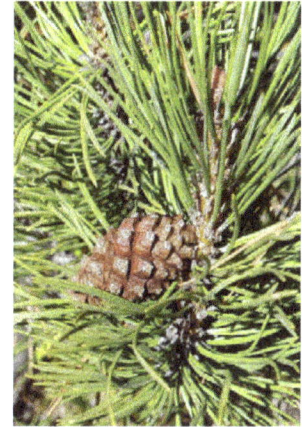

7.187 Cone of *Pinus mugo*.

Stem cross sections

7.188 *Picea abies* without heartwood.

7.189 *Pinus sylvestris* with heartwood and sapwood.

7.190 Radial stem strip with 840 annual rings of *Juniperus sibirica*.

Secondary growth

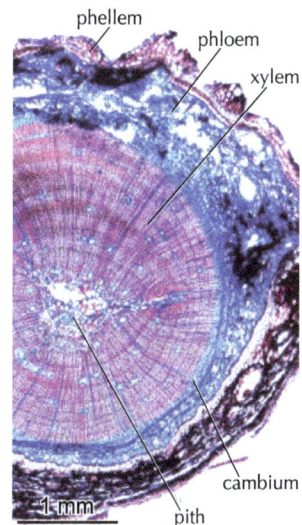

7.191 Twig of *Pinus sylvestris*.

Variable ring distinctness and presence or absence of resin ducts

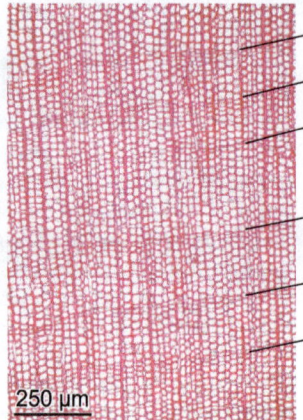

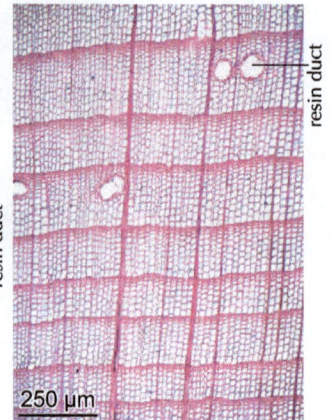

7.192 Distinct ring boundaries, without resin ducts, in *Fitzroya cupressoides*, South America.

7.193 Weak ring boundaries, without resin ducts, in *Podocarpus lambertii*, subtropical South America.

7.194 Distinct ring boundaries and resin ducts in *Pinus banksiana*, boreal North America.

7.195 Distinct ring boundaries and resin ducts in *Picea obovata*, boreal Siberia.

Pits and helical thickenings on tracheids

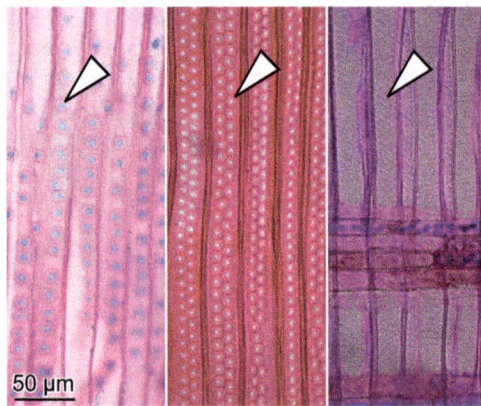

Ray pitting on central and border cells

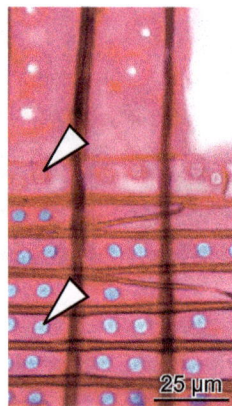

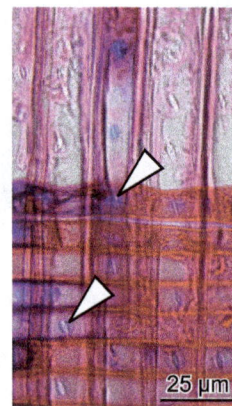

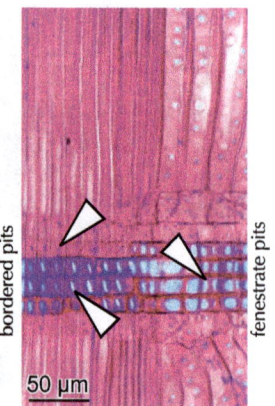

7.196 Left: Uniseriate pits in *Pinus banksiana*. Middle: Biseriate pits in *Araucaria angustifolia*. Right: Helical thickenings in earlywood tracheids in *Pseudotsuga menziesii*.

7.197 Uniform pitting: tracheid and ray pits with round apertures in *Metasequoia glyptostroboides*.

7.198 Uniform pitting: tracheid and ray pits with slit-like apertures in *Podocarpus falcatus*.

7.199 Heterogeneous pitting: fenestrate pits on ray-parenchyma cells and bordered pits with round apertures on ray tracheids and axial tracheids in *Pinus sylvestris*.

Structure of the phloem

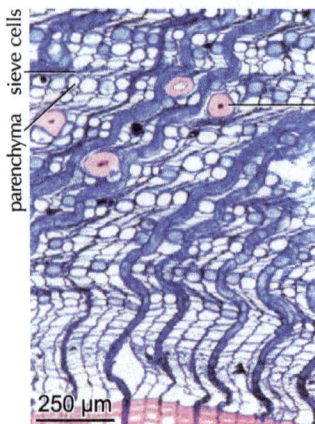

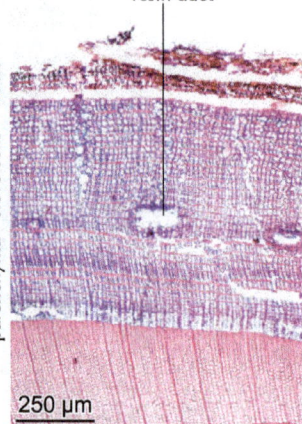

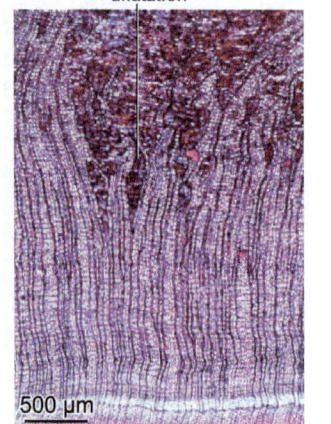

7.200 Alternating rows of sieve cells and parenchyma cells with a few isolated sclereids in *Larix decidua*.

7.201 Alternating rows of sieve cells and parenchyma cells and fibers in *Metasequoia glyptostroboides*.

7.202 Resin ducts in the phloem of *Juniperus communis*.

7.203 Ray dilatation in the phloem of *Podocarpus falcatus*.

7.4.4 Gnetales

Ephedraceae

Ephedra is the only genus within the family of Ephedraceae. All 30–45 leaf-less species grow on dry sites, mostly in arid regions. Their growth forms vary from dwarf shrubs and shrubs to lianas (*Ephedra campylopoda*).

Secondary growth is characteristic for all *Ephedra* species, and growth rings are generally distinct. The xylem is composed of vessels, tracheids, "fiber tracheids" and rays. Foraminate perforation plates with distinct borders characterize vessels. Cell walls of vessels and tracheids frequently contain helical thickenings and bordered pits with round apertures and distinct, unlignified tori. "Fiber tracheids" are hybrids between parenchyma cells and

tracheids: pits are simple (parenchyma-like), but horizontal walls are absent (tracheid-like). Classical parenchyma cells were not observed. Crystal sand is frequently present in most unlignified parts of the xylem and phloem. Dark-stained substances occur mainly in the pith.

The phloem is characterized by parenchyma cells, fibers, sclereids and sieve elements (sieve cells) with sieve fields on radial walls. Companion cells are absent. All large rays are dilated. Older stems contain a distinct, multilayered rhytidome.

See also Carlquist 1992.

Macroscopic aspect

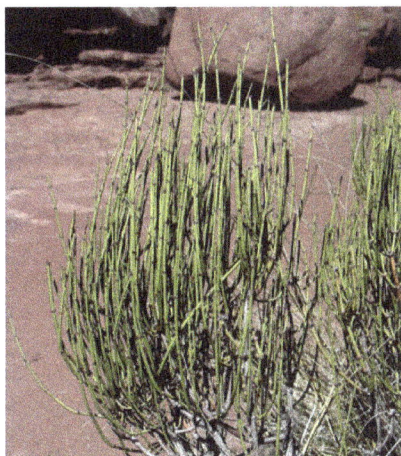

7.204 *Ephedra* sp. on a dry site in southwestern North America.

7.205 *Ephedra* sp. with flowers. Photo: A. Moehl.

7.206 *Ephedra distachya* ssp. *helvetica* with fruits. Photo: A. Moehl.

Cross sections

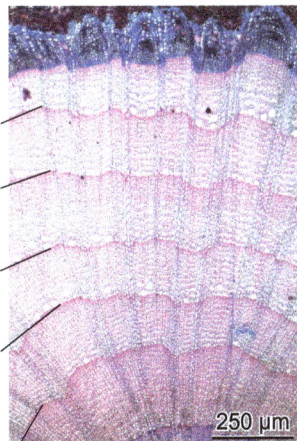

7.207 Dwarf shrub *Ephedra nebrodensis* with distinct annual rings.

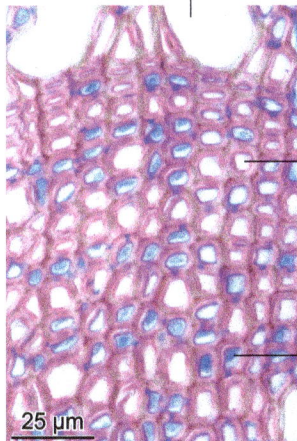

7.208 Latewood of *Ephedra trifurcata* with tracheids, "fiber tracheids" and vessels.

7.209 Liana *Ephedra campylopoda* with irregular radial growth.

7.210 Shrub *Ephedra gerardiana* on a dry site at high altitude, Ladakh, India, 4,400 m a.s.l.

Structure of conducting elements

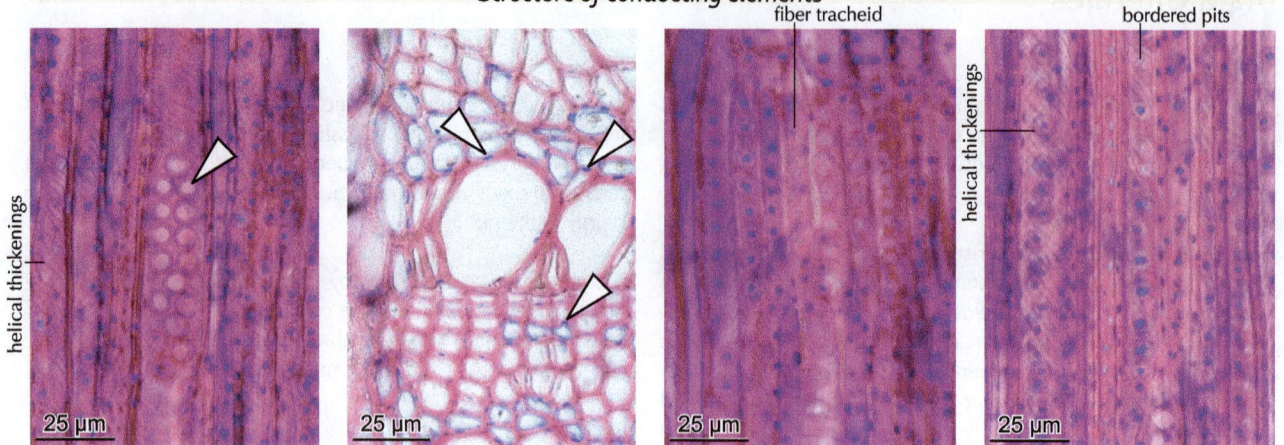

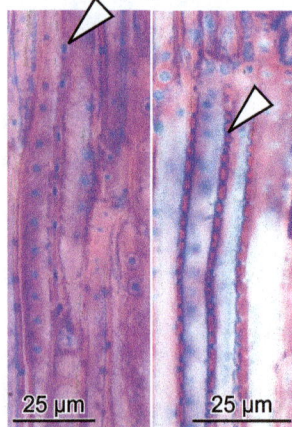

7.211 Vessel with a foraminate perforation in *Ephedra viridis*.

7.212 Pits with unlignified tori in "fiber tracheids" in *Ephedra viridis*.

7.213 Vessel and tracheids with bordered pits and a "fiber tracheid" with simple pits in *Ephedra viridis*.

7.214 Tracheids with bordered pits and helical thickenings in *Ephedra distachya*.

Conducting elements

7.215 Longitudinal section of *Ephedra trifurcata*. Left: Tracheids with bordered pits. Right: "Fiber tracheids" with simple pits.

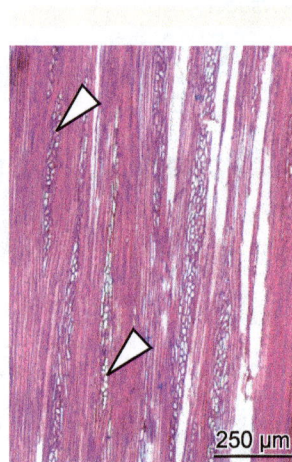

7.216 Tri- to six-seriate rays with irregularly formed cells in *Ephedra viridis*.

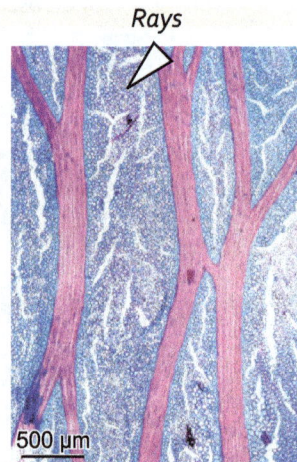

Rays

7.217 Extremely large, unlignified rays between lignified tracheid/vessel strips in *Ephedra gerardiana*.

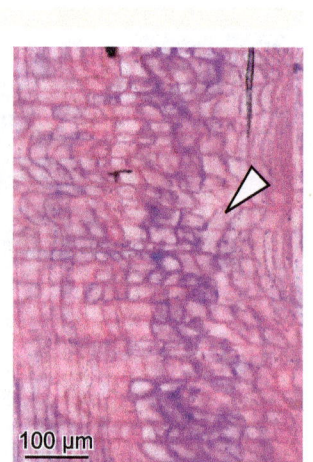

7.218 Prostrate and square ray cells in *Ephedra trifurcata*.

Ray pitting

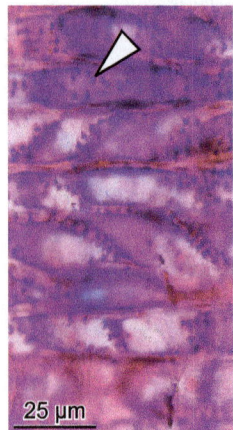

7.219 Lateral walls with simple pits in *Ephedra viridis*.

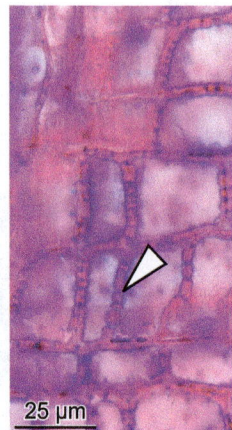

7.220 Axial and radial walls with simple pits in *Ephedra viridis*.

Crystal sand

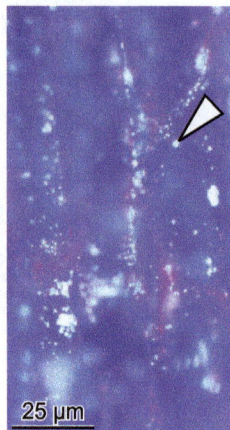

7.221 Crystal sand in *Ephedra gerardiana*.

Bark

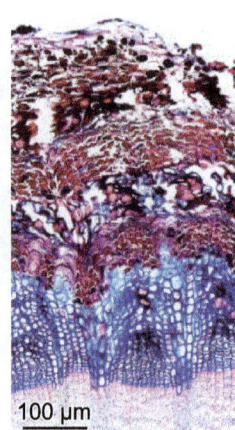

7.222 Living phloem and dead phellem/phloem parts (rhytidome) in *Ephedra nebrodensis*.

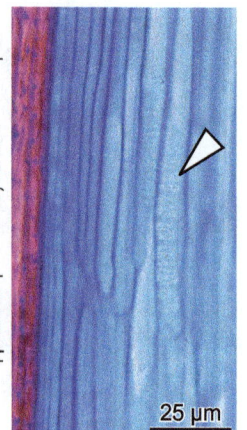

7.223 Sieve fields on radial walls of sieve cells in *Ephedra viridis*.

Welwitschiaceae

Welwitschia mirabilis is the only species within the family of Welwitschiaceae. The below-ground stem/root and the above-ground continuously growing two leaves are characteristic. The plant grows in the arid zone of Namibia and Angola. The stem/root anatomy was described in detail by Carlquist & Gowans 1995.

Anatomy of the leaf

A layer of palisade cells is situated between anatomically undifferentiated epidermal surfaces. Open collateral vascular bundles are located in the central parenchymatic tissue.

Anatomy of the female flower stalk

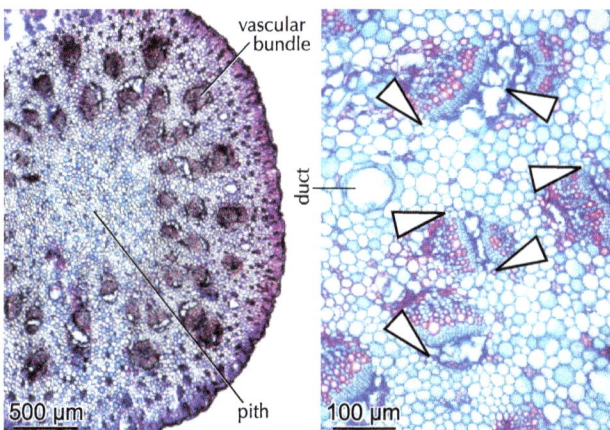

Collateral open vascular bundles are embedded in a parenchymatic tissue and a few ducts. The irregularly distributed bundles are arranged around a pith. The bundles consist of tracheids with annular thickenings and round, bordered pith-like structures. Phloem cells expand shortly after their formation and remain as collapsed structures.

Stem/root

The arrangement of the xylem and phloem is normally chaotic, however, in central parts of the stem successive cambia produce several layers of xylem/phloem zones. A circular, closed xylem/phloem is absent, a lateral vascular cambium seems to be absent, therefore the form of vascular bundles remains. Thin-walled, unlignified rays separate the bundles. The xylem consists of thick-walled vessels with simple perforation plates, tracheids and thin-walled, unlignified parenchyma cells. Radial vessel and tracheid walls contain bordered pits without tori and with round to slit-like apertures. Pits are often arranged in alternate position. Parenchyma is pervasive. In parenchymatic zones, including rays, there are very thick-walled fibers, containing a mantle of small prismatic crystals. Crystal sand occurs in most parenchyma cells.

The phloem consists primarily of very thick-walled gelatinous fibers. A few thin-walled sieve elements occur between them.

Morphology of the plant

7.224 *Welwitschia mirabilis* in the Namibian desert. Photo: P. Poschlod.

Anatomy of leaves

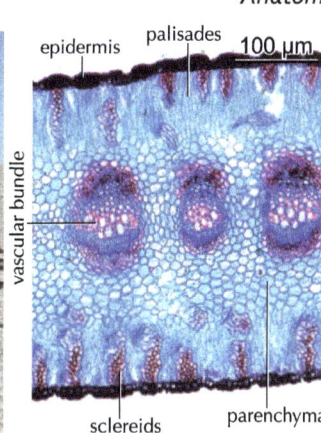

7.225 Leaf cross section.

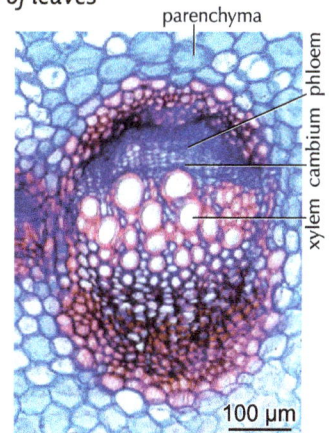

7.226 Open collateral vascular bundle.

Anatomy of the female flower stalk

7.227 Distribution of vascular bundles.

7.228 Randomly oriented vascular bundles in a parenchymatic tissue containing ducts.

Anatomy of vascular bundles in flower stalks

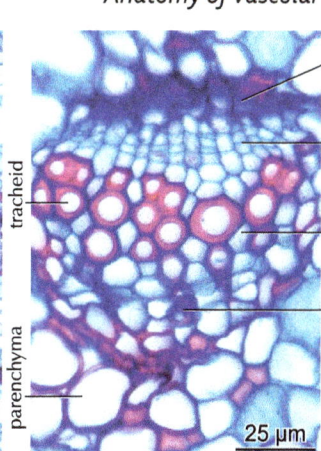

7.229 Cambium between the xylem and the phloem.

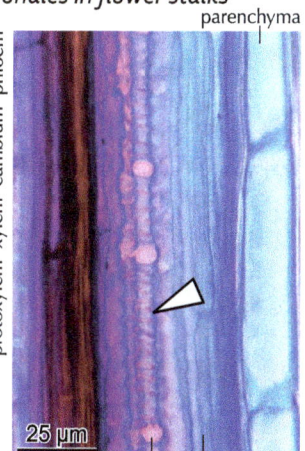

7.230 Annular thickenings in tracheids.

Anatomy of the stem/root

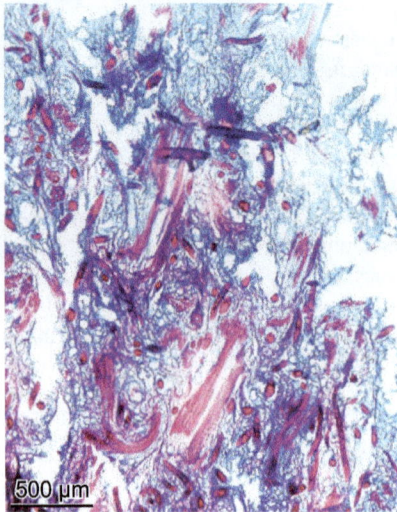

7.231 Chaotic orientation of the tissue in external parts of stems.

Anatomy of the stem

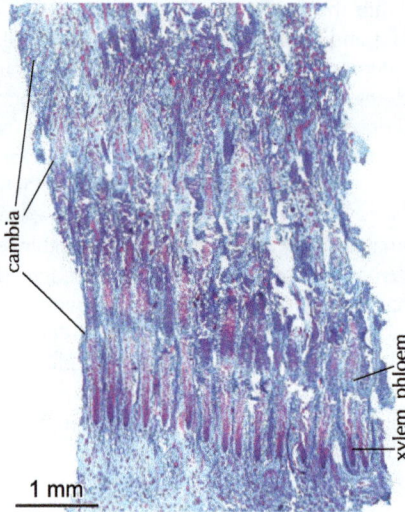

7.232 Successive cambia between radially elongated vascular bundles in internal parts of stems.

Anatomy of vascular bundles

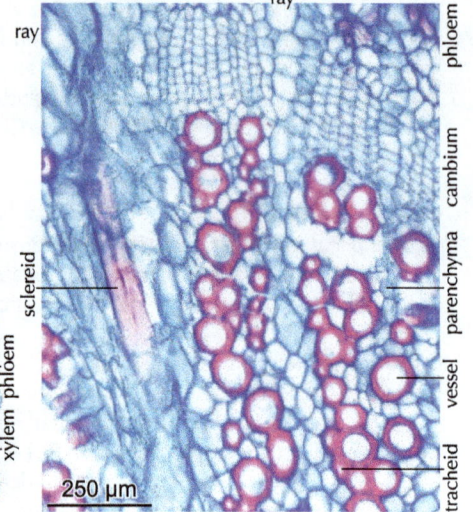

7.233 Cambial zone of an open collateral bundle.

Anatomy of vascular bundles in the stem/root

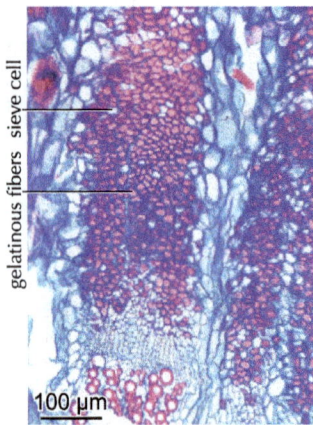

7.234 Phloem with thick-walled gelatinous fibers.

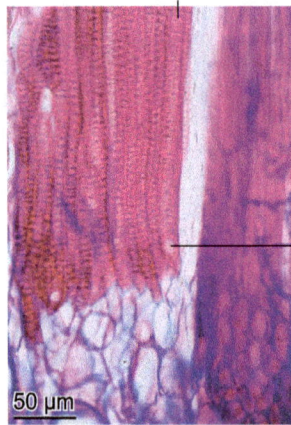

7.235 Vessels with perforation plates and tracheids.

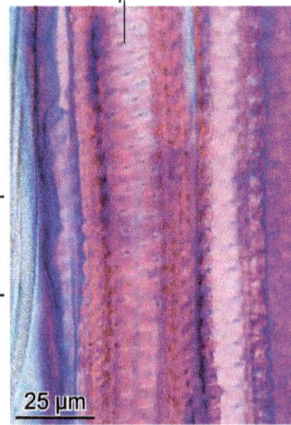

7.236 Biseriate bordered pits.

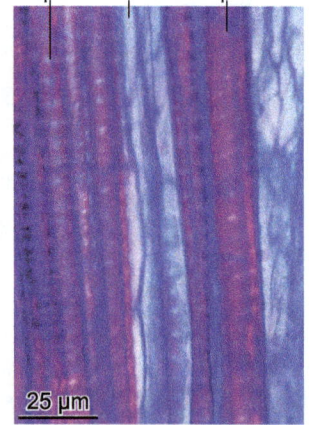

7.237 Uniseriate bordered pits.

Sclereids and crystals

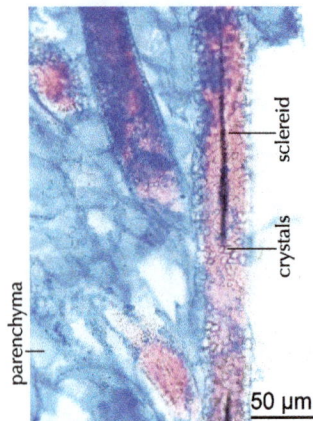

7.238 Elongated sclereids with small prismatic crystals.

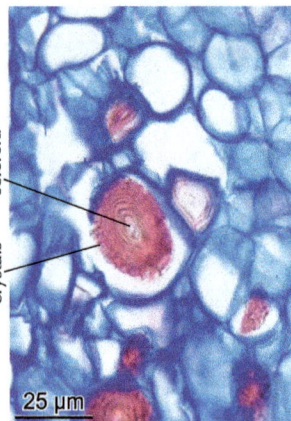

7.239 Layered sclereids surrounded by crystals.

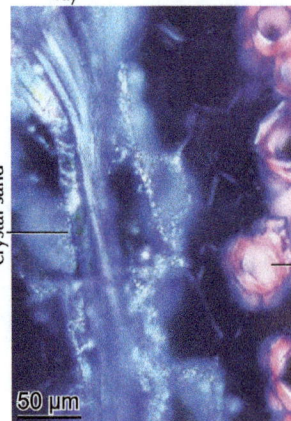

7.240 Crystal sand in thin-walled parenchyma cells.

Anatomy of the cortex

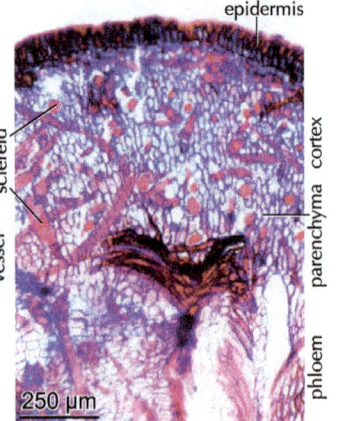

7.241 Thin-walled parenchyma and sclereids.

Gnetaceae

Gnetum is the only genus within the family of Gnetaceae. All 30 species, lianas and small trees with broad leaves, grow in the tropics. Described here is the small tree *Gnetum gnemon* from the Philippines. Its anatomy is described in detail by Carlquist 1996.

Secondary growth is characteristic. Growth rings are generally absent, however, density variations indicate intra-annual differences in climatic growing conditions. The xylem is composed of vessels, tracheids, septate fibers, axial parenchyma cells and rays. Vessels are solitary. Simple and foraminate perforation plates with distinct borders characterize vessels, with both types occurring within the same individual. Vessel walls contain small, vestured pits with oblique apertures. Radial walls of septate tracheids are perforated by large bordered pits. Round apertures and distinct unlignified tori are characteristic. Helical thickenings occur, but are rare. Horizontal walls of septate fibers are not lignified. Axial parenchyma is dominantly vasicentric paratracheal and occasionally apotracheal. The width of the homocellular rays with prostrate cells varies between one and five rows. Ray pits are slightly to distinctly bordered. Many small prismatic crystals are deposited in ray and pith cells.

The phloem is characterized by a few parenchyma and sieve elements (sieve cells) with sieve fields on radial walls. Companion cells are absent. Sieve cells collapse soon after formation. Rays are dilated. Gelatinous fibers and small crystals occur in the cortex and the phloem. Sclereids form a band between the cortex and the phloem. The phellem is interrupted by lenticels.

Morphology of the plant

7.242 Broad, thin leaves of *Gnetum gnemon*.

7.243 Breaking zone of a twig (apoptosis).

Cross sections

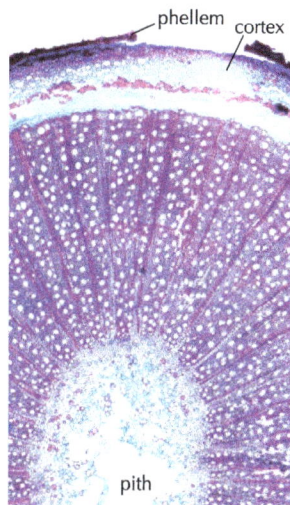

7.244 Pith, xylem and bark.

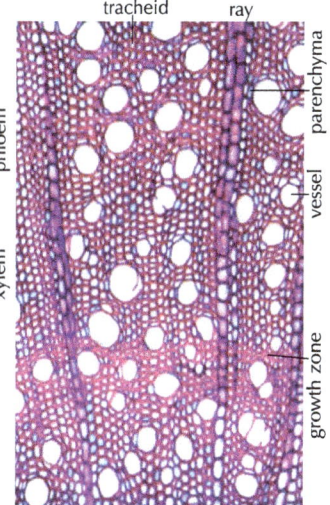

7.245 Xylem with a growth zone.

Perforation plates and pits

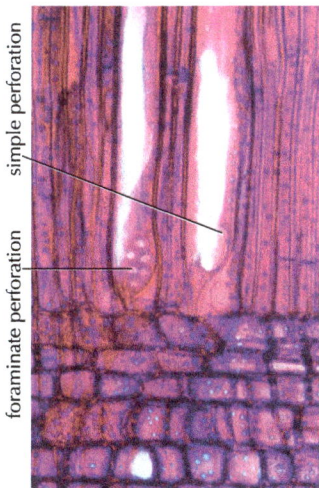

7.246 Perforation plates.

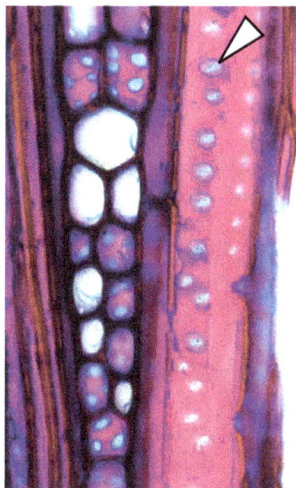

7.247 Vestured intervessel pits.

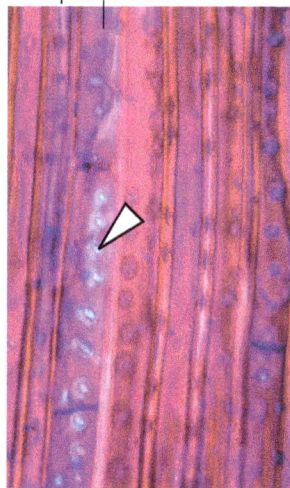

7.248 Bordered pits on septate tracheids.

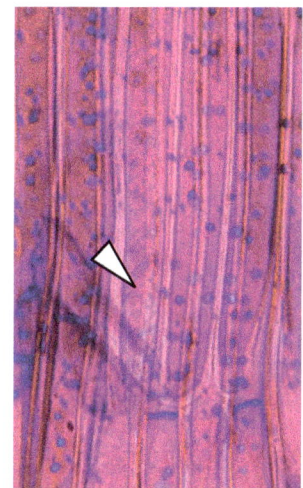

7.249 Thin helical thickenings in tracheids.

Rays

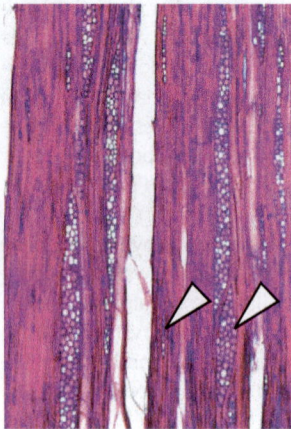

7.250 Uni- to five-seriate homocellular rays.

7.251 Homocellular ray with prostrate and square cells.

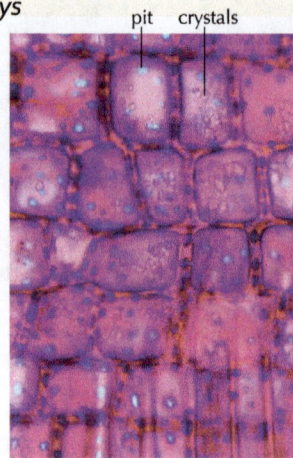

7.252 Ray cells with pits and small prismatic crystals.

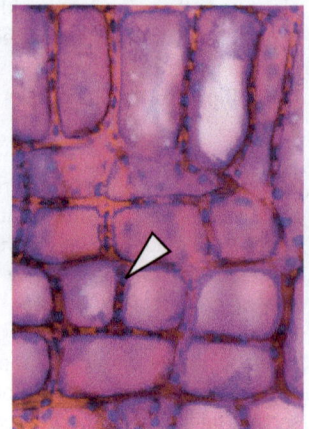

7.253 Ray cells with bordered pits.

Bark

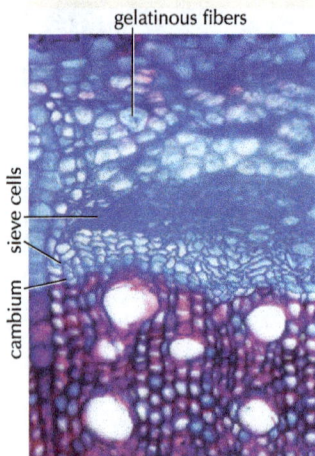

7.254 Conducting and collapsed sieve elements and gelatinous fibers.

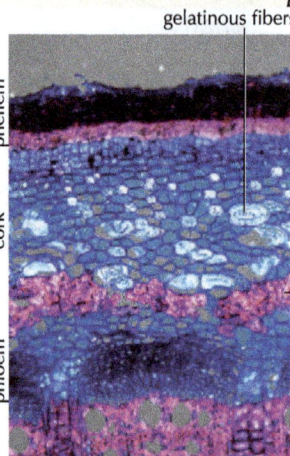

7.255 Phloem, cortex with sclereids and gelatinous fibers and periderm.

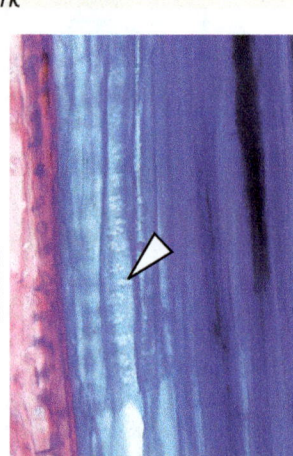

7.256 Sieve fields in sieve cells.

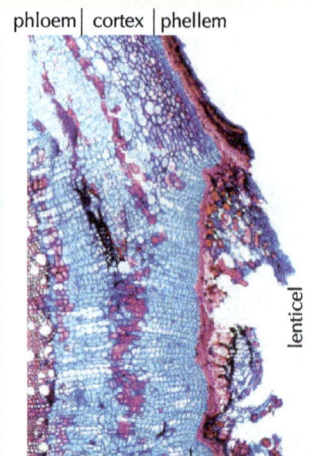

7.257 Lenticel.

Gnetales: Conifers or Angiosperms?

Gnetales are seed plants (spermatophyta) that are taxonomically related to conifers. Bordered pits on tracheids in the xylem confirm that. However, many stem-anatomical features show relations to angiosperms. A few features are common to all three families: the presence of vessels, tracheids with large bordered pits and small crystals in parenchyma cells and sclereids in the bark. Each genus has its own "specialty".

Ephedra mainly forms leafless shrubs and prefers arid climates or dry sites in the Northern and Southern Hemisphere. The xylem has no axial parenchyma but "fiber tracheids" with simple pits. All perforation plates are foraminate.

Welwitschia mirabilis forms a subterranean stem and two continuously growing leaves. The plant is geographically isolated and occurs only in the desert of Namibia. The xylem has successive cambia and pervasive parenchyma. Collateral vascular bundles remain and do not grow together into a compact belt of xylem.

Gnetum grows as lianas and small trees with broad leaves. All species grow in moist tropical climates. The xylem consists of solitary vessels, tracheids and septate tracheids and paratracheal parenchyma. Perforations are simple and foraminate. A special feature are vestured vessel pits and bordered ray pits.

In common for *Welwitschia* and *Gnetum* is the presence of gelatinous fibers in the phloem and the cortex.

In conclusion: Anatomical features in the xylem relate all the three families to the angiosperms. Solely the bordered pits indicate a relation to the conifers. It remains unclear to what extent site differences and geographical isolation drove stem evolution in such different directions.

7.4.5 Angiosperms: Monocotyledons and their growth forms

The monocotyledons are extremely manifold. There are approximately 60,000 species within 11 orders. Monocotyledons grow in all vegetation zones from tropical to arid and arctic zones, and in all habitats from extremely dry to submersed sites. Growth forms vary from little herbs to lianas and large trees (palms).

Presented are exemplarily some few species and growth forms of different taxonomic groups in various habitats. This selection gives an impression of the anatomical variability within the monocots. So far, there are hardly any stem anatomical features observed that are common for the entire group.

Palms (Arecaceae)

Within the Arecaceae family, there are approximately 2,600 species. They occur preferentially in tropical regions. A common characteristic for all species is the absence of secondary growth. Described here is the date palm, *Phoenix dactylifera*. Vascular bundles are irregularly distributed over the whole stem cross section (atactostele). Closed collateral vascular bundles contain many small vessels with scalariform pits. The bundles are surrounded by thick-walled sheaths of fibers. See also Tomlinson *et al.* 2011.

Macroscopic aspect

7.258 Date palm (*Phoenix dactylifera*) in an oasis of the Sahara desert.

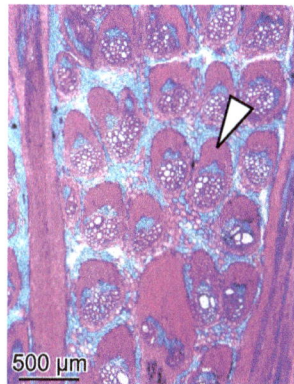

7.259 Irregular distribution of vascular bundles (atactostele) in *Phoenix dactylifera*.

Microscopic aspect

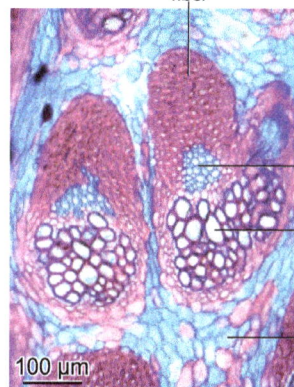

7.260 Closed collateral vascular bundles with dense sheaths in *Phoenix dactylifera*.

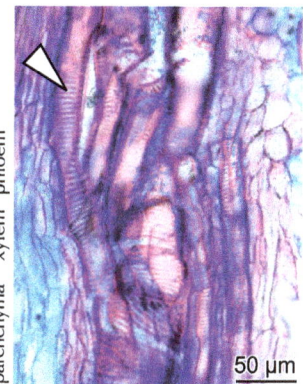

7.261 Vessels with scalariform pits in *Phoenix dactylifera*.

Bamboo (Poaceae)

The more than 1,000 bamboo species occur primarily in southern Asia and South America. Most of the species form heavily lignified, straight stems. Here, the species *Phyllostachys bambusoides* is anatomically described. Vascular bundles are irregularly distributed over the whole stem cross-section (atactostele). Closed collateral vascular bundles are composed of two large vessels and an external group of phloem. Vessels are characterized by a large perforation plate in almost horizontal position, many small pits perforate lateral vessels. Sieve plates in sieve tubes occur as transversely perforated walls. A thick-walled fiber sheath surrounds vascular bundles. More information in Grosser & Liese 1971.

Macroscopic aspect

7.262 Bamboo (*Chusquea* sp.) in the Asian tropics, 20 m tall.

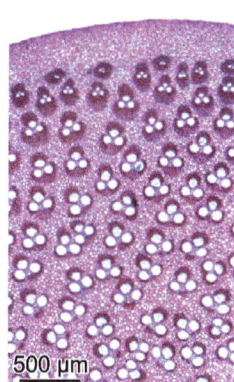

7.263 Irregular distribution of vascular bundles (atactostele) in *Phyllostachys bambusoides*.

Microscopic aspect

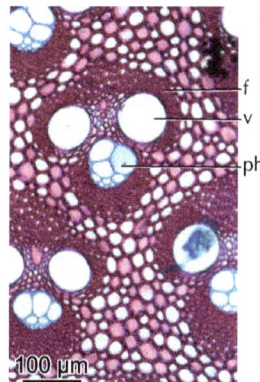

7.264 Closed collateral vascular bundles in *Phyllostachys bambusoides*.

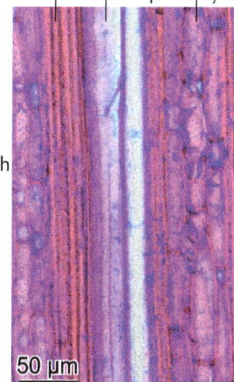

7.265 Vessel with small pits, fibers and parenchyma cells in *Phyllostachys bambusoides*.

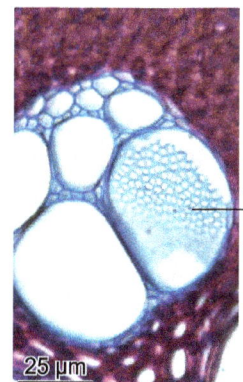

7.266 Phloem with very large sieve tubes in *Phyllostachys bambusoides*.

Grass-like terrestrial herbs (Cyperaceae)

More than 5,000 species occur from the tropics to the arctic on extremely dry sites as well as on lake shores. The outlines of culms are triangular, but often roundish. Presented here are one species from a wet site (*Carex foetida*) and another from a dry, alpine site (*Kobresia simpliciuscula*).

Secondary growth is absent. The closed collateral vascular bundles are composed of a xylem with a few enlarged vessels and a round group of phloem. The bundles are often surrounded by a belt of thick-walled fibers. See also Metcalfe 1971.

Macroscopic aspect

Cross sections of culms

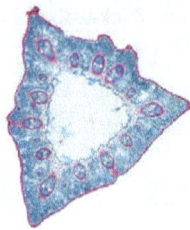

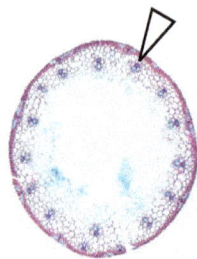

7.267 *Carex pendula*.

7.268 Triangular culm of the 10 cm-tall grass-like herbaceous *Carex foetida* on an alpine moist site.

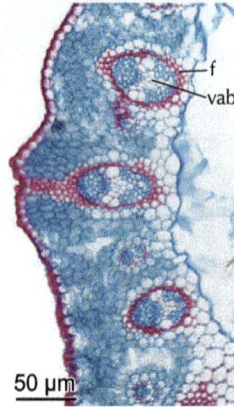

7.269 A belt of thick-walled, lignified fibers surrounds the closed vascular bundles in *Carex foetida*.

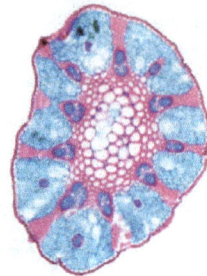

7.270 Round culm of the 10 cm-tall grass-like herbaceous *Kobresia simpliciuscula* on an alpine, dry meadow.

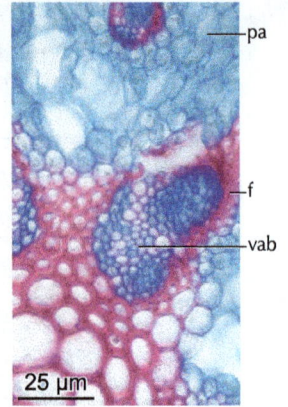

7.271 A belt of thick-walled, lignified fibers surrounds the closed vascular bundles in *Kobresia simpliciuscula*.

Terrestrial grasses (Poaceae)

More than 10,000 grass species occur from the tropics to the arctic on extremely dry sites as well as wet sites like lake shores. The outlines of culms are mostly roundish. Presented here are a 30 cm-tall grass species from a ruderal site (*Hordeum murinum*) and a 4 m-tall species from a moist Mediterranean site (*Arundo donax*). Secondary growth is absent. The collateral closed vascular bundles are composed of a xylem with a few enlarged vessels and a round group of phloem. A belt of thick-walled fibers often surrounds the bundles. See also Schweingruber & Berger 2017.

Macroscopic aspect

Cross sections of culms

7.272 *Hordeum vulgare*.

7.273 Round culm of *Hordeum murinum*, a 30 cm-tall grass. Vascular bundles are circularly arranged (siphonostele).

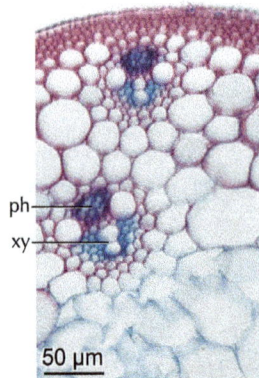

7.274 A belt of thin-walled, lignified fibers surrounds the closed vascular bundles in *Hordeum murinum*.

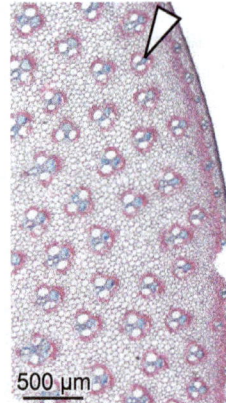

7.275 *Arundo donax*, a 4 m-tall reed. Vascular bundles are arranged in a Fibonacci spiral pattern (atactostele).

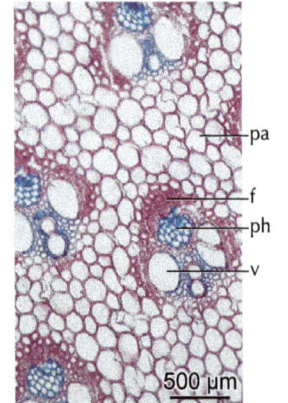

7.276 A belt of thick-walled, lignified fibers surrounds the closed vascular bundles in *Arundo donax*.

Orchids (Orchidaceae)

Approximately 20,000 autotroph and parasitic species occur from the tropics to the arctic on dry as well as on wet sites. The outlines of culms are mostly roundish. Presented here are a 10 cm-tall upright species from a dry site (*Spiranthes spiralis*) and a 40 cm-tall species from a moist site (*Epipactis atrorubens*). The closed collateral vascular bundles are composed of a xylem with a few small vessels and phloem. See also Stern 2014.

Macroscopic aspect

Cross sections of a culm

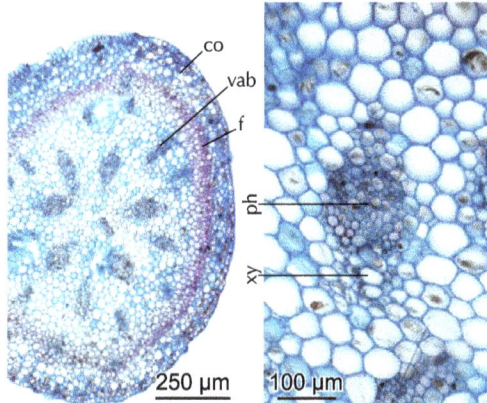

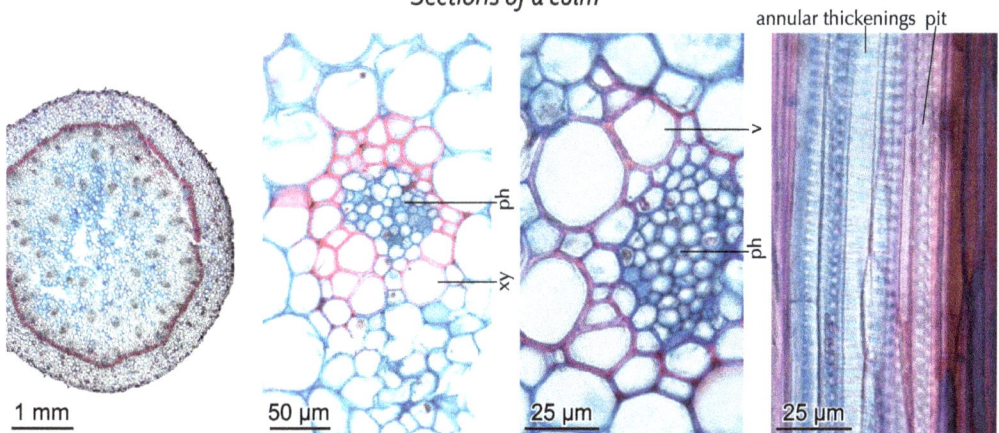

Cross sections of a bulb

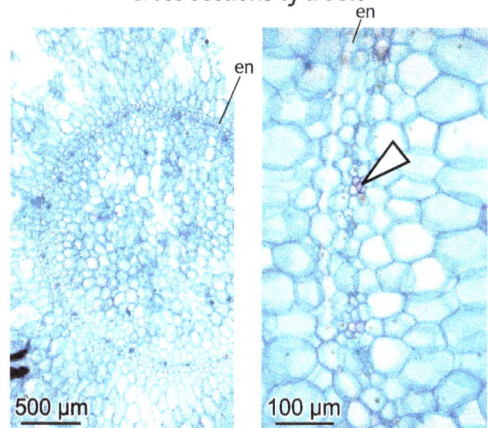

7.277 *Spiranthes spiralis*, a 10 cm-tall species of a dry site in the temperate zone.

7.278 Vascular bundles are irregularly distributed across the whole stem of *Spiranthes spiralis*.

7.279 Xylem and phloem are anatomically not distinctly separated in the vascular bundle of *Spiranthes spiralis*.

7.280 An endodermis and circularly arranged vascular bundles separate the pith and the cortex in *Spiranthes spiralis*.

7.281 Rudimentary vascular bundles inside an endodermis in the bulb of *Spiranthes spiralis*.

Macroscopic aspect

Sections of a culm

7.282 *Epipactis atrorubens*, a 40 cm-tall species of a wet site in the temperate zone. Photo: L.B. Tettenborn, Wikimedia Commons, CC BY-SA 3.0.

7.283 Vascular bundles are irregularly distributed across the whole stem of *Epipactis atrorubens*.

7.284 The xylem surrounds the phloem in the vascular bundle of *Epipactis atrorubens*.

7.285 Phloem and vessels of *Epipactis atrorubens*. Sieve tubes and companion cells are not differentiated in the phloem.

7.286 Annular thickenings in large vessels, and bordered pits in small vessels of *Epipactis atrorubens*.

Lianas

Monocotyledonous lianas occur in various families, e.g. in the Dioscoreaceae, Asparagaceae and Poaceae. Some species are perennial, e.g. the Mediterranean spiny *Smilax aspera*, or the tropical bamboo *Chusquea cumingii*. *Dioscorea communis* or *D. caucasica* have permanent subterranean bulbs and annual liana-like shoots. Characteristic for all liana-like species are collateral vascular bundles with large vessels (150–260 µm in radial diameter).

Macroscopic aspect   *Cross sections of culms/shoots*

Annual lianas

7.287 *Dioscorea communis*, an annual liana.

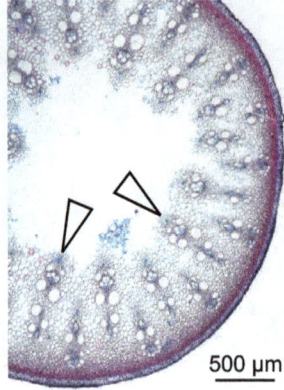

7.288 Annual shoot of *Dioscorea caucasica*, with circular arranged vascular bundles.

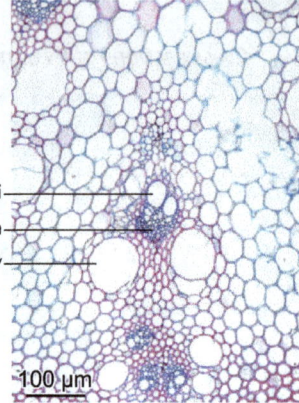

7.289 Closed collateral vascular bundle with large vessels and large sieve tubes in *Dioscorea caucasica*.

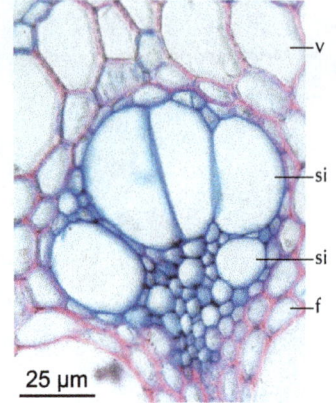

7.290 Phloem with very large and small sieve tubes and small companion cells in *Dioscorea caucasica*.

Perennial lianas

7.291 *Smilax azorica*. Photo: JCapelo via Wikimedia Commons.

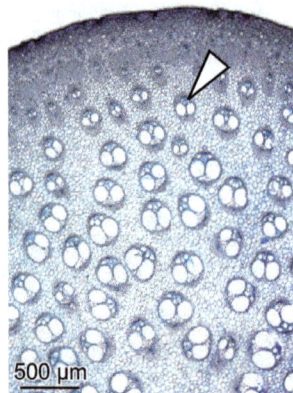

7.292 Irregular distribution of vascular bundles in *Smilax azorica*.

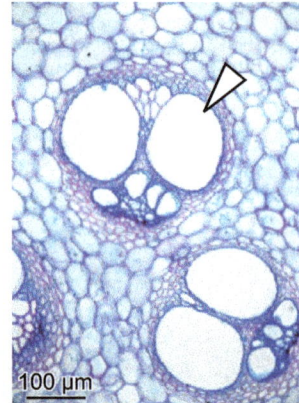

7.293 Xylem of *Smilax azorica* with very large vessels.

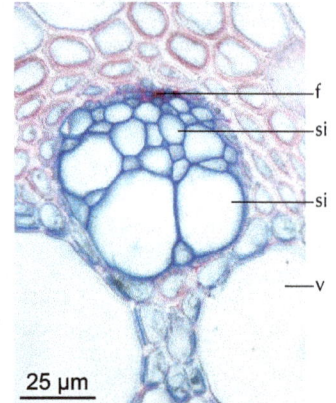

7.294 Phloem of *Smilax azorica* with very large sieve tubes.

Bamboo

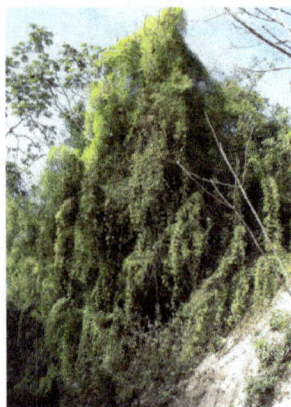

7.295 *Chusquea cumingii*, a tropical bamboo species.

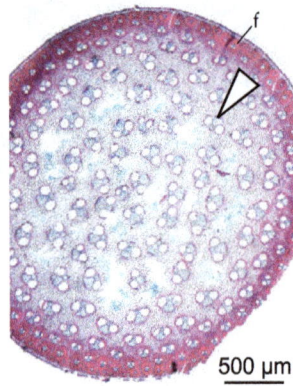

7.296 Irregularly distributed vascular bundles inside of a dense belt of fibers in *Chusquea cumingii*.

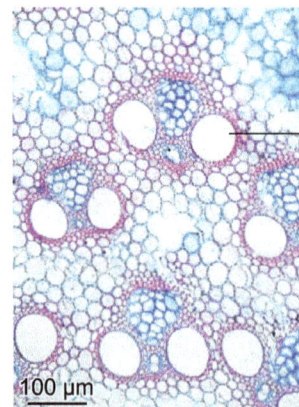

7.297 Xylem of *Chusquea cumingii* with very large vessels.

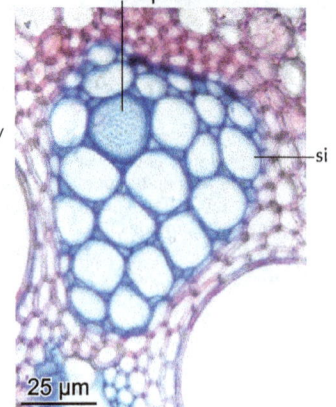

7.298 Phloem of *Chusquea cumingii* with very large sieve tubes.

Hydrophytes

There are species of various families in this life form. Most of them are distributed worldwide in fresh and marine aquatic environments. Presented here are two submerse species from the Potamogetonaceae and Zosteraceae and two floating species from the Hydrocharitaceae and Lemnaceae (today included in the Araceae family).

The submerse species have a central strand of conducting tissues inside of a more or less distinct endodermis. Phloem and xylem are difficult to differentiate (use polarized light). Characteristic for all species is the presence of poorly lignified or unlignified aerenchymatic tissues.

Potamogetonaceae are mainly submerse and floating plants in fresh water. The family includes approximately 120 species. Zosteraceae are submerse marine plants. The family includes two species. Hydrocharitaceae are aquatic plants in both fresh water and marine habitats. The family includes approximately 120 species. Lemnaceae are floating hydrophytes. The exact number of species is unknown.

Potamogetonaceae

Macroscopic aspect

7.299 *Potamogeton pectinatus*, a 40 cm-long submerse aquatic plant.

Cross sections of a shoot

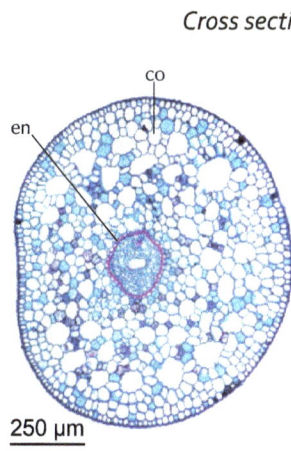

250 µm

7.300 Stem of *Potamogeton pectinatus* with a large aerenchymatic cortex and a central strand which is surrounded by a thick-walled endodermis.

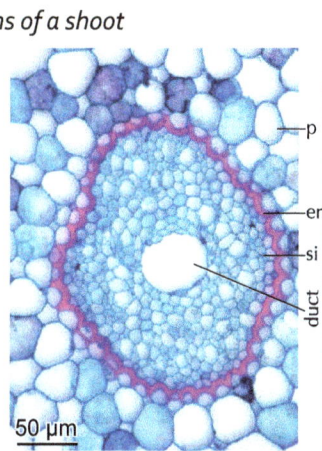

50 µm

7.301 Central strand of *Potamogeton pectinatus* consisting of a central air conducting canal, surrounding sieve tubes and parenchyma cells. The unlignified vessels are difficult to recognize (use polarized light).

Longitudinal section

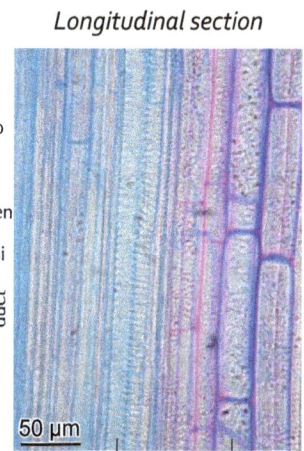

50 µm

7.302 Vessels with helical thickenings in *Potamogeton natans*.

Zosteraceae

Macroscopic aspect

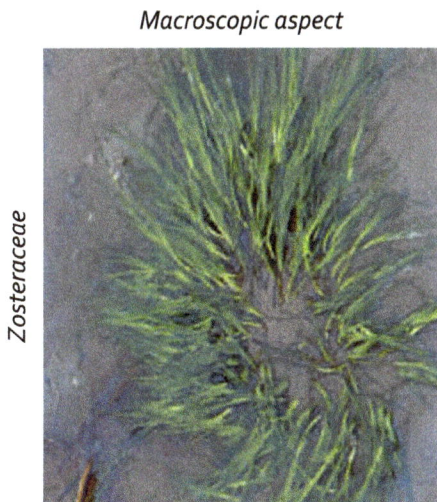

7.303 *Zostera marina*, a submerse marine plant.

Cross sections of a shoot

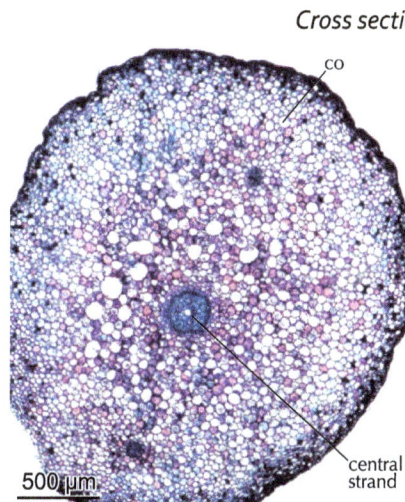

500 µm

7.304 Stem of *Zostera marina* with a large parenchymatic cortex and a central strand which is surrounded by a thin-walled endodermis.

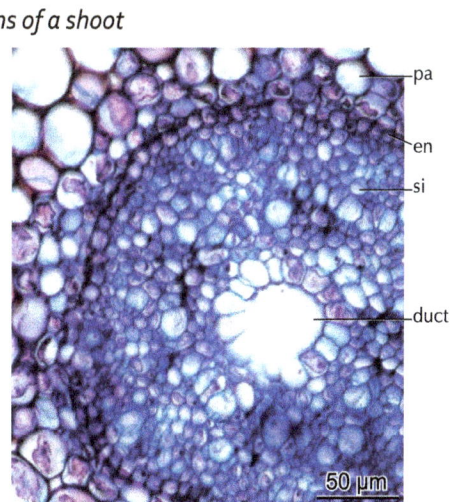

50 µm

7.305 Central strand of *Zostera marina*, consisting of a central air-conducting duct with surrounding sieve tubes and parenchyma cells.

Hydrocharitaceae

Macroscopic aspect

Cross sections of a culm

7.306 *Stratiotes aloides*, a floating aquatic plant.

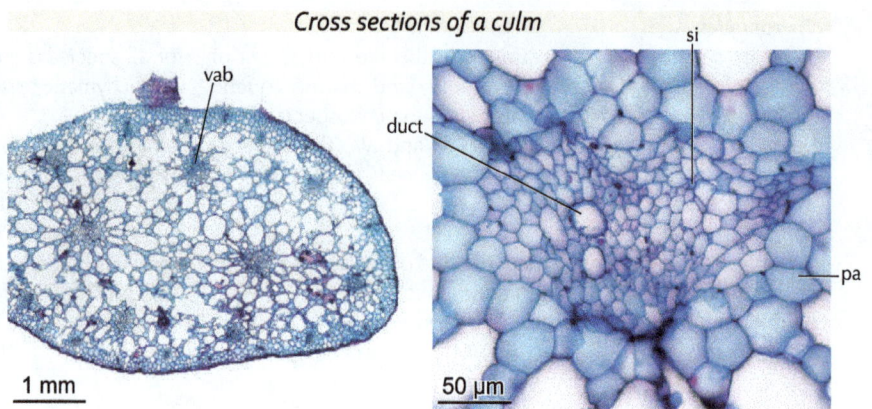

7.307 Culm of *Stratiotes aloides*. Vascular bundles are irregularly distributed within an aerenchymatic tissue.

7.308 Vascular bundles of *Stratiotes aloides* consist of air-conducting tubes with surrounding sieve tubes and parenchyma cells.

Lemnaceae (now Araceae)

Macroscopic aspect

Cross sections of the plant body

7.309 *Lemna minor*, a floating aquatic plant.

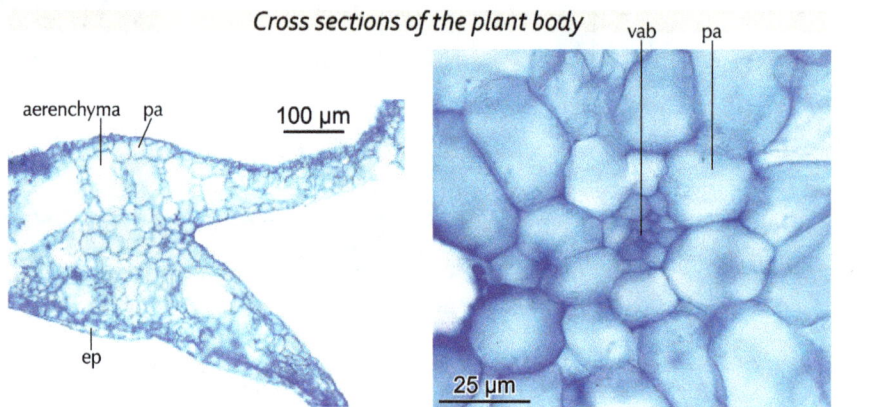

7.310 Plant body of *Lemna minor*, consisting of a thin-walled aerenchymatic tissue.

7.311 Heavily reduced vascular bundle in *Lemna minor*. The central cells probably represent sieve tubes.

Trees and shrubs with secondary growth (Dracaena, Aloe)

Secondary growth is rare in monocots, but it occurs in a few families, e.g. in the Asparagaceae. Secondary growth is different than in dicots. The cambium is located outside of the conducting tissue for water and assimilates. Towards the inside it produces concentric vascular bundles with a phloem in the center, towards the outside it produces a uniform parenchymatic cortex with a periderm at the periphery. Vessels have round, simple pits. See also Chapter 5.2.

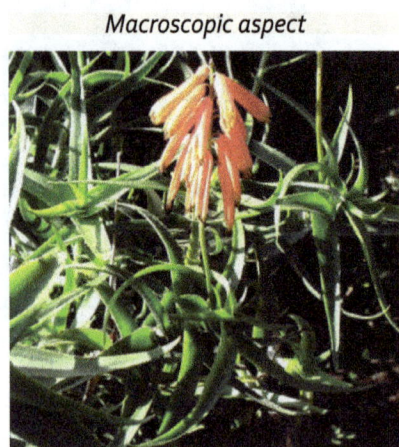

Macroscopic aspect

Cross sections of a stem

7.312 *Aloe* sp., a monocotyledonous plant with secondary growth.

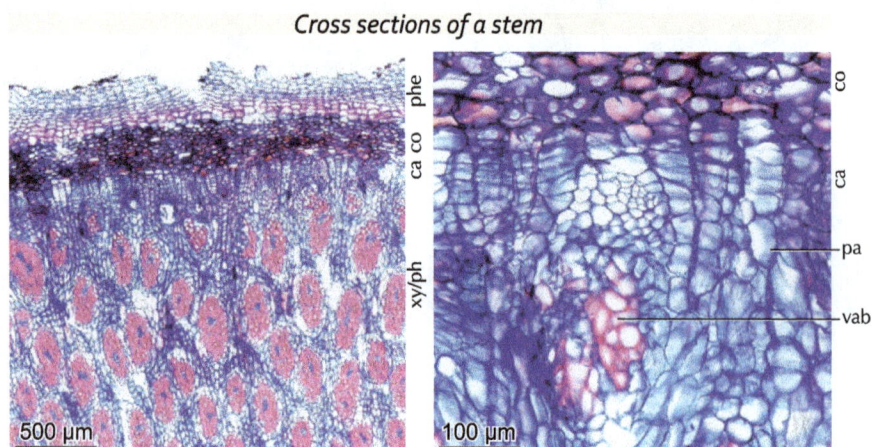

7.313 Xylem, phloem and periderm of *Aloe dhufarensis*.

7.314 Formation of a vascular bundle within the cambial zone in *Aloe dhufarensis*.

7.4.6 Angiosperms: Dicotyledons and their growth forms

The dicotyledons include approximately 210,000 species in the basal orders Magnolids and Eudicotyledons (Rosids and Asterids; Christenhusz & Byng 2016). Growth forms and life forms cover a wide range (see Chapter 3), and vary from little herbs to lianas and large trees. Dicotyledons grow in all vegetation zones from tropical to arid and arctic zones and in all habitats from extremely dry to submerse sites.

Exemplarily presented here are some few species of different growth or life forms in various habitats. Excluded are species with successive cambia (for those see Chapter 6.3). The following short presentation will give an impression of the anatomical variability within the dicotyledons of the temperate zone. Taxonomic characteristics on the level of families are presented in Schweingruber *et al.* 2011 and 2013, and Crivellaro & Schweingruber 2015.

Annual herbs (therophytes)

The height of annual herbs can vary from 3 cm to more than 4 m. They grow during one vegetation season. However, the growing time within this season differs. Some species grow in very early spring and fulfill their entire life cycle within a few weeks, e.g. *Erophila verna*, others grow late in the season and last only for one or two months, and some have a longer life span within one year, e.g. *Helianthus annuus*. The spectrum of the xylem structure varies. It can be very light, with a density of

0.3 g cm^{-3}, thin-walled and hardly lignified (*Erophila*) or heavy, with a density of 0.1 g cm^{-3}, thick-walled and intensively lignified (*Euphrasia*). The term "annual" can be misleading: not a single species has a life span of an entire year. Common for all annuals is the presence of just one growth ring, which is formed in the limited period of one astronomical year. Presented here are one very small and one very large annual plant.

Macroscopic aspect

Small annual plant

7.315 *Erophila verna*, a 3 cm-tall plant growing in early spring, with a life span of about four weeks.

Cross section of a root

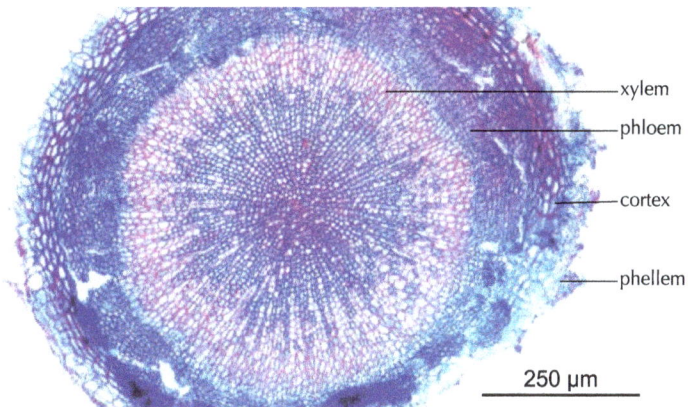

7.316 Root of *Erophila verna* with poorly lignified cells in the center and lignified cells at the periphery of the xylem. Vessels are extremely small. Rays are absent.

Macroscopic aspect

Very tall annual plant

7.317 *Helianthus annuus*, a 200 cm-tall plant growing in late summer and fall, with a life span of about three months.

Cross section of a shoot

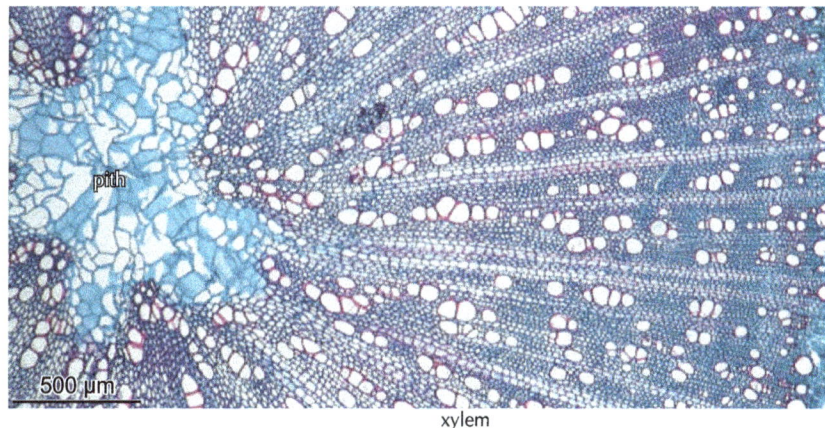

7.318 Xylem of *Helianthus annuus* with thin- to thick-walled fibers, large vessels and very distinct rays.

Perennial herbs (hemicryptophytes and geophytes)

As for annual herbs, the height of perennial herbs can vary from 3 cm to 4 m. Perennial herbs grow over several vegetation periods. The term "herb" is anatomically not clearly defined. Following the common floras (e.g. Aeschimann *et al.* 2004), included in this term are plants with soft, poorly lignified stems or with short, intensively lignified stems. Transitions from herbs to dwarf shrubs are morphologically continuous. Most perennial

herbs fulfill their life cycle during several vegetation periods. Their growth is interrupted during cold or dry seasons (dormant periods). The life span varies from two to approximately 50 years. Annual plants with two growth rings, which germinate in fall, stay dormant during winter and flower in spring (winter annuals), are an exception in temperate regions. The variability in stem structure is as large as in annual plants.

Herb with long rhizomes

Herb with a short rhizome

Perennial herbs

7.319 *Duchesnea indica*

7.320 Rhizome of *Fragaria viridis* with a diffuse-porous xylem and four annual rings.

7.321 *Geranium sanguineum*

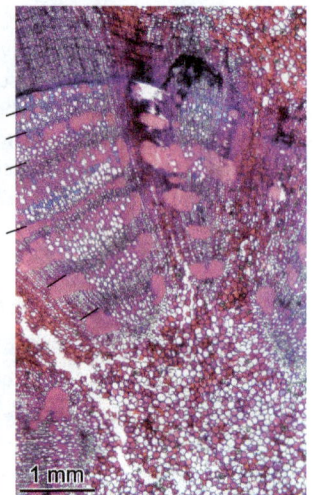

7.322 Rhizome of *Geranium sanguineum* with a semi-ring-porous xylem with seven annual rings.

Cushion plants with a tap root

Perennial herbs

7.323 *Paronychia argentea*

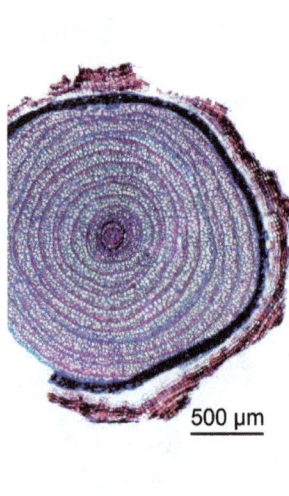

7.324 Taproot of *Paronychia argentea* with a semi-ring-porous xylem with 15 annual rings.

7.325 *Antennaria dioica*, a small plant of colder climates.

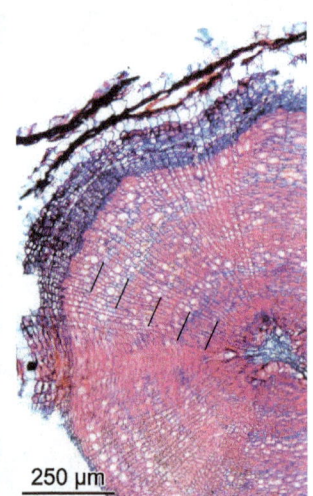

7.326 Taproot of *Antennaria alpina* with a dense, semi-ring-porous xylem with six annual rings.

Dwarf shrubs (chamaephytes and nanophanerophytes)

Included are approximately 5–50 cm-tall, largely branched, perennial plants with hard, intensively lignified (woody) stems. Rhizomes and roots live for approximately five to 250 years.

Dwarf shrubs

Small costal shrub with a taproot

Prostrate dwarf shrub

7.327 *Frankenia ericoides.*

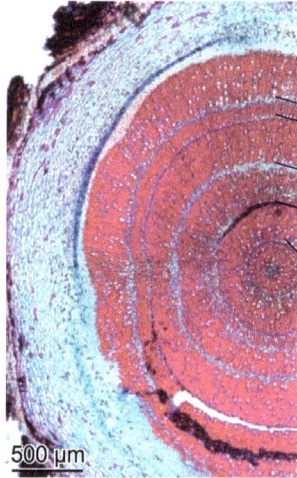

7.328 Taproot of *Frankenia pulverulenta* with a very dense, diffuse-porous xylem with six annual rings.

7.329 *Rhamnus pumila.*

7.330 Stem of *Rhamnus pumila* with a dense, semi-ring-porous xylem.

Shrubs (nanophanerophytes)

Included are intensively branched, perennial, 50–400 cm-tall plants with hard, intensively lignified (woody) stems. The oldest known individuals reach ages of 800 years (*Juniperus sibirica*).

Shrubs

One meter-tall shrubs

7.331 *Paeonia lutea.*

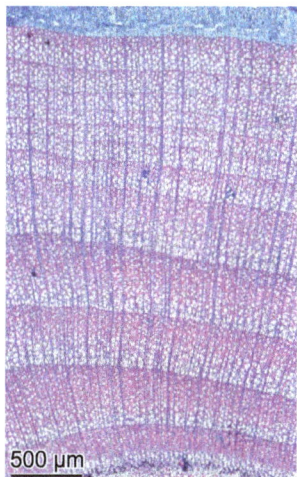

7.332 Stem of *Paeonia suffruticosa* with a semi-ring-porous xylem.

7.333 *Ribes rubrum.*

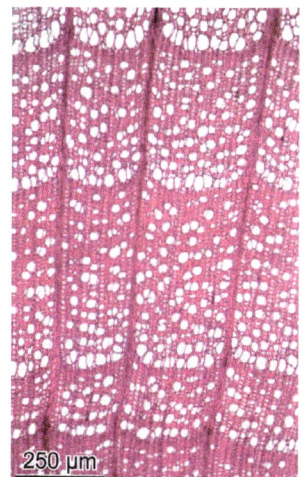

7.334 Stem of *Ribes rubrum* with a semi-ring-porous xylem.

Trees (phanerophytes)

Included are perennial plants with one basal stem, more than 4 m height, with hard, intensively lignified (woody) stems. They can reach ages up to 5,000 years (*Pinus longaeva*).

20–30 m-tall trees (temperate zone)

7.335 *Acer pseudoplatanus*

7.336 Diffuse-porous xylem with small vessels in *Acer campestre*.

7.337 *Quercus robur*

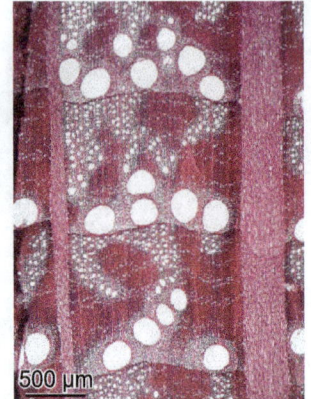

7.338 Semi-ring-porous xylem in *Quercus robur*.

Lianas

Included here are annual and perennial plants which need support from other plants to grow upwards. Characteristic for all lianas is the presence of large vessels. However, the real demand for water conductance is related to the occurrence of transpiration stress.

Annual lianas

7.339 *Calystegia arvensis* on a cereal stem.

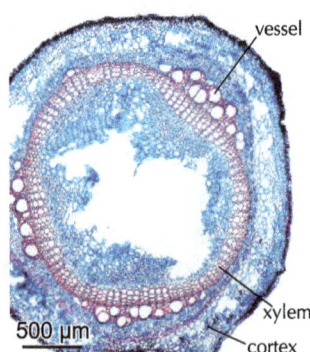

7.340 Partial ring with large vessels in *Calystegia arvensis*. It is formed in the same year as the closed ring.

7.341 *Citrullus colocynthis*, a prostrate liana in the extreme desert.

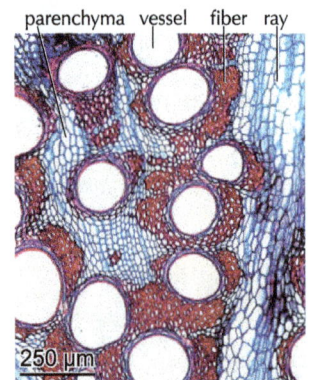

7.342 Heterogeneous tissue with large vessels (active only in the short rainy periods) in *Citrullus colocynthis*.

Perennial lianas

7.343 *Celastrus flagellaris* on a tree stem in the temperate zone.

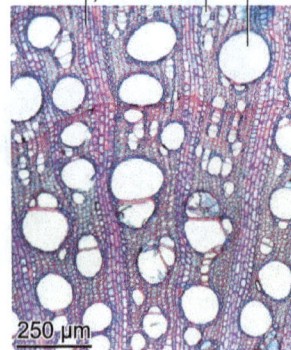

7.344 Diffuse-porous xylem with large and small vessels in *Celastrus flagellaris*.

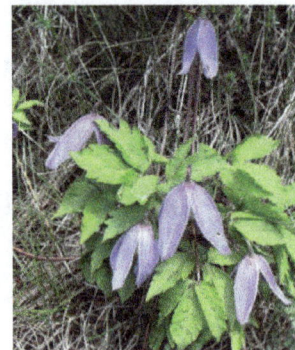

7.345 *Clematis alpina*, one of the few lianas in subalpine and boreal/arctic environments.

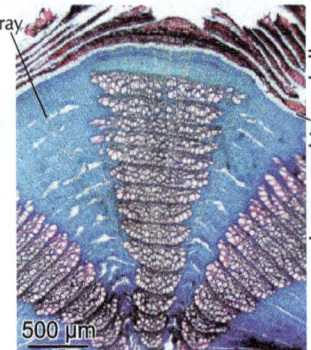

7.346 Semi-ring-porous xylem in *Clematis alpina*, with a large earlywood containing many vessels.

Succulents

Included are annual and perennial terrestrial plants with water-storing stems, growing mostly on dry sites. Common for all observed species is the extended water-storing tissue, be it in the pith, the xylem or the bark. Succulent plants occur in many taxonomic units. They are dominant in hot, arid regions, but are also frequent at dry sites in most other biomes.

Annuals

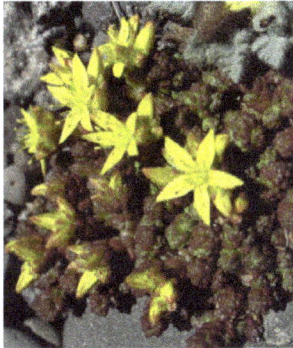

7.347 *Sedum annuum*, a 3 cm-tall herb with succulent leaves and stems, from the temperate zone at low to high altitudes.

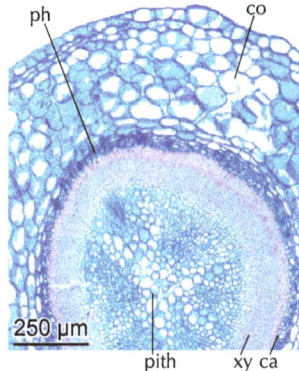

7.348 In *Sedum annuum*, a dense xylem is surrounded by a small phloem and a very large cortex, consisting of large, parenchymatic cells.

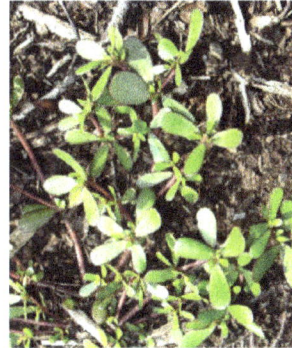

7.349 *Portulaca oleracea*, a prostrate herb with fleshy leaves and a succulent stem, from the temperate zone at low altitudes and dry sites.

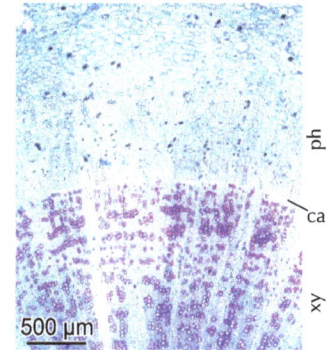

7.350 In *Portulaca oleracea*, a thin-walled, parenchyma-dominated xylem is surrounded by an unlignified phloem and cortex.

Perennials

7.351 *Sempervivum montanum*, an alpine herb with a basic rosette and sporadic flowers, from the temperate zone at high altitudes.

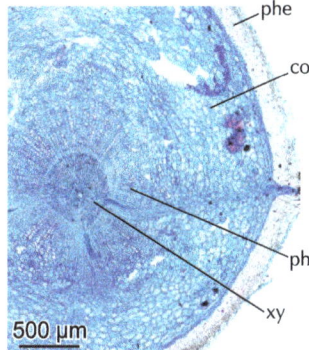

7.352 Root collar of *Sempervivum montanum* with a small, unlignified xylem, a large phloem and a very large cortex.

7.353 *Euphorbia canariensis*, a cactus-like plant, from the subtropical zone at dry sites.

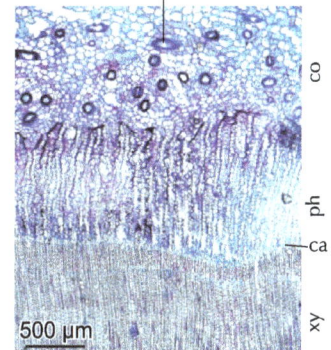

7.354 In *Euphorbia canariensis*, a xylem with thin-walled fibers and few vessels is surrounded by a very large unlignified phloem and cortex.

Parasites

Included are annual and perennial terrestrial plants. Growth forms are extremely different, but all of them maintain or complete their metabolism with nutrients from host plants.

Common for all observed species is the large, water-storing cortex.

Annual parasite

7.355 The thin, annual shoots of *Cuscuta epithymum* are not self-supporting and attach to photosynthetically active host plants with haustoria.

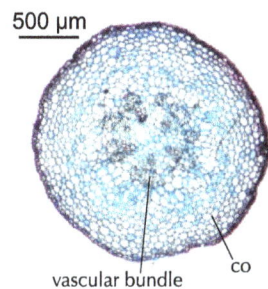

7.356 Circular arranged vascular bundles are surrounded by a large thin-walled parenchymatic cortex in *Cuscuta* sp.

Semiparasite

7.357 The perennial *Viscum album* on apple trees. This semiparasite connects to the xylem of the host plant by haustoria (see p. 78).

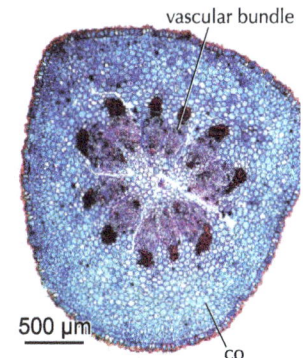

7.358 Annual shoot of *Viscum album*. Vascular bundles are surrounded by unlignified parenchymatic cortex cells.

Parasites

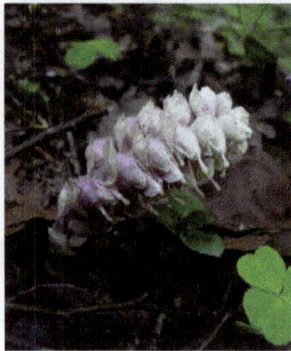

7.359 The chlorophyll-free *Lathraea squamaria* connects to the xylem of alder roots by haustoria.

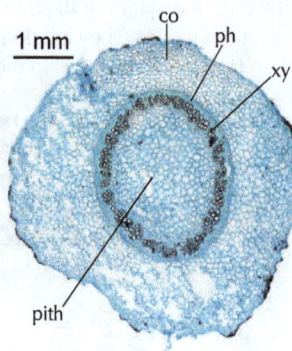

7.360 In *Lathraea squamaria*, a small ring of lignified vessels surrounds a large pith, and is itself surrounded by a small phloem and a large cortex. This is a typical succulent structure.

7.361 The chlorophyll-free *Orobanche alba* is a parasite on several Lamiaceae, and connects to the phloem of the photosynthetically active host plant.

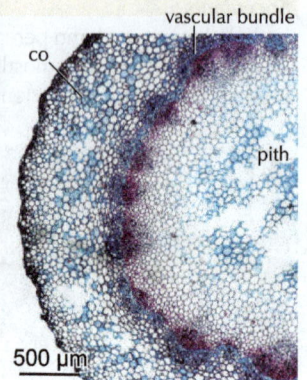

7.362 In *Orobanche alba*, a ring of vascular bundles surrounds a large pith and is surrounded by a large cortex. This is a typical succulent structure.

Hydrophytes and helophytes

Included are annual shoots of plants which grow under water (hydrophytes) or are rooted in permanently wet soils (helophytes). Common for all hydrophytes and helophytes are unlignified aerenchymatic tissues. There are no principal anatomical differences between the two growth forms. Despite the homogeneous, stressful anaerobic environment, specific anatomical differences between species are related to taxonomy.

Hydrophytes

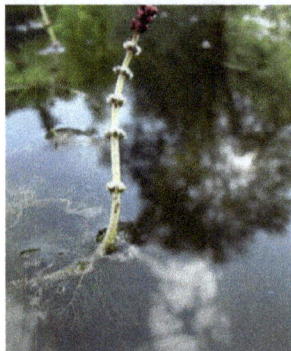

7.363 Flower of *Myriophyllum spicatum* above the water table. The major part of the plant is submersed.

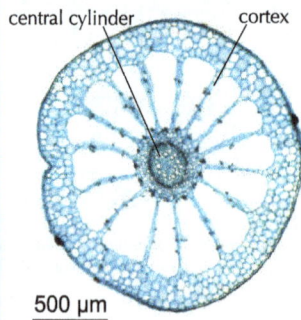

7.364 The central cylinder in *Myriophyllum spicatum* is surrounded by a cortex with large, uniseriate aerenchymatic tubes.

7.365 Submersed shoot of *Ceratophyllum demersum*.

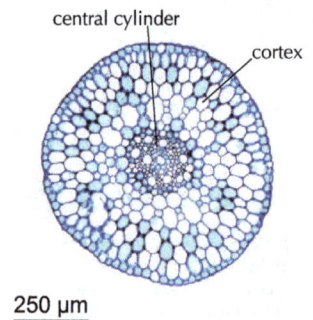

7.366 The central cylinder in *Ceratophyllum demersum* is surrounded by a cortex with small aerenchymatic canals between large parenchymatic cells.

Hydrophyte

7.367 The floating leaves and flowers of *Nuphar lutea* root deep in the ground below the water table.

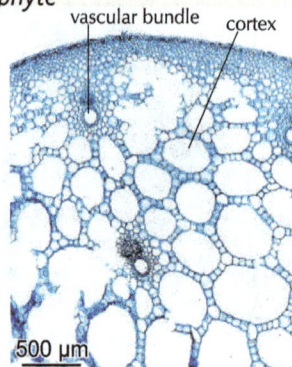

7.368 Single vascular bundles are surrounded by an aerenchymatic tissue in *Nuphar lutea*.

Helophyte

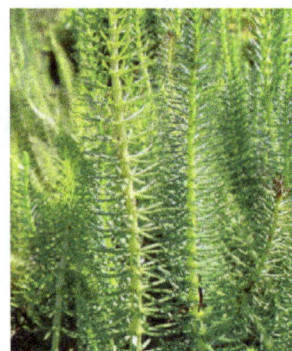

7.369 Annual shoots of *Hippuris vulgaris*.

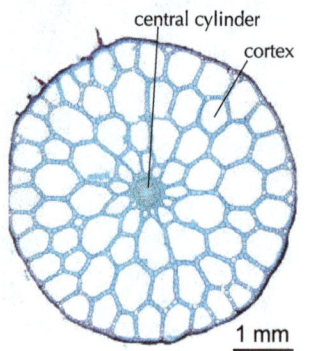

7.370 In *Hippuris vulgaris*, the small central cylinder is surrounded by a cortex with large aerenchymatic tubes.

Stem Evolution: An Overview

8

8.1 Paleobotanic evidence of stems

Writing a comprehensive story about stem evolution is difficult because evidence of only few early plants has been preserved over millions of years. Samples exist of plants which were fossilized under anaerobic conditions, enclosed in resin (amber) or carbonized. The majority of plants from sites with aerobic soil conditions are not preserved. Presented here are some evolutionary events of major importance, primarily based on petrified and carbonized plant remains. Not discussed are the fossil ferns, the living seed ferns, and the ginkgos and angiosperms. For their structure refer to the description of their living representatives in Chapter 7.

Early plant evolution

A four-billion-years-lasting evolutionary process of water organisms lacking a nucleus (prokaryota) set the basis for life on land. First, marine microbial mats, the stromatolites, increased the amount of oxygen in the atmosphere through photosynthesis. Next, the marine algae developed polarity by forming long, geo- and heliocentrically oriented, axially ramified cellular strands.

The fact that ancient plant structures have survived for 400 million years is demonstrated by one fossil and few living green algae (chlorophyta). The modern green algae genus *Chara* is a typical living fossil.

Paleontologists assume that early fossil algae in tides zones had root-like structures. Shown here is a modern *Laminaria* species with rhizoids. Rhizoids are basal filaments which attach the whole plant to a stable ground. Water transport in aquatic algae occurs by diffusion, therefore vessels have not been developed.

Fossil and recent green algae   *Recent brown algae*

500 μm

8.1 *Heterocladus waukaheshaensis*, a 5 cm-tall Silurian green alga with a central stem and many whorls. Reprinted from Taylor *et al.* 2009.

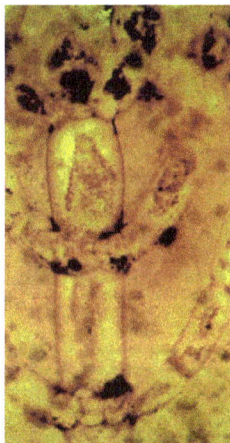

8.2 Stem of *Palaeonitella cranii*, a Devonian green alga with a central stem and three whorls. Reprinted from Taylor *et al.* 2009.

8.3 Stem of *Chara* sp., a comparable recent green alga showing four whorls.

8.4 *Laminaria* sp., a modern brown alga with rhizoids.

8.5 Cross section of a modern aquatic brown alga stem without liquid-conducting structures. The whole plant consists of parenchyma cells. Small, fairly thick-walled cells surround the periphery.

The move to the land

First land plants are known from the Upper Silurian and Upper Devonian period. Over the course of 60 million years, approximately from 420 to 360 million years ago, plants developed mechanisms to survive in an atmosphere hostile to life. The first terrestrial plants featured several new traits in comparison to preceding aquatic plants. Lignified elements guaranteed stability, tracheids and sieve cells allowed the transport of liquids, an epidermis prevented dehydration, stomata enabled gas exchange, and secondary growth provided longevity. All these features are so fundamental that they survived in all existing land plants until the present day.

Presented here are a few paleobotanical reconstructions. For further information the well-illustrated book from Taylor *et al.* 2009 is recommended.

The earliest small land plants

First land plants developed over the course of 20 million years in the late Silurian period. They reached a height of approximately 20 cm, were leaf-less, and had a simple stem structure. Characteristic is a central cylinder consisting of a round or star-shaped xylem with tracheids and a phloem (protostele or actinostele). Tracheids have annular thickenings.

Rhynia

8.6 Reconstruction of the Devonian *Rhynia gwynne-vaughanii*. Reprinted from Hirmer 1927.

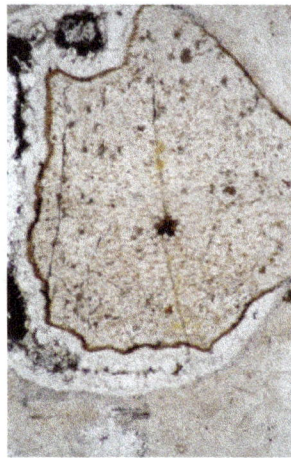

8.7 Cross section of a stem of *Rhynia gwynne-vaughanii* with a small xylem and a large cortex, surrounded by an epidermis. The cortex is divided into an inner and outer part.

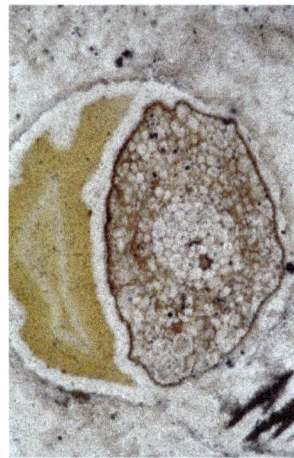

8.8 Cross section of a branch of *Rhynia gwynne-vaughanii* with an extremely small xylem and a large cortex, surrounded by an epidermis.

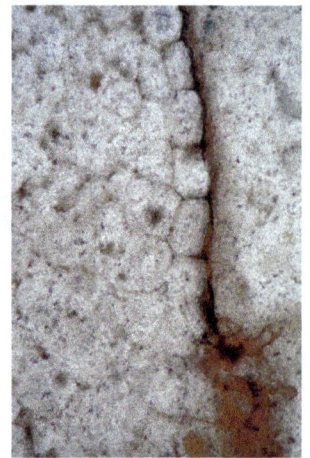

8.9 Epidermis of *Rhynia gwynne-vaughanii*.

Asteroxyleon

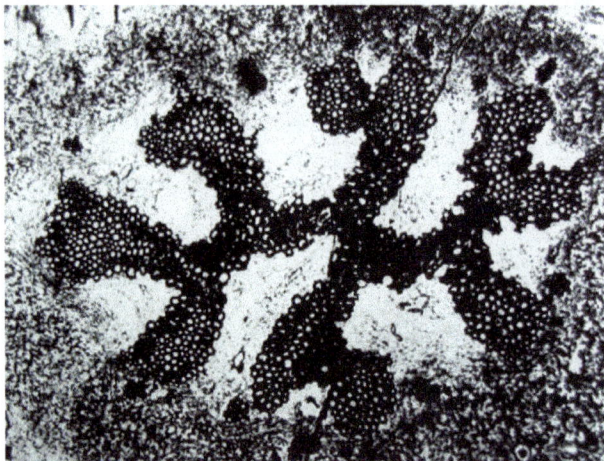

8.10 Cross-section of a stem of the Devonian *Asteroxyleon mackiei* with a star-shaped xylem (actinostele). Reprinted from Zimmermann 1959.

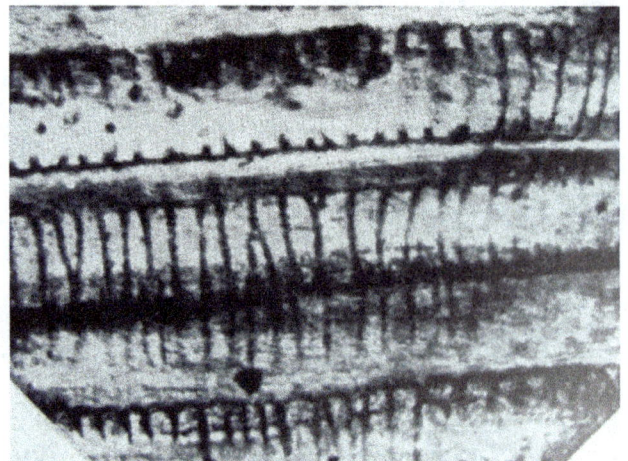

8.11 Tracheids of *Asteroxyleon mackiei* with annular thickenings. Reprinted from Henes 1959.

The earliest trees

First trees with a modern habit appeared in the Late Devonian period and lasted until the Early Carboniferous period. *Archeopteris* trees (syn. *Callixylon*) were widely distributed approximately 380 to 360 million years ago. With the presence of a coniferous wood structure, but still reproducing by spores, these trees are classified as progymnosperms. Characteristic are tracheids with bordered pits, produced by a cambium (secondary growth).

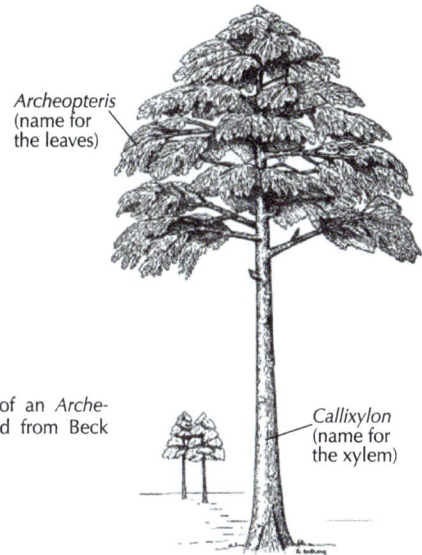

8.12 Reconstruction of an *Archeopteris* tree. Reprinted from Beck 1962.

Archeopteris (name for the leaves)

Callixylon (name for the xylem)

The Carboniferous clubmoss and horsetail forests (Lycophyta and Sphenophytes)

A huge diversity of the spore-reproducing horsetail and clubmoss trees dominated the Late Devonian and the Carboniferous period approximately from 390 to 300 million years ago. Widely distributed were the genera *Lepidodendron* and *Sigillaria* (Lycophyta) and *Calamites* (Equisetales).

The structure of clubmoss trees is principally similar to modern trees. However, the proportions between xylem, phloem, cortex and periderm greatly differ. *Lepidodendron* trees are "bark trees"—a small xylem/phloem center is surrounded by an extremely large primary cortex and a very large periderm. The major part of the xylem is produced by a cambium (secondary growth). The arrangement of tracheids and rays is similar to conifers, however, cell walls are scalariform perforated. The absence of annual rings leads to the assumption that these trees grew in a tropical environment.

Calamites trees have a principally similar structure, but the bark is much smaller and the pith is enlarged. *Calamites* are "pith trees". The presence of carinal canals in some species indicates a relation to living herbaceous horsetails. Radial pitting on tracheids varies from scalariform to circular bordered.

Clubmoss and horsetail trees

Lepidodendron

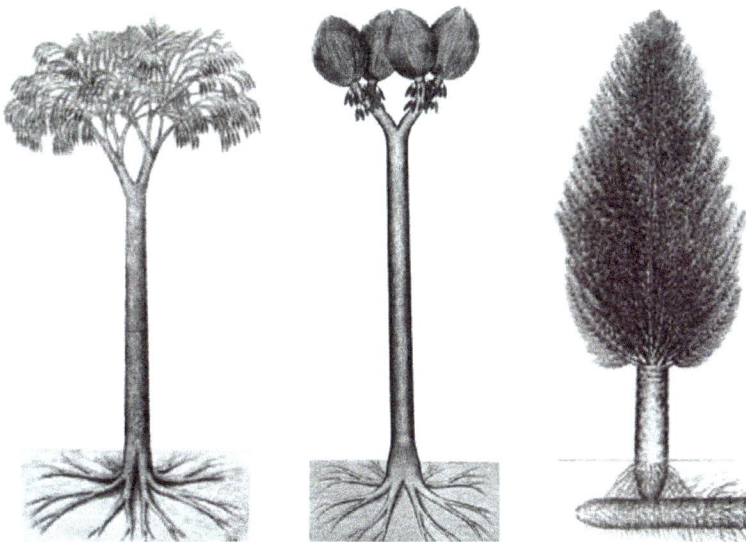

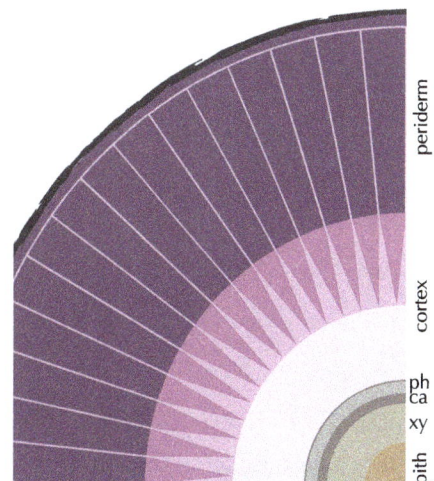

periderm

cortex

ph
ca
xy

pith

8.13 Reconstruction of clubmoss trees *Lepidodendron* sp. and *Sigillaria* sp., and the arborescent horsetail *Calamites* sp. Reprinted from Hirmer 1927.

8.14 Stem reconstruction of the clubmoss tree *Lepidodendron*. One third of the stem radius consists of pith, xylem and phloem and two thirds of cortex and periderm (after Hirmer 1927).

Sigillaria		*Stigmaria*	*Calamites*

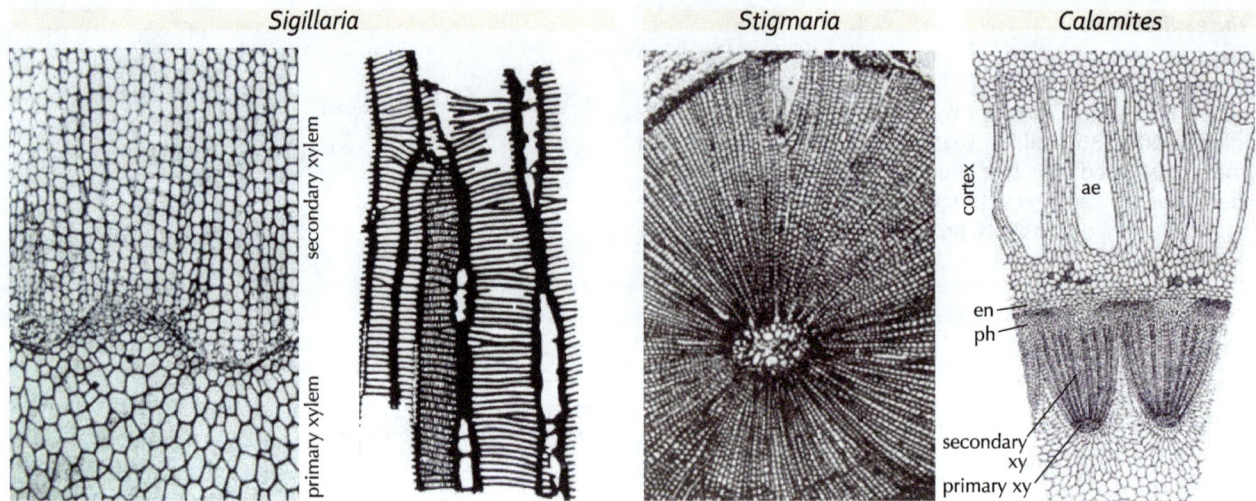

8.15 Cross section of the xylem of *Sigillaria saulii*. Tracheids of the secondary xylem are radially arranged. Annual rings are absent. Reprinted from Henes 1959.

8.16 Structure of tracheids of *Sigillaria saulii*. Characteristic for all Lepidphytales are scalariform wall structures with longitudinal thin strips. Reprinted from Hirmer 1927.

8.17 Cross section of a root of *Stigmaria* sp., Lycopodiales, from the Lower Carboniferous. The secondary xylem consists of tracheae and rays. Reprinted from Hirmer 1927.

8.18 Cross section of a root of *Calamites* sp. from the Lower Carboniferous, consisting of a secondary xylem, and a cortex with aerenchyma. Reprinted from Hirmer 1927.

The first gymnosperms (Cordaitales)

Shown here is the anatomical structure of *Dadoxylon* sp. (syn. *Araucarioxylon*), one representative of an extinct group of conifer-like trees with secondary growth. The Cordaitales existed in the Carboniferous and were seed plants with a xylem similar in structure to conifers. The only difference to living conifers is the existence of bi- to triseriate rays. Multiseriate bordered pits are very similar to the modern Araucariaceae. The plants with large leaves have a similar aspect to some living Podocarpaceae. The absence of annual rings indicates that *Dadoxylon* trees grew in a tropical climate.

Cordaitales

8.19 Leaves of a species of the Cordaitales. Reprinted from Taylor *et al*. 2009.

8.20 Cross section of *Dadoxylon* sp. without annual rings.

8.21 Radial section of *Dadoxylon* sp. with multiseriate bordered pits on tracheid walls.

8.22 Tangential section of *Dadoxylon* sp. with bi- to triseriate rays.

The development of angiosperms

Fossil stem remains of angiosperms are rare, but pollen records from the early Cretaceous period indicate a large taxonomic diversity 140 million years ago. Wood remains from the late Cretaceous already show an anatomical diversity comparable to that at present. Here, it is presented in Chapter 7 on modern material starting with palm ferns (Chapter 7.4.1) and ending with monocots (Chapter 7.4.5).

8.2 Evolution and homoplasy of wood anatomical traits

It makes sense to consult fossilized stems to analyze structural developments in stems. In Table 1 below, a few anatomical features are set in relation to fossil evidence and in a phylogeny-based taxonomic system. Already at first glance it becomes obvious that there was no straightforward evolution of most features. Most of the features are homoplastic—they developed or disappeared in different taxa (convergent developments). Spicer & Groover 2010 and Carlquist 2012 have demonstrated this on the basis of anatomical traits of fossil and living trees. In contrast to previous studies, herbs are included in the analysis here, and angiosperms are grouped by growth form (trees, shrubs, lianas, large and small terrestrial herbs and hydrophytic herbs) instead of by phylogenetic units. Some results are therefore different to the aforementioned studies.

In Table 1, the appearance and occurrence of secondary growth and perforation plates of vessels are highlighted over a period of 400 million years.

Secondary growth is present in Devonian and Carboniferous Pteridophyta. It was lost in most living Pteridophytes. The presence of secondary growth in small herbs of modern Ophioglossaceae can be interpreted as a relict or a reinvention (Chapter 7.3.2). Radial growth, including radial growth with successive cambia, occurs in the majority of living seed plants (gymnosperms, dicotyledons). Secondary growth is principally absent in monocotyledons, however a reinvention began in Dracaenaceae and Agavaceae with a new mode of formation (Chapter 7.4.5). Secondary growth is absent in many dicotyledonous hydrophytes e.g. in Nymphaeceae, Ranunculaceae, Droseraceae and others. Since secondary growth exists in the majority of genetical taxa its absence can be interpreted as a retrogressive development.

Vessels are a characteristic of dicotyledons. They are absent in all Paleozoic taxa and fossil and modern gymnosperms. Vessels appeared parallel to the development of angiosperms in the Lower Cretaceous period. The occurrence of vessels in living Equisetaceae and in ferns of Ophioglossaceae seems to be a modern (Cretaceous?) development. A few taxa without vessels exist within the angiosperms. Phylogenetic studies lead to the assumption that at least *Amborella* is a relict of Pteridophytes, however Trochodendronaceae and Winteraceae are either relicts or retrogressive taxa. The following pictures clearly show that the presence of scalariform pits and pits with slit-like apertures are relictic, and large rays and ray dimorphism are modern features. Paleozoic and Cretaceous developments meet in a very few shrubs and trees.

Table 8.1 Occurrence or absence of vessels and secondary growth in relation to major taxonomic units from Paleozoic times to the present.

Group	taxonomic units	reinvented? / lost	vessels present	vessels absent	perforations simple	perforations sclariform normal	perforations sclariform aberrant	secondary growth absent	with secondary growth	successive cambia present
Pteridophyta		*Rhynia* herb, fossil		x				x		
		Lycopodiaceae herb		x				x		
		Lycopodiaceae tree, fossil		x					x	
		Selaginellaceae herb		x				x		
		Isoetaceae		x				x		
		Psilotaceae herb		x				x		
		Leptosporangiatae (ferns) tree		x				x		
		Leptosporangiatae (ferns) herb		x				x		
		Leptosporangiatae (ferns) hydrophyte		x				x		
		Ophioglossaceae herb	x		x				x	
	Equisetopsida	*Calamites*, fossil							x	
	Equisetopsida	Equisetaceae herb	x		x			x		
Seed plants	Gnetales	Spermatophytes *Ephedra*	x			x			x	
	Gnetales	Spermatophytes *Welwitschia*	x		x				x	x
	Gnetales	Spermatophytes *Gnetum*	x		x		x		x	
	Gymnosperms	*Dadoxylon* sp., fossil							x	
	Gymnosperm conifers	Araucariaceae tree							x	
	Gymnosperm conifers	Juniperaceae s.lt tree, shrub							x	
	Ginkgopsida	*Ginkgo* tree							x	
	Cycadopsida	*Cycas* tree-like	r						x	x
	Dicots	Amborellaceae shrub		x					x	
	Dicots	Trochodendron tree		x					x	
	Dicots	Winteraceae tree		x					x	
	Dicots other families	Dicots trees & shrubs	x		x	x			x	x
	Dicots other families	Dicots lianas	x		x				x	x
	Dicots other families	Dicots hemicryptophytes <20 cm	x		x				x	x
	Dicots other families	Dicots hemicryptophytes >20 cm	x		x		x		x	x
	Dicots other families	Dicots hydrophytes	x		x				x	
	Monocots	Monocots Dracaenaceae	x		x				x	
	Monocots	Monocots all others	x		x			x		
	Dicots few Asteraceae	Dicots hemicryptophytes <20 cm					x			

Amborellaceae

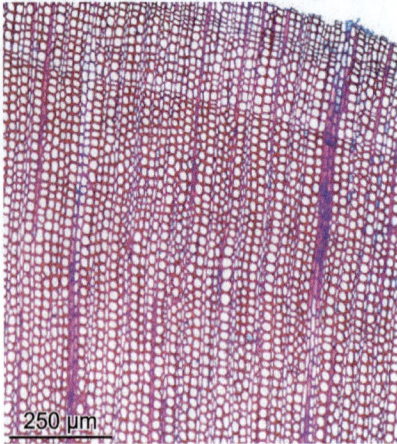

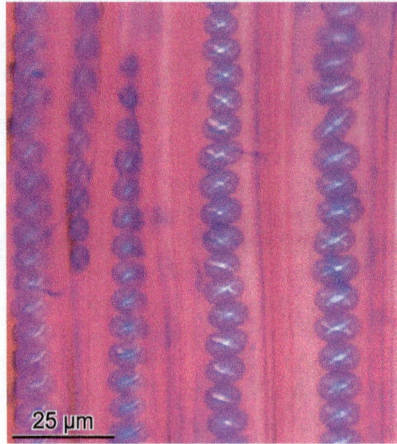

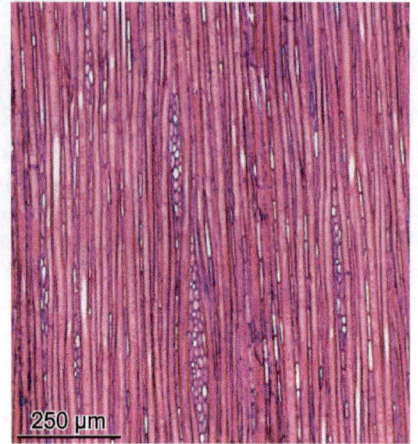

8.23 Radial arrangement of tracheids, no annual rings in *Amborella trichocarpa*.

8.24 Tracheid pits of *Amborella trichocarpa* with slit-like apertures. Tori seem to be absent.

8.25 Uniseriate and bi- to four-seriate heterogeneous rays in *Amborella trichocarpa*.

Trochodendraceae

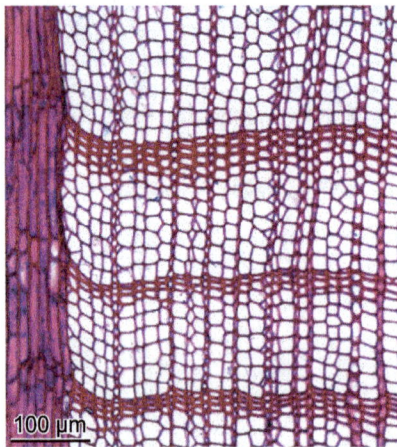

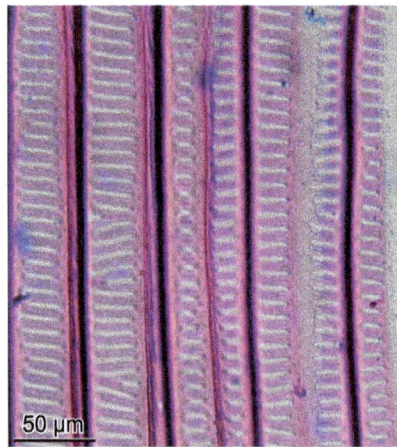

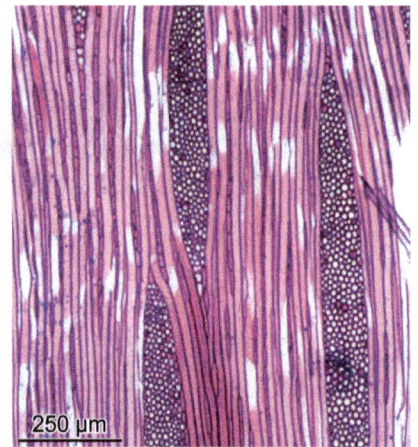

8.26 Radial arrangement of tracheids. Distinct early- and latewood in *Trochodendron aralioides*.

8.27 Scalariform tracheid pits in *Trochodendron aralioides*. Tori seem to be absent.

8.28 Uniseriate and bi- to multiseriate heterogeneous rays in *Trochodendron aralioides*.

Winteraceae

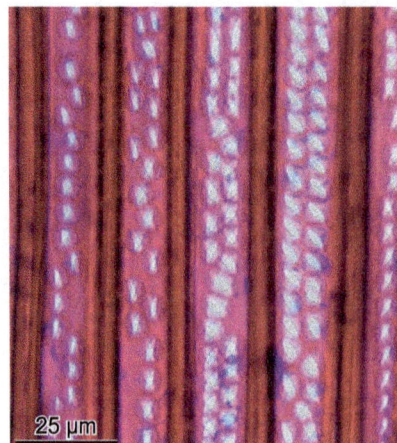

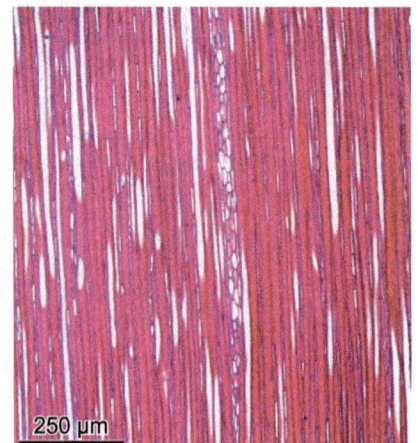

8.29 Radial arrangement of tracheids, no annual rings in *Drimys winteri*.

8.30 Tracheid pits with slit-like apertures in *Drimys winteri*. Tori seem to be absent.

8.31 Uniseriate and bi- to four-seriate heterogeneous rays in *Drimys winteri*. Distinct sheath cells.

Homoplasy and evolution

Evolutionary trends in xylem evolution are difficult to recognize because functional adaptation to hydraulic and mechanic needs shaped anatomical structures. This is demonstrated here by relating a few anatomical characteristics to plant size. The following graphs express probabilities of occurrence, and are based on observations of 3,347 dicotyledons from seasonal climates in the Northern Hemisphere. The general trends shown in the figures below were confirmed with detailed analyses of the Apiaceae, Asteraceae, Brassicacea, Fabaceae, Rosacea and Lamiaceae (Schweingruber *et al.* 2013).

Vessel diameters <20 µm primarily occur in plants <10 cm in height. Such plants often grow in alpine zones. Vessel diameters >50 µm occur mainly in tall trees.

Porosity types correspond with plant size. Ring porosity occurs mainly in trees. It is absent in very small plants.

Scalariform perforations are rare. They are almost absent in small plants and occur only in approximately 10% of the analyzed specimens. They are absent in plants of the arid zone. *Simple perforations* dominate in most taxa of any size and in any climate type.

Fibers and parenchyma. Plant sizes shape the anatomy. Fibers dominate the xylem of all large plants. They are often absent in very small plants or in hydrophytes. In contrast, parenchyma dominates the anatomy of small plants. Pervasive parenchyma is absent in large plants, and very frequent in very small plants.

Crystals. Plant size also influences physiological processes. Prismatic crystals in the xylem and the bark are much more frequent in large than in small plants.

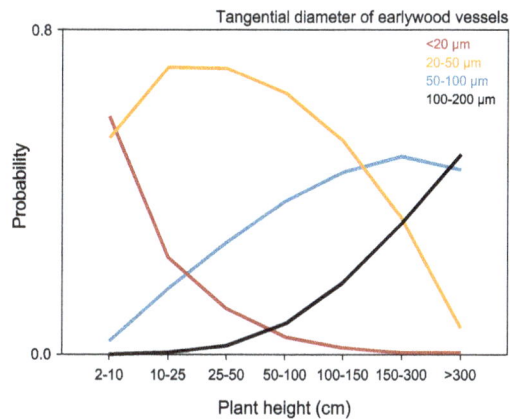

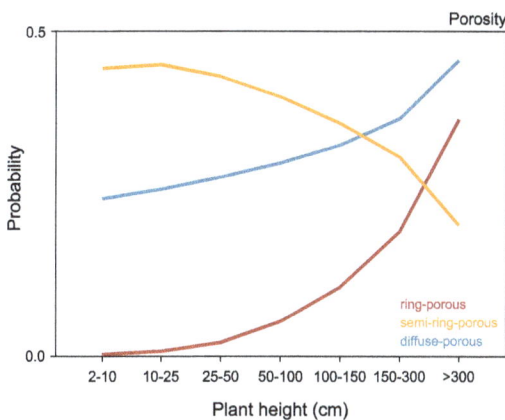

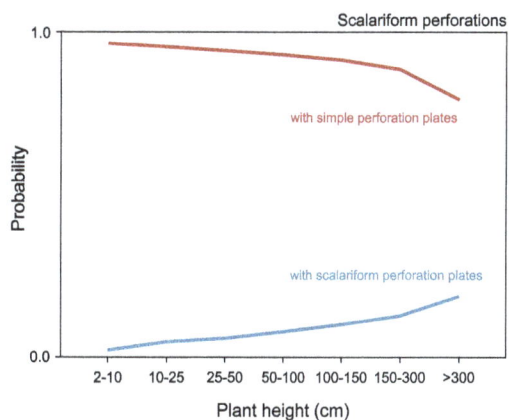

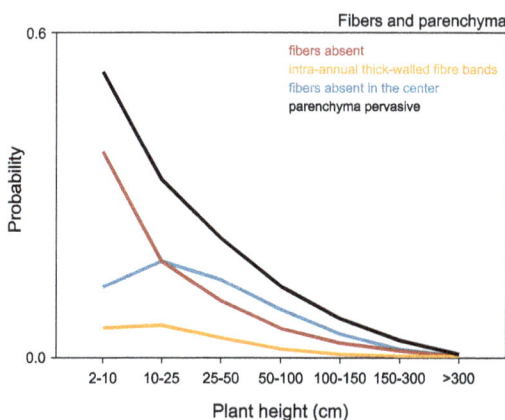

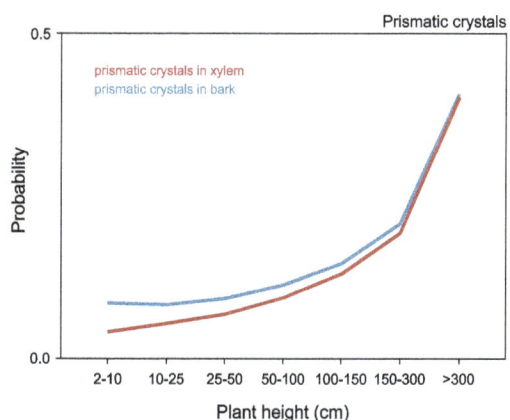

8.32 Probability of occurrence of anatomical characteristics in relation to plant size, based on 3,347 dicotyledons from seasonal climates in the Northern Hemisphere. Reprinted from Schweingruber *et al.* 2013.

8.3 Parallel evolution of macroscopic and microscopic traits

The question is raised here of how local environmental factors influence the macroscopic and microscopic traits of stems. Exemplarily presented and compared are the macroscopic aspect (plant formation), the life form of plants, the floristic composition and the anatomical structure of stems of a subalpine forest, an oceanic shrub community, and an alpine herb community.

A suite of specific environmental factors enables plants with similar physiological qualities and surviving strategies to grow together.

Plant formations are defined by their physiognomic character, e.g. forests are plant formations with a dominance of trees. Heathlands are dominated by dwarf shrubs and meadows by herbs. **Plant communities** are defined by their floristic composition, e.g. the tree storey of subalpine forests in the Alps consists of conifers, the shrub layer of heathlands consists of *Calluna vulgaris* and *Genista* sp., and alpine meadows are dominated by herbs. An ecophysilogical and taxonomic interplay determines formations and communities.

The term parallel evolution is related to taxonomic groups. If parallel evolved taxa are genetically far apart, e.g. conifers and

dicotyledons, parallel developments are obvious, e.g. *Abies* and *Sorbus* are both trees. Parallel developments are less obvious for genetically closer related taxa, e.g. two genera within the conifers such as *Pinus* and *Abies* are both Pinaceae and always grow as trees, or *Vaccinium myrtillus* and *Vaccinium uliginosum* are both Ericaceae and always grow as dwarf shrubs.

Forest

8.33 The tree layer in the fir forest in Emmental, Switzerland (left), is dominated by *Abies alba*, accompanied by *Sorbus aucuparia* (right).

Atlantic heathland

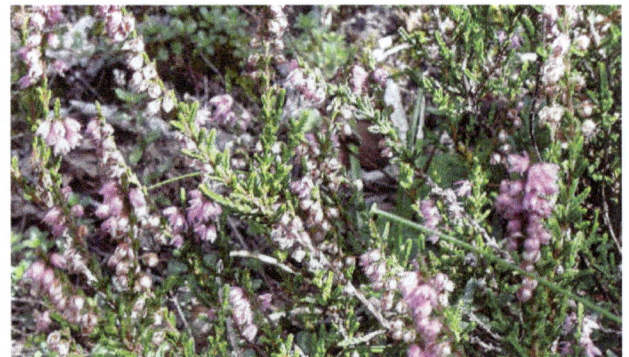

8.34 Heathland at the North Atlantic coast, Cornwall, England (left). Dwarf shrub cushions are dominated by *Erica* sp., *Calluna vulgaris* (right), *Ulex* sp. and *Genista* sp.

Alpine meadow

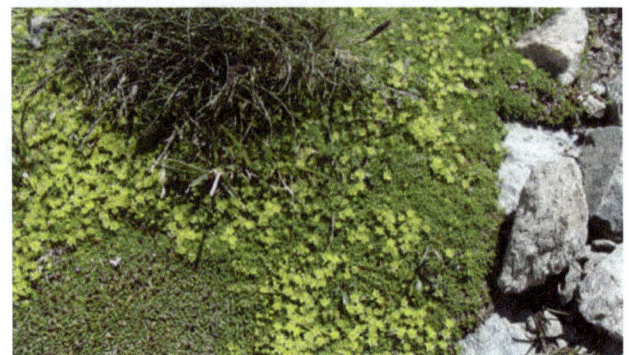

8.35 Alpine meadow at 2,900 m a.s.l. in the European Alps with *Carex curvula* and cushion plant *Minuartia sedoides*.

8.36 *Minuartia sedoides*, a perennial alpine cushion plant with 85 annual rings in the taproot.

Mesic European fir forest (Adenostylo-Abietetum)

Conifers, few bushes and tall herbs are characteristic for forests at the northern slopes of the Alps. Firs (*Abies alba*) prefer sites with high precipitation (>1,500 mm/year), long winters with longer frost periods and nutrient rich soils at altitudes between 1,300–1,700 m a.s.l. The tree layer is dominated by *Abies alba* and accompanied by *Sorbus aucuparia*. The shrub layer consists of *Lonicera nigra* and *Rosa pendulina*, and tall herbs such as *Adenostyles alliariae*, *Petasites albus*, *Prenanthes purpurea* and *Athyrium filix-femina* dominate the herb layer.

Species within the tree, shrub and herb layers evolved parallel to adopt the same growth form. However, the anatomical stem structures of all occurring plants preserved their genetic heritage.

The xylem of the conifer *Abies alba* consists of tracheids; the anatomical structure expresses its coniferous heritage. Vessels of all dicotyledonous species in the tree, shrub and herb layers are part of parallel evolution. High soil water content and intensive water transport triggered the formation of large earlywood vessels with a diameter of 40–70 μm. However, the vessel/fiber/parenchyma patterns are still species-specific. Environmental factors influenced the capacity of the conduction system but not the arrangement of cell types. See also Keller *et al*. 1998.

Isolated evolution of conifer

Parallel evolution of vessel diameters in the tree and shrub layers

8.37 The xylem of *Abies alba*, Pinaceae, consists of tracheids.

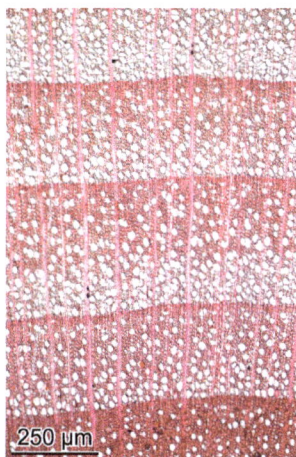

8.38 Xylem of the 5 m-tall *Sorbus aucuparia*, Rosaceae, with large earlywood vessels.

8.39 Xylem of the 1.5 m-tall *Lonicera nigra*, Caprifoliaceae, with large earlywood vessels.

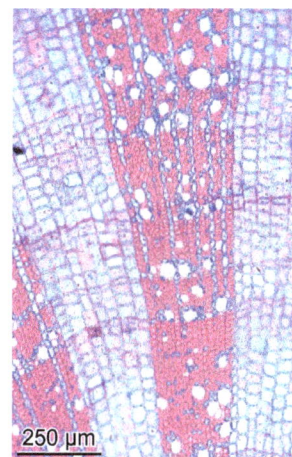

8.40 Xylem of the 1 m-tall *Rosa pendulina*, Rosaceae, with large earlywood vessels.

Parallel evolution of vessel diameters in rhizomes of herbs

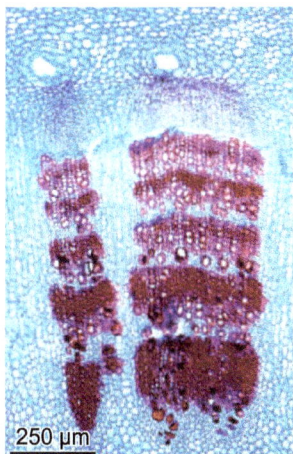

8.41 Vessels in the rhizome of the 40 cm-tall *Adenostyles alliarae*, Asteraceae.

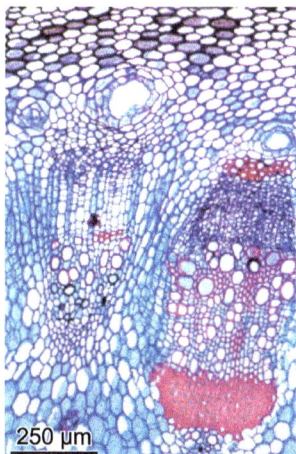

8.42 Vessels in the rhizome of the 50 cm-tall *Petasites albus*, Asteraceae.

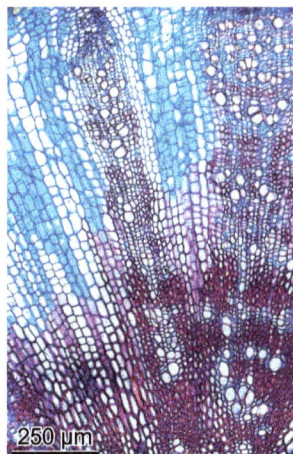

8.43 Vessels in the rhizome of the 80 cm-tall *Prenanthes purpurea*, Asteraceae.

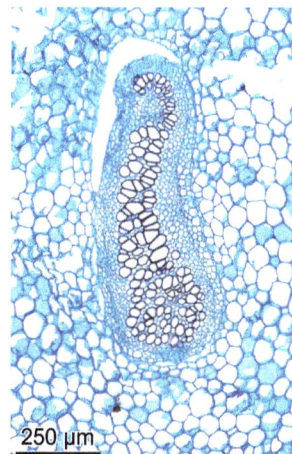

8.44 Vessels in the rhizome of the 80 cm-tall *Athyrium filix-femina*, Polypodiaceae.

Heathlands along the European North Atlantic coast

Dwarf shrubs and small bushes are characteristic for the vegetation along the northwestern European coasts. This coastal region is characterized by high precipitation (>1,000 mm/year), mild temperatures without frosts (e.g. annual mean temperature approx. 10°C in Cardiff, England), strong winds, granitic bedrock and thick, acidic, often wet organic soils.

The plant formation, the heath, is dominated by dwarf shrubs of different Ericaceae (genera *Calluna*, *Erica*) and Fabaceae (genera *Ulex*, *Genista*, *Cytisus*; Gorissen 2004).

Environmental factors have influenced the morphological aspect of the plant formation (small shrubs) and the diameter of earlywood vessels (30–50 µm) in the xylem, but distribution patterns of vessels/fibers/parenchyma reflect the genetic heritage of the taxa. *Erica* species are diffuse-porous and Fabaceae species are semi-ring-porous, with vessels arranged in dendritic patterns.

The morphological aspect as well as the anatomical stem structure of the accompanying herbs vary greatly. All species preserved their external and internal taxonomical characteristics. Their specific anatomical structures enable them to resist extreme local environmental conditions. Evolutionary processes therefore did not drive the basic anatomical structure of herbs in a similar direction.

Parallel evolution of vessel/fiber distribution patterns within taxonomic units

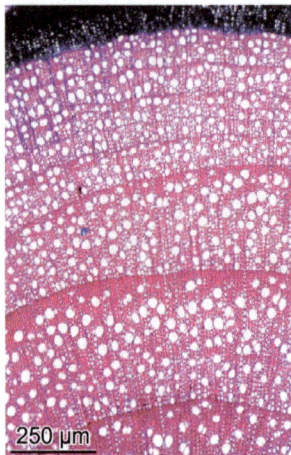

8.45 Xylem of *Calluna vulgaris*, Ericaceae.

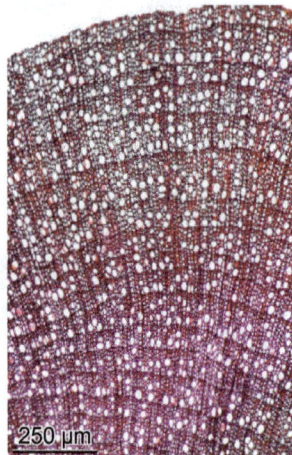

8.46 Xylem of *Erica tetralix*, Ericaceae.

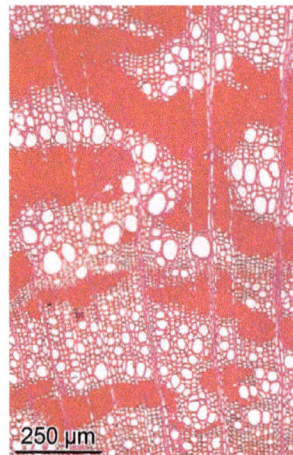

8.47 Xylem of *Ulex parviflora*, Fabaceae.

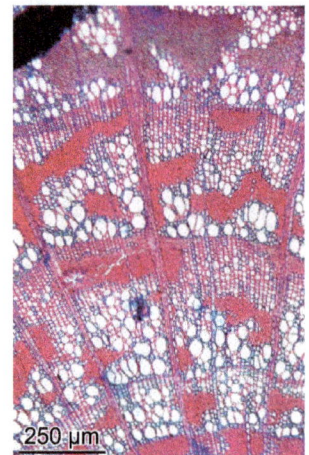

8.48 Xylem of *Genista cinerea*, Fabaceae.

Divergent evolution of the anatomy of accompanying species

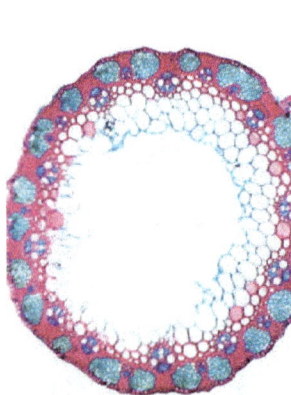

8.49 Cross section of a culm of the monocotyledonous *Agrostis stolonifera*, Poaceae.

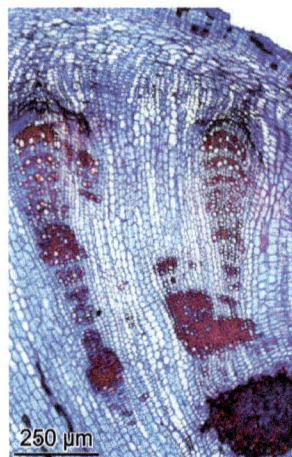

8.50 Cross section of the stem of the dicotyledonous *Potentilla erecta*, Rosaceae.

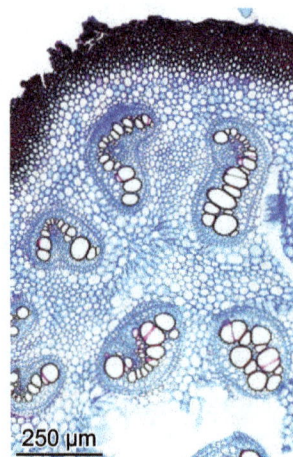

8.51 Cross section of the petiole of the fern *Pteridium aquilinum*, Polypodiaceae.

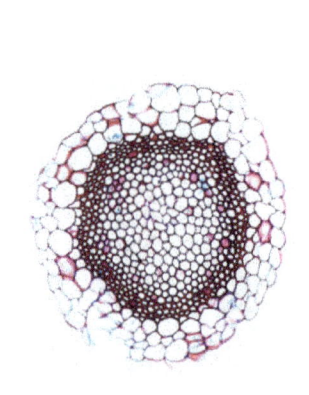

8.52 Cross section of the stem of the moss *Sphagnum subnitens*, Sphagnaceae.

Alpine meadows (Caricetum curvulae)

Small, 3–10 cm-tall grasses and dicotyledonous herbs characterize meadows on gentle slopes on granitic bedrocks in alpine zones over 2,500 m a.s.l. in the Alps. The snow cover is reduced due to snow drift and vegetation periods are hardly longer than three months. *Carex curvula* and *Silene acaulis* and other species form dense cushions on dry, acidic soils, with an extended root system and minimal aboveground biomass. A few other grasses and dicots persist between those cushions (Klötzli *et al.* 2010).

Parallel evolution is expressed by small plant height and vessels with small diameters (15–30 μm). However, the general anatomical stem structure of all species present preserved their genetic heritage.

Parallel evolution of vessel diameters of cushion plants

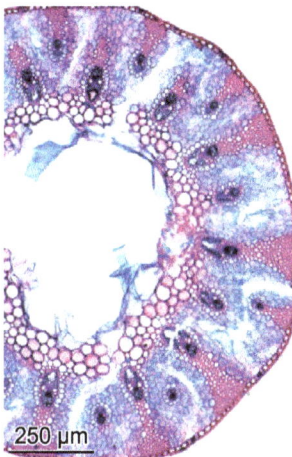

8.53 Cross section of the culm of the monocotyledonous *Carex curvula*, Cyperaceae.

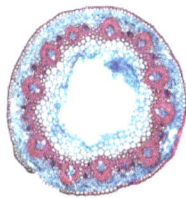

8.54 Cross section of the culm of the monocotyledonous *Juncus trifidus*, Juncaceae.

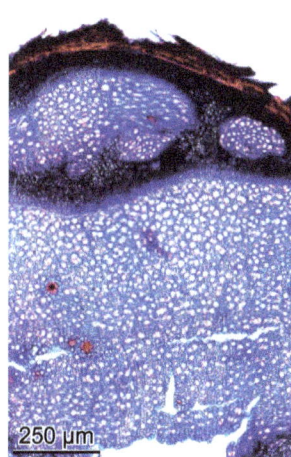

8.55 Cross section of the stem of the dicotyledonous *Silene acaulis*, Caryophyllaceae.

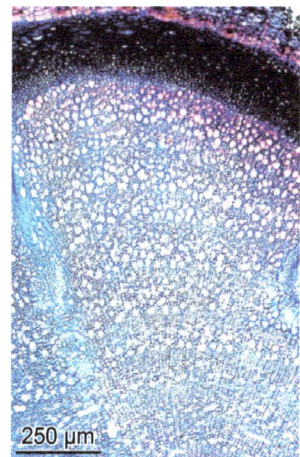

8.56 Cross section of the stem of the dicotyledonous *Minuartia recurva*, Caryophyllaceae.

Parallel evolution of non-cushion-forming herbs

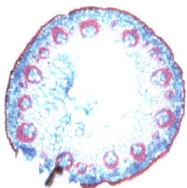

8.57 Cross section of a culm of the monocotyledonous *Luzula alpinopilosa*, Juncaceae.

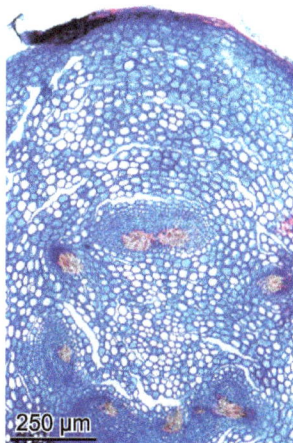

8.58 Cross section of the stem of the dicotyledonous *Primula minima*, Primulaceae, growing isolated or as a small cushion.

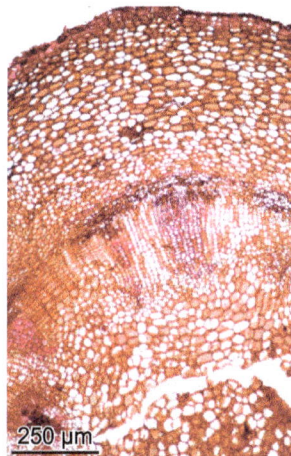

8.59 Cross section of the stem of the dicotyledonous *Leontodon helveticus*, Asteraceae.

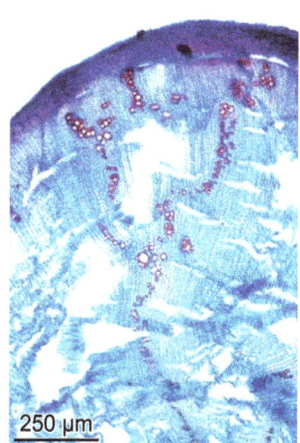

8.60 Cross section of the stem of the dicotyledonous *Phyteuma hemisphaericum*, Campanulaceae.

Adaptation to Permanently Changed Environments

9.1 Anatomical and morphological plasticity of species

All wood anatomy classification books describe the structure of so-called "normally grown" specimens. Nature, however, also produces a few giants, and many dwarfs. Here the question is raised to which degree local environmental factors modify the morphology and anatomy of stems of individual species. Exemplarily presented are the xylem structures of a few large (giants) and small individuals (dwarfs) of trees and herbs. Larger-than-average individuals have experienced mostly favorable growing conditions when they were young, and during their lifetime escaped the effects of extreme growth-limiting factors such as lack of nutrients, water or light, frost, or injury. In contrast, limiting and extreme factors negatively affected growth in small individuals. Giants are the winners, and dwarfs are the losers of competition.

From the following figures can be concluded:
○ Trees, shrubs and herbs follow the same reaction mechanisms to environmental influences.

○ The surface area of the assimilating tissue, the height of the plant, and the size of water-conducting area (vessel diameter, number of vessels) are correlated. Large transpiring crowns need efficient water-conducting tissues.

○ The intensity of mechanical stress and the amount of stabilizing elements (fiber-wall thickness, ray width) are correlated. Fibers in large plants are normally thicker-walled than those in small plants.

○ Vessels and rays are not a compulsory element of plants. Fibers and parenchyma in plants growing under extremely poor conditions are capable to conduct water, and to store photosynthetic products, e.g. starch.

○ Anatomical structural differences between extremely large and extremely small plants can be obvious (e.g. in *Fraxinus* or *Carpinus*) or minor (e.g. in *Alliaria* or *Arabidopsis*). The production rates are constant—a high annual production in large, and a low production in small plants.

Anatomical characteristics of genetically large and small species

9.1 *Fraxinus excelsior.* Top: 20 m-tall as part of a riparian forest. Bottom: a 60 cm-tall browsed bush in a meadow.

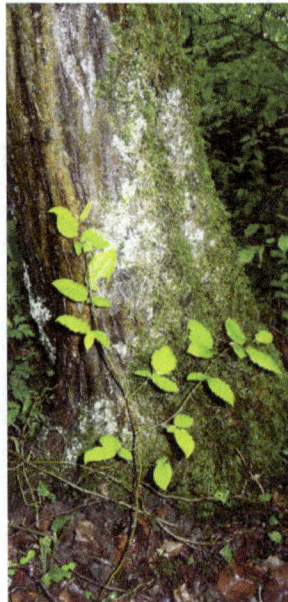

9.2 *Carpinus betulus.* Stem of a 10 m-tall dominant tree and a 50 cm-tall suppressed sapling.

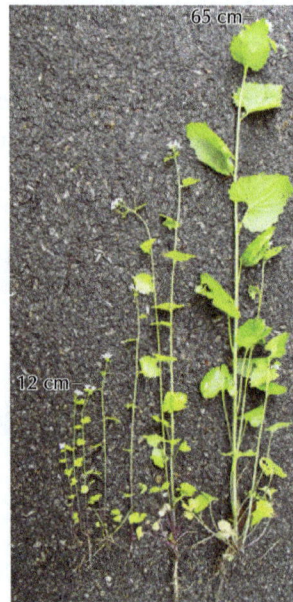

9.3 *Alliaria petiolata.* Bi-annual plants on a ruderal site, 12–65 cm in height.

9.4 *Arabidopsis thaliana.* Annual plants on a ruderal site, 6–35 cm in height.

Fraxinus excelsior

Tall old trees of common ashes (*Fraxinus excelsior*) with large crowns and large stem diameters grow mostly on nutrient rich, moist soils. The xylem anatomy of dominant trees and suppressed individuals is extremely different. Characteristic for large trees are ring porosity, large dense latewood and large rays. Small individuals with small crowns and thin stems grow under the canopy and are periodically exposed to grazing. They survive periodic crown damages and can get old. Characteristic for them are tissues consisting of thin-walled fibers, few vessels and small rays. A large water-conducting earlywood in trees is in accordance with a large assimilating tissue and extensive sap flow. In contrast, the few vessels in stems of suppressed or crippled trees reflect the presence of a reduced leaf mass and indicate a reduced sap flow. A major function of large rays in large trees is probably to enhance radial stem stability.

Carpinus betulus

Hornbeam (*Carpinus betulus*) grows mostly on shallow soils. Trees form the 10 m-tall canopy and suppressed individuals (saplings) are in the herb layer in the shadow. The xylem anatomy of dominant trees and suppressed individuals is fairly similar.

Common are vessels in radial rows and radial vessel-free zones. Vessels in trees are slightly larger than those in suppressed trees. Rays in trees are distinctly distinguished from the ground tissue and those in suppressed trees are similar to the axial parenchyma cells. A functional explanation of the xylem structure between trees and saplings is difficult.

giant

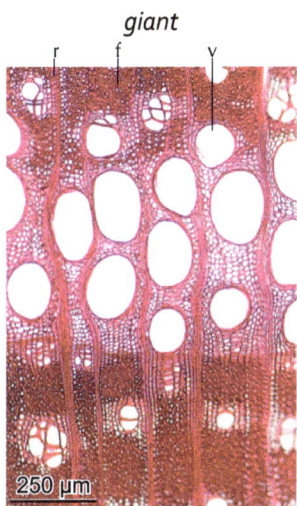

9.5 Ring-porous xylem with a large earlywood zone and many large vessels in the cross section of a dominant tree of *Fraxinus excelsior*.

dwarf

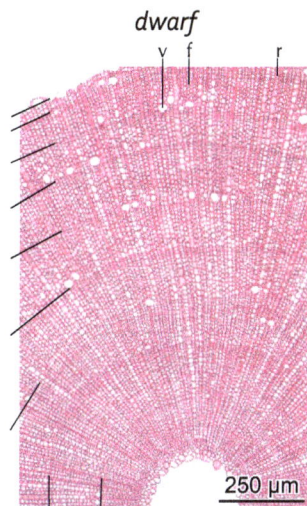

9.6 Few small vessels in a fibrous tissue in the cross section of the xylem of a browsed sapling of *Fraxinus excelsior*.

giant

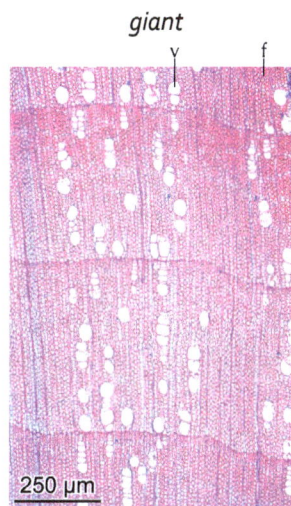

9.7 Diffuse-porous xylem with vessels in radial rows between thin- to thick-walled fibers in the cross section of a dominant tree of *Carpinus betulus*.

dwarf

9.8 Diffuse-porous xylem with few vessels in radial rows between thin-walled fibers in the cross section of a suppressed sapling of *Carpinus betulus*.

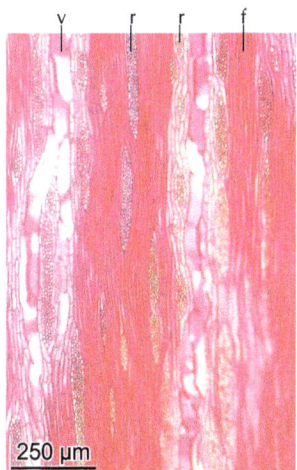

9.9 Tri- to five-seriate homogeneous rays between fibers, parenchyma cells and vessels in the tangential section of a dominant tree of *Fraxinus excelsior*.

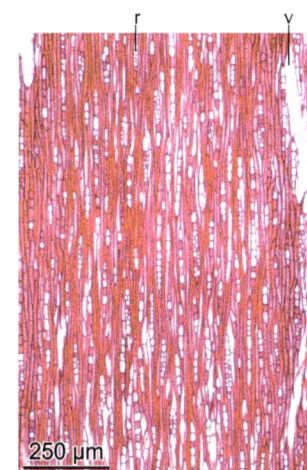

9.10 Uni- to biseriate homogeneous and heterogeneous rays between fibers and axial parenchyma cells in the tangential section of a browsed sapling of *Fraxinus excelsior*.

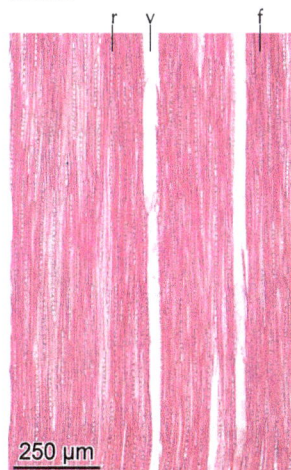

9.11 Slender uni- and biseriate homogeneous rays in the tangential section of a dominant tree of *Carpinus betulus*.

9.12 Wide uni- and biseriate rays in the tangential section of a suppressed sapling of *Carpinus betulus*.

Alliaria petiolata

Garlic mustard (*Alliaria petiolata*) is a biannual herb growing on deep, rich soils. Dense stands are composed of 15–100 cm-tall individuals. The xylem anatomy of dominant and suppressed individuals is very different. The second ring in large individuals contains 60–80 µm wide, radially arranged vessels within a thin-walled fiber tissue. Rays are similar to axial parenchyma cells, some are in confluent groups. The second ring in small individuals contains a few vessels with a diameter of 20–30 µm within a thin- to tick-walled fiber tissue. Rays are similar to the ones in large individuals but the fibers are smaller. A large water-conducting earlywood with vessels in large individuals is in accordance with a large assimilating tissue and extensive sap flow. In contrast, the few vessels in stems of the small individuals reflect a reduced leaf mass and indicate a reduced sap flow. The dense fiber tissue guarantees stability of the very thin stem. Dense stands suggest that root competition is the major growth-limiting factor.

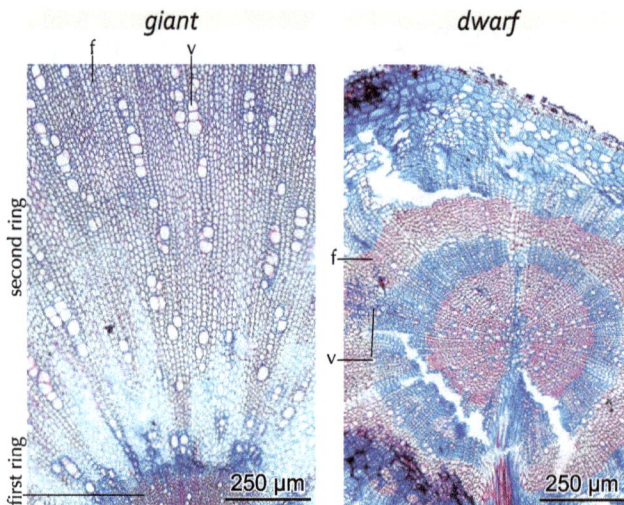

Arabidopsis thaliana

Thale cress (*Arabidopsis thaliana*) is an annual herb and grows on dry to medium moist sites with rich soils. Stands are composed of 5–50 cm-tall individuals. The center of large and small individuals has no fibers; it consists mainly of unlignified parenchyma cells. Rays are absent in large and in small individuals. Very small vessels with a diameter of 15–25 µm are arranged in long radial rows. Vessels are almost absent in small individuals. Small vessels guarantee sap flow in large individuals. Water conductance in small individuals occurs probably through fibers. Different root spaces are growth-limiting.

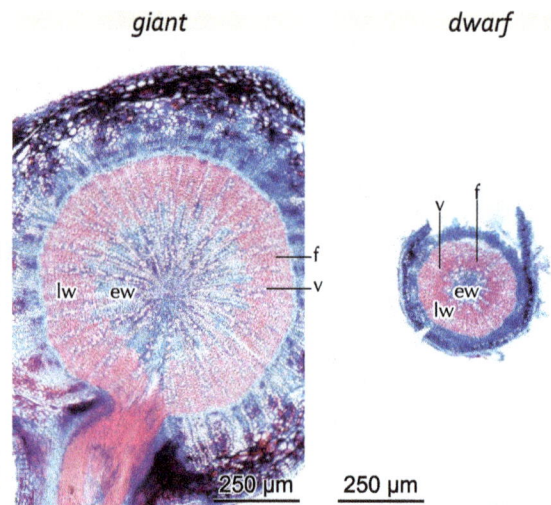

giant

dwarf

giant

dwarf

9.13 Large radially arranged vessels within a thin-walled fiber tissue in the cross section of a large individual of *Alliaria petiolata*.

9.14 Few small vessels within a thin- to thick-walled fiber tissue in the cross section of a small individual of *Alliaria petiolata*.

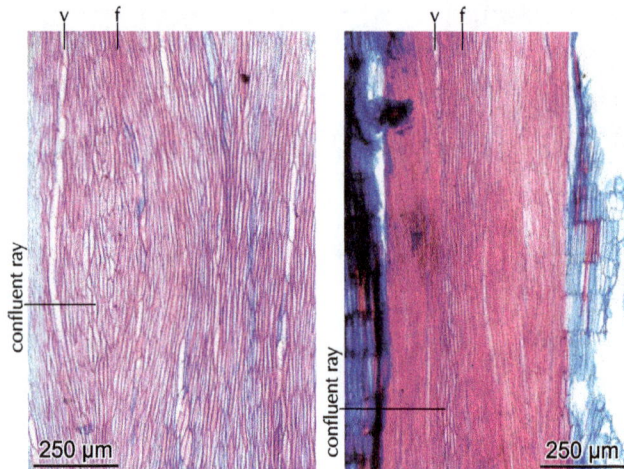

9.15 Small, radially arranged vessels within a thin-walled fiber tissue in the cross section of a large individual of *Arabidopsis thaliana*.

9.16 Very few small vessels within a thin- to thick-walled fiber tissue in the cross section of a small individual of *Arabidopsis thaliana*.

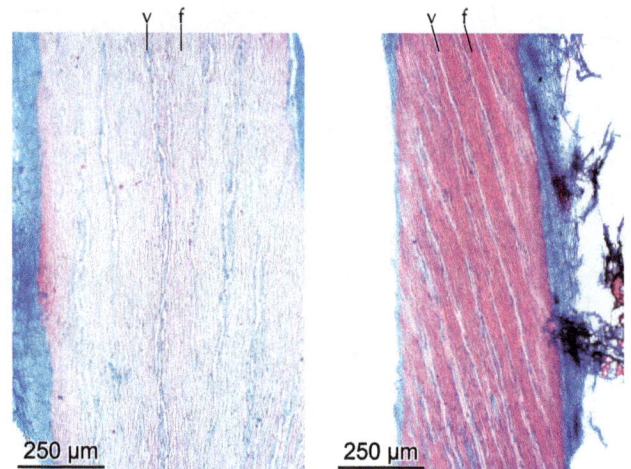

9.17 Rays, axial fibers and parenchyma cells are not or difficult to distinguish in the tangential section of a large individual of *Alliaria petiolata*.

9.18 Rays, axial fibers and parenchyma cells are not or difficult to distinguish in the tangential section of a small individual of *Alliaria petiolata*.

9.19 Rays, axial fibers and parenchyma cells are indistinguishable in the tangential section of a large individual of *Arabidopsis thaliana*.

9.20 Rays, axial fibers and parenchyma cells are indistinguishable in the tangential section of a small individual of *Arabidopsis thaliana*.

9.2 Anatomical and morphological adaptation to different climates

Vegetation zones are the result of climatic influences. Zones are defined by climatic conditions and the physiognomy of the vegetation, e.g. the constantly wet tropical zone is dominated by evergreen trees, epiphytes and lianas, or the dry cold-temperate (boreal) zone is dominated by conifers and evergreen or summer-green dwarf shrubs (Pfandenhauer & Klötzli 2014).

Here the question is raised to which degree zonal climatic conditions affect the physiognomy and the anatomy of stems. Wheeler *et al.* 2007 analyzed the xylem of almost 6,000 trees from the tropics to the boreal zones around the globe. They found that the xylem in trees of different vegetation zones is

expressed in a modest way by ring distinctness and the water conductivity system (vessels). Based on their results, presented here are frequently occurring anatomical structures of trees and shrubs from different latitudinal and altitudinal zones.

A) Several tree species from the tropical rain forest, the European cold, moist temperate zone, and the Eurasian boreal forest.

B) Shrubs from the tropical desert (Sahara), the European dry thermo-Mediterranean zone and the arctic zone.

9.2.1 Trees in the tropics, the temperate and the boreal zone

Tropical rain forest
Trees with adaptations to a persistently warm, aseasonal climate with more than 3,000 mm of annual precipitation. Water transport through the tall stems with heights up to 50 m to the

large, evergreen crowns occurs through a few large vessels. Most trees in lower altitudes of tropical rain forests do not form annual ring boundaries, making cross-dating impossible.

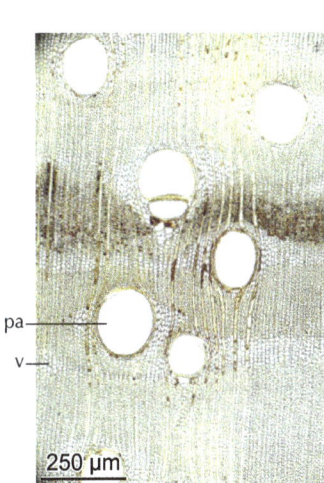

9.21 *Microberlinia brazzavillensis*, Fabaceae, a 40 m-tall evergreen tree. Diffuse-porous wood with 5–10 vessels/mm², >200 μm vessel diameter and paratracheal parenchyma.

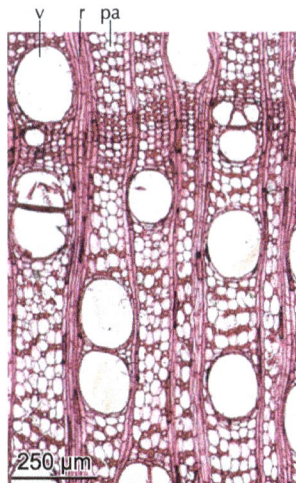

9.22 *Pseudobombax munguba*, Malvaceae, a 30 m-tall deciduous tree. Diffuse-porous wood with approx. 5–10 vessels/mm², >200 μm vessel diameter, very thick-walled fibers, few vessels and a lot of parenchyma.

9.23 *Diospyros ebenum*, Ebenaceae, a 25 cm-tall evergreen tree. Diffuse-porous wood with black heartwood, approx. 5 vessels/mm², >200 μm vessel diameter, and parenchyma around vessels and in bands.

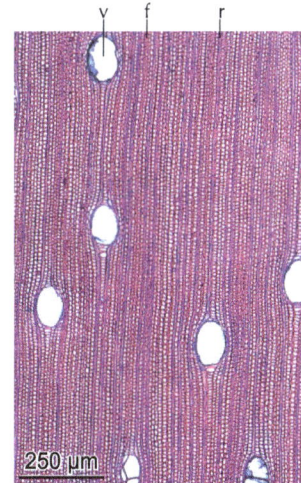

9.24 *Sarcotheca* sp., Oxalidaceae, an evergreen tree. Diffuse-porous wood with approx. 5 vessels/mm², 200 μm vessel diameter, many fibers and few axial parenchyma cells.

Temperate forest

Trees with adaptations to a temperate seasonal climate with 1,000 mm of annual precipitation. Water transport through the tall stems with heights up to 35 m to the large, deciduous crowns occurs mostly through many small vessels (diffuse-porous wood), or in a few species through large vessels in the earlywood and small vessels in the latewood (ring-porous wood). The anatomical character of the earlywood reflects a tropical and that one in the latewood a temperate climate. Principally all trees and shrubs in temperate climates form annual rings. Cross-dating is therefore possible. About 80% of perennial dicotyledonous herbs also produce annual rings.

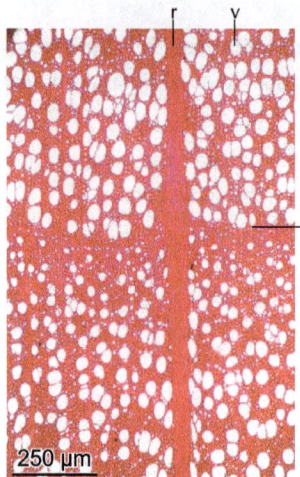

9.25 *Fagus sylvatica*, Fagaceae, a 25 m-tall deciduous tree. Diffuse-porous wood with approx. 300 vessels/mm², 100 µm earlywood vessel diameter.

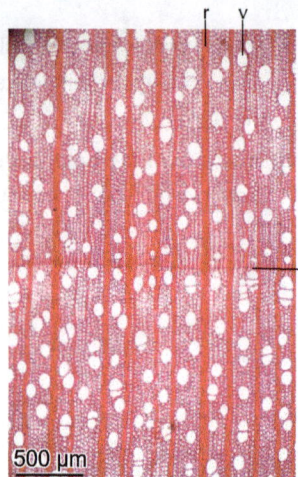

9.26 *Acer pseudoplatanus*, Sapindaceae, a 25 m-tall deciduous tree. Diffuse-porous wood with approx. 150 vessels/mm², 100 µm vessel diameter.

9.27 *Pyrus communis*, Rosaceae, a 15 m-tall deciduous tree. Diffuse-porous wood with approx. 300 vessels/mm² and 100 µm vessel diameter.

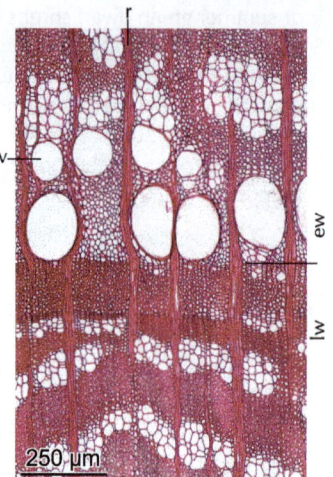

9.28 *Ulmus glabra*, Ulmaceae, a 25 m-tall deciduous tree. Ring-porous wood, latewood vessels in tangential bands.

Boreal forest

Trees with adaptations to a seasonal climate with warm short summers and cold winters in the conifer belt of the boreal zone, dominated by spruces (*Picea* sp.), larches (*Larix* sp.) and pines (*Pinus* sp.). Birch trees (*Betula pendula*) and mountain-ash (*Sorbus aucuparia*) are common accompanying tree species. Water transport through the tall stems with heights up to 25 m to the evergreen crowns occurs primarily through the earlywood tracheids. The anatomically diffuse-porous structure of the accompanying deciduous species is like in the temperate zone. All trees, shrubs and perennial herbs form annual rings. Cross-dating is possible in individuals with more than 20 rings.

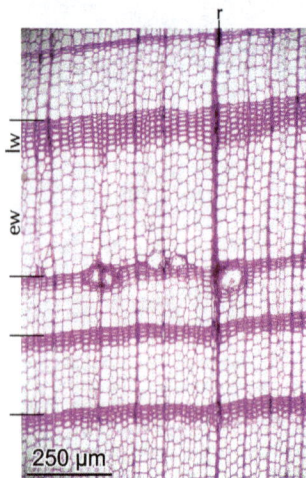

9.29 *Larix sibirica*, Pinaceae, a 15 m-tall deciduous tree.

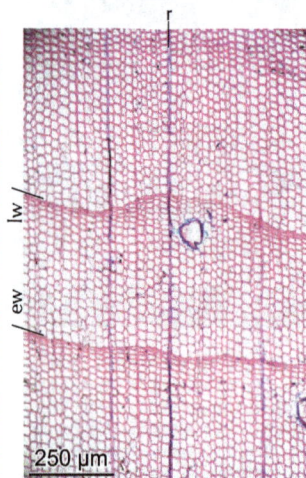

9.30 *Pinus sibirica*, Pinaceae, a 25 m-tall evergreen tree.

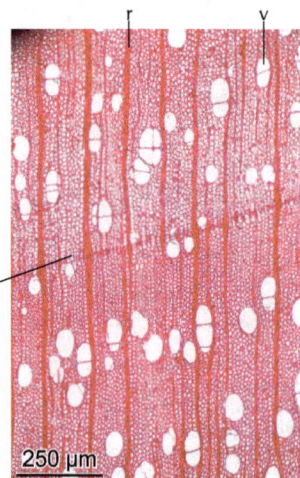

9.31 *Betula pendula*, Betulaceae, a 10 m-tall deciduous tree. Diffuse-porous wood with approx. 150 vessels/mm² and 100 µm vessel diameter.

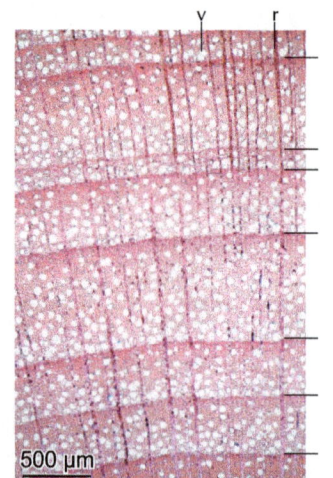

9.32 *Sorbus aucuparia*, Rosaceae, 10 m-tall deciduous tree. Diffuse-porous wood with approx. 300 vessels/mm² and 50 µm vessel diameter.

9.2.2 Shrubs in the tropics, the Mediterranean and arctic zone

Subtropical African dry climate, Sahara

Shrubs with adaptations to persistent drought in a tropical desert (Sahara) with less than 5 mm of annual precipitation. Very few shrubs and dwarf shrubs survive the extremely long drought periods and often also impact by grazing. Very intensive growth occurs for very short periods after rainfall events. Most of the species have no visible growth zones, and if there are any they correlate to rain events rather than regular seasons. Cross-dating is therefore impossible.

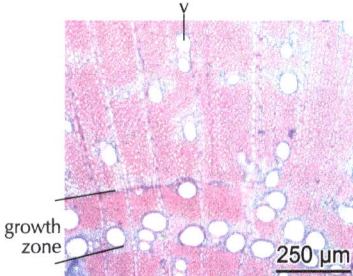

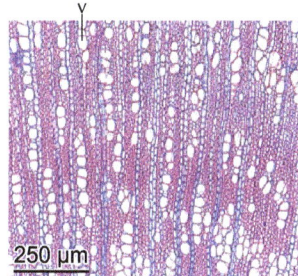

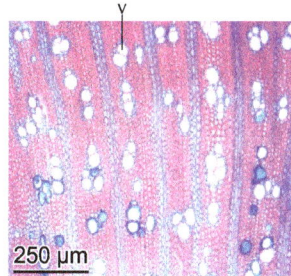

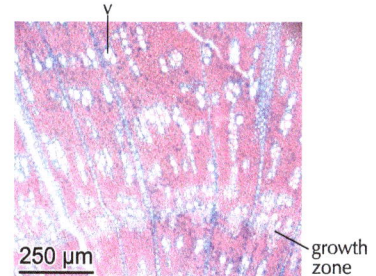

9.33 *Calligonum azel*, Polygonaceae, a 1 m-tall shrub. Diffuse-porous wood with approx. 40 vessels/mm² and 100 µm vessel diameter.

9.34 *Euphorbia calyptrata*, Euphorbiaceae, a 30 cm-tall dwarf shrub. Growth zones absent. With approx. 200 vessels/mm² and 40 µm vessel diameter.

9.35 *Zilla spinosa*, Brassicaceae, a 40 cm-tall dwarf shrub. Growth zones absent. With approx. 150 vessels/mm² and 50 µm vessel diameter.

9.36 *Farsetia aegyptia*, Brassicaceae, a 40 cm-tall dwarf shrub. Distinct growth zones. With approx. 200 vessels/mm² and 400 µm vessel diameter.

European thermo-Mediterranean zone

Shrubs and dwarf shrubs with adaptations to winter rain and summer drought, with less than 200 mm of annual precipitation. Many shrubs of numerous families profit from the winter rains and survive the extreme summer droughts. Most of them form true annual rings. Radial growth greatly varies. Plants on shallow soils form small rings (<1 mm width), those on deep soils produce larger rings (2–3 mm). Cross-dating is principally possible, but disturbances (fire, grazing) dominate climatic effects.

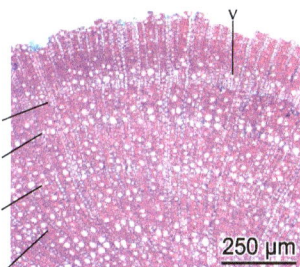

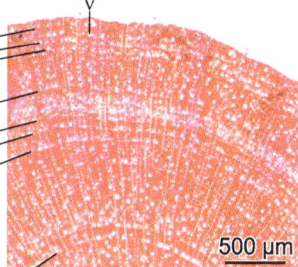

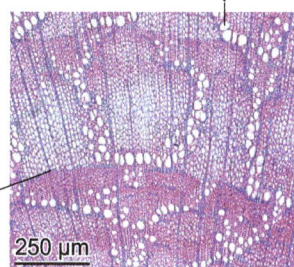

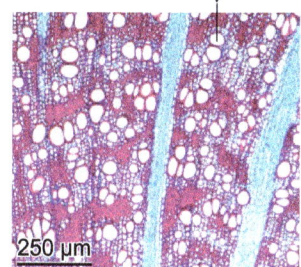

9.37 *Teucrium chamaepitys*, Lamiaceae, a 20 cm-tall dwarf shrub. Semi-ring-porous wood with approx. 400 vessels/mm² and 25 µm vessel diameter.

9.38 *Fumana ericoides*, Cistaceae, a 10 cm-tall dwarf shrub. Diffuse-porous wood with approx. 400 vessels/mm² and 30 µm earlywood vessel diameter.

9.39 *Daphne gnidium*, Thymelaeaceae, a 60 cm-tall dwarf shrub. Diffuse-porous wood with approx. 250 vessels/mm² and 40 µm vessel diameter.

9.40 *Launaea lanifera*, Asteraceae, a 30 cm-tall dwarf shrub. Without growth rings. With approx. 200 vessels/mm² and 50 µm vessel diameter.

Arctic zone

Dwarf shrubs with adaptations to persistently cold temperatures and short vegetation periods. Dwarf shrubs from a few families (Ericaceae, Salicacea and Betulaceae) grow only during the one or two summer months. They form mostly distinct, but extremely small annual rings (0.05–0.5 mm). Cross-dating is difficult, but possible.

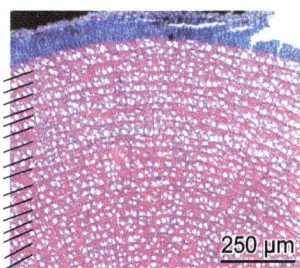

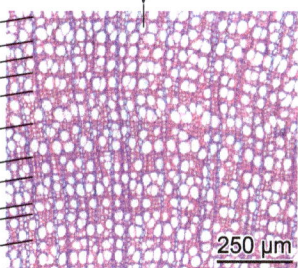

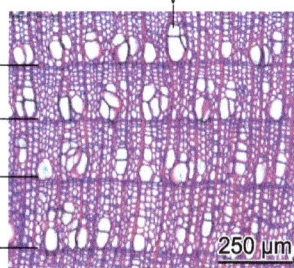

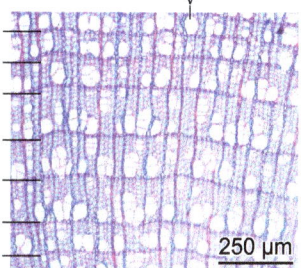

9.41 *Cassiope tetragona*, Ericaceae, a 20 cm-tall dwarf shrub. Diffuse-porous wood with >500 vessels/mm² and 20 µm vessel diameter.

9.42 *Empetrum nigrum*, Ericaceae, a 30 cm prostrate dwarf shrub. Diffuse-porous wood with approx. 500 vessels/mm² and 35 µm vessel diameter.

9.43 *Betula nana*, Betulaceae, a 50 cm prostrate dwarf shrub. Diffuse-porous wood with approx. 200 vessels/mm² and 40 µm vessel diameter.

9.44 *Salix arctica*, Salicaceae, a 40 cm-tall dwarf shrub. Diffuse-porous wood with approx. 300 vessels/mm² and 50 µm vessel diameter.

Adaptation to Temporarily Changed Environments

Two principal capacities characterize seed plants. Primarily, cell formation pathways determine the basic structure of plant bodies, and the formation of different cell types. This is shown in the following figure.

Node ① determines the formation of fusiform mother cells, vessels and axial parenchyma mother cells.

Node ② determines the formation of fibers, tracheids and vessels.

Node ③ determines the formation of parenchyma cells or vessels.

Node ④ guides living parenchyma cells in various directions.

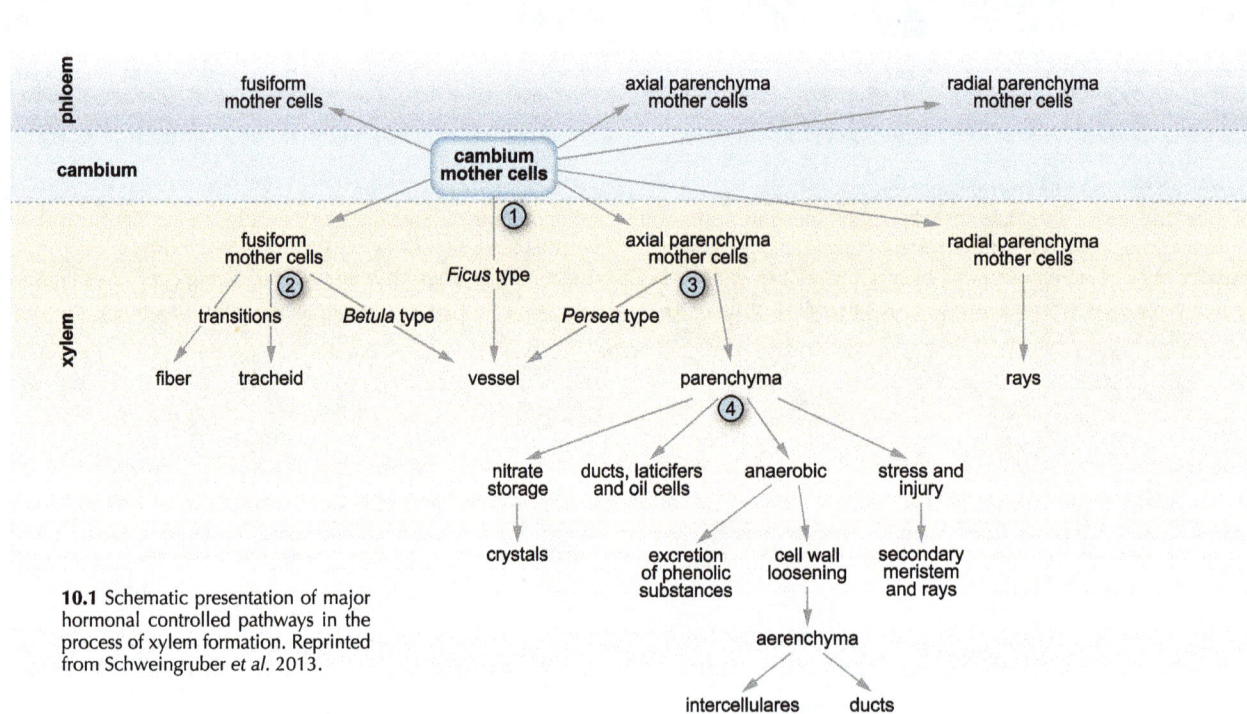

10.1 Schematic presentation of major hormonal controlled pathways in the process of xylem formation. Reprinted from Schweingruber *et al.* 2013.

Secondarily, the cambium and cells are able to react to short-term environmental changes with a reduction or an enhancement of their activity. This is described in detail in Larson 1994, Timell 1986, Schweingruber 2007 and Fromm 2013. In the following chapter the anatomical reactions to various environmental influences are demonstrated in a condensed form.

Two basic processes determine temporal anatomical reactions in all growth forms—from herbs over shrubs to trees—in all climatic zones and sites:

- Xylem mother cells in the cambial zone produce different amounts of fibers, vessels and axial and radial (ray) parenchyma. The consequence of intensive stress is structural change.
- Cambial cells and living parenchyma cells react to minor environmental influences with cell-wall expansion,

secondary-cell-wall formation and lignification, and to intensive stress with the formation of callus tissue or ducts, and the excretion of various substances.

Short-term environmental conditions determine the anatomy of the xylem. However, the structures greatly vary because modifications occur based on genetic, taxon-specific information, and based on the location of the environmental impact (tropism). Cambial and cellular reactions are mostly combined. Reactions to the impact occur immediately or with a time lag. Anatomical reactions are mostly not unique to specific environmental influences.

In the following pages, the full range of variability has to be limited to showing a few examples for conifers and deciduous trees.

10.1 Anatomical effect of short-term environmental changes during the vegetation period

10.1.1 Individual small and large annual rings and missing rings

"Large" and "small" are relative terms, and are always related to the mean values of width within a ring sequence. Extremely small or large rings express cambial reactions to short-term events in the leaf or root area of plants. A few extremely small rings can be the consequence of extreme droughts, cold days, short vegetation periods, or crown damage by insects. These are **negative pointer years**.

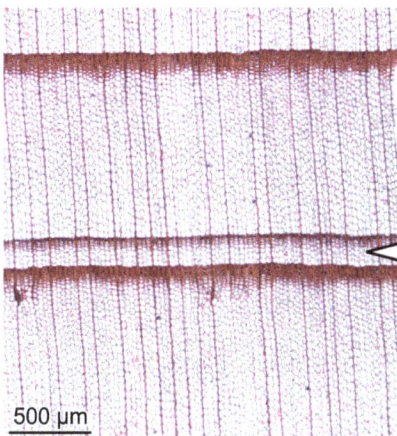

If environmental conditions are too extreme, annual rings can be missing completely; cambial and cellular activities are interrupted for at least one vegetation period. Missing rings can only be detected by cross-dating with other samples.

In contrast, above-average precipitation or warm days can lead to the formation of a few extremely large rings. These are **positive pointer years**.

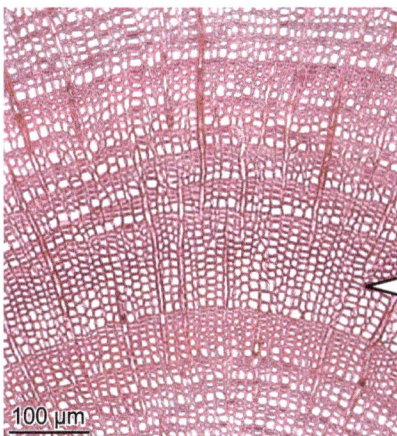

Negative pointer years

10.2 Extremely small ring with very few earlywood tracheids in the conifer *Metasequoia glyptostroboides*. A severe summer drought reduced the cambial activity mainly at the beginning of the growing season.

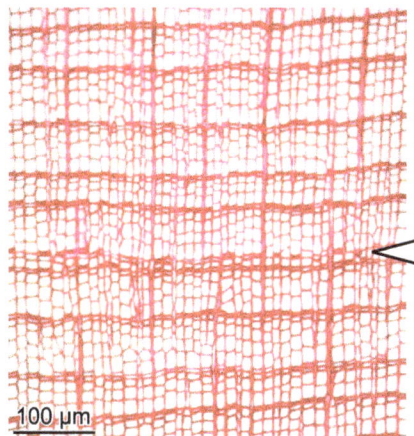

10.3 Extremely small ring with one earlywood tracheid in the conifer *Juniperus communis*. A very short and cold vegetation period in the subarctic environment reduced the cambial activity at the beginning of the growing season.

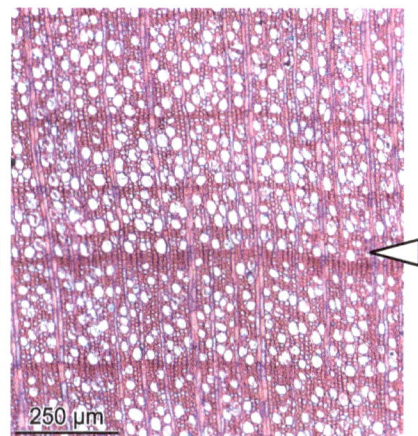

10.4 Extremely small ring in the deciduous tree *Pyrus pyraster*. The reason for this growth reduction is unknown.

Positive pointer years

10.5 Extremely large ring in the conifer *Larix dahurica*. An unusually rainy and warm period probably induced accelerated growth at the northern timberline in eastern Siberia.

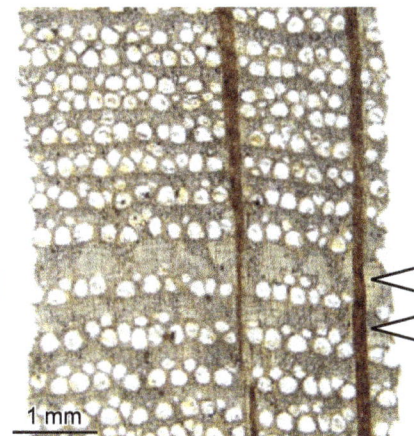

10.6 Two rings with large latewood zones in a sequence of smaller rings in a subfossil *Quercus* sp. from a bog with usually high water table in northern Germany. Larger rings may be the result of a larger aerobic root zone in drier years.

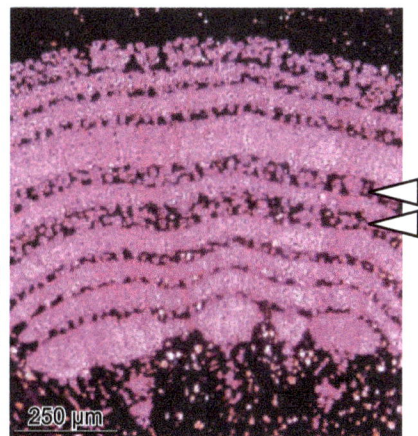

10.7 Two rings with large earlywood zones in the root collar of the herb *Silene vulgaris* in a meadow of the temperate zone, polarized light.

10.1.2 Discontinuous growth – Wedging rings

The presence of wedging rings is an indication of a principal plant physiological law. Each cambial cell and each living differentiated cell reacts autonomously to external mechanical stresses. This capacity permits an optimal reaction to internal and external influencing factors and consequently, the physiological optimization of plants (Schweingruber 2007).

Wedging rings in increment cores or discs only provide a limited insight into the actual presence of rings within the whole plant. Nogler 1981 demonstrated that wedging or missing rings are only locally missing in stems. Ring wedging occurs in all growth forms and taxonomic units with secondary growth. Individual or several rings wedge out in short or long periods of suffering from stress, after position changes, or just in relation to the formation of fluted stems.

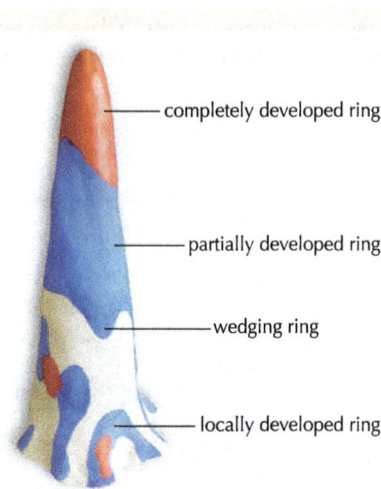

Occurrence of a few wedging rings

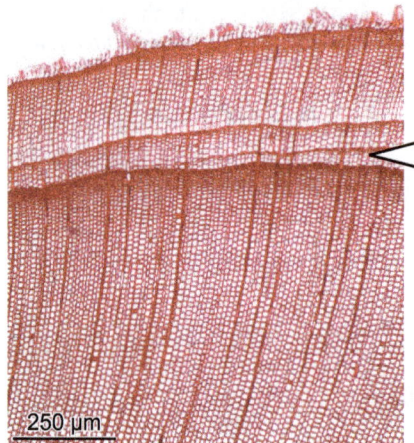

completely developed ring

partially developed ring

wedging ring

locally developed ring

10.8 Schematic diagram of a locally absent, wedging or missing annual ring. Model: Nogler 1981.

250 µm

10.9 Individual wedging ring in the conifer *Juniperus nana*.

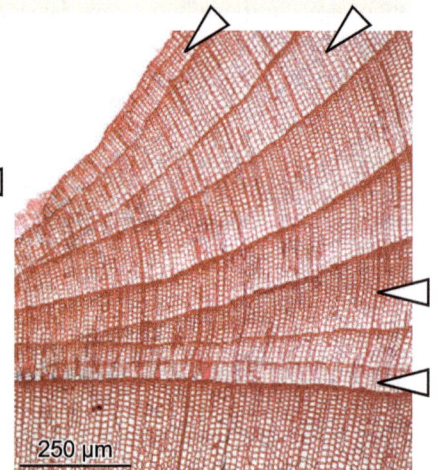

250 µm

10.10 Many wedging rings in a zone of fluting of the stem of the conifer *Juniperus nana*.

Occurrence of wedging ring zones

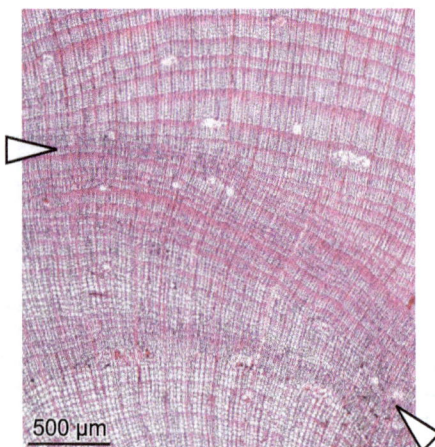

500 µm

10.11 Wedging areas in a branch of *Pinus mugo* which changed its position after rock fall events.

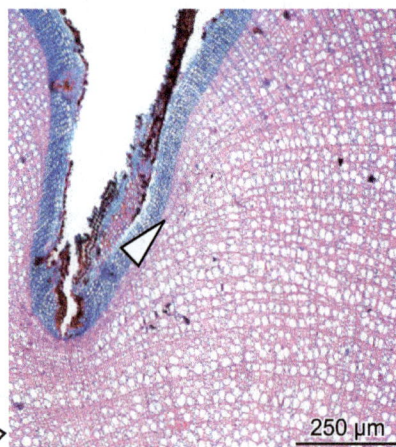

250 µm

10.12 Wedging rings in a zone of fluting of the dwarf shrub *Empetrum nigrum*.

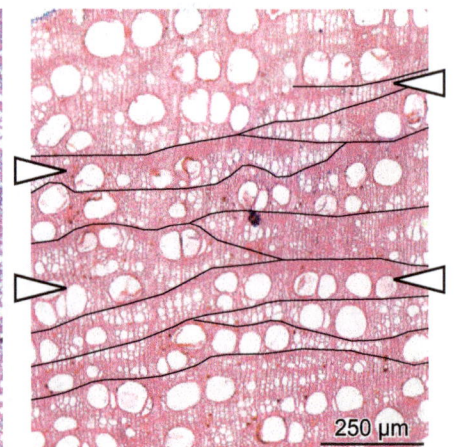

250 µm

10.13 Multiple wedging rings in the ring-porous wood of *Ulmus glabra* after intensive browsing.

10.1.3 Individual small and large latewood zones and latewood zones with thin- or thick-walled tracheids

Noticeable latewood zones in conifers are related to reduced or enhanced cambial activity or cell-wall formation activity of living tracheids. If the cell walls remain small they are called light years (low density; Kaennel & Schweingruber 1995). Reduced cell wall growth is a result of low temperatures, insufficient water supply in the late summer or a lack of photosynthetic products after crown or root damages. Reduced cambial and cell activity often occurs together.

Extremely light or dense latewood zones

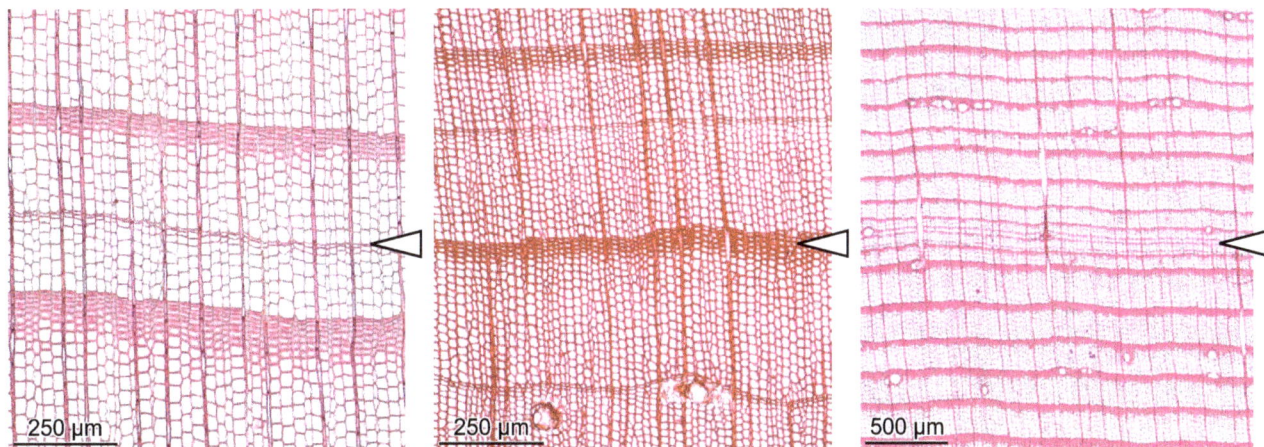

10.14 Individual ring with small latewood consisting of thin-walled tracheids in *Larix decidua*.

10.15 Individual ring with large latewood consisting of thick-walled tracheids in *Pinus sylvestris*.

10.16 Sequence of small rings with small latewood zones consisting of thin-walled tracheids in *Larix decidua*.

10.1.4 Individual, not fully lignified latewood zones ("blue rings")

Lignification is the last step in the process of cell formation. If this process is interrupted, and the cells are not fully lignified, they appear blue in slides stained with Astrablue/Safranin. These so-called "blue rings" were found exclusively in conifers of the northern high latitudes. It is believed that low temperatures in fall induce this feature. See Piermattei *et al.* 2015.

Blue rings in conifers

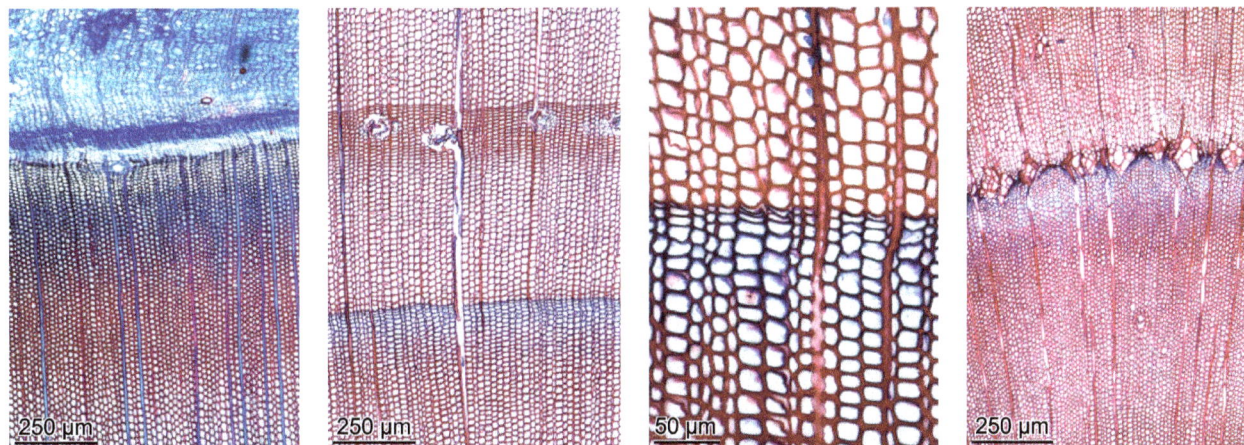

10.17 Latewood in a ring of *Larix decidua* growing at the treeline. The last section of the latewood tracheids is still in formation and the unlignified cells appear unstained or blue.

10.18 Unlignified latewood zone in *Pinus sylvestris* growing at the northern tree line.

10.19 Completely formed latewood zone in *Pinus sylvestris* growing at the northern tree line. Primary walls are lignified (red), secondary walls are unlignified (blue).

10.20 Frost ring and "blue ring" in the latewood of *Larix decidua* growing at the tree line. Extreme frost in early fall damaged the cambium cells and interrupted cell-wall formation. Callus tissue is formed and the secondary walls remain unlignified.

10.1.5 False rings and density variations

False rings are normally formed rings with distinct latewood borders and cannot be anatomically distinguished from true annual rings. Their intra-annual occurrence can be identified only by cross-dating. The term density variation is based on x-ray analysis of conifers and indicates intra-annual structural variations without distinct tangential borders. The difference between the two types is gradual.

False rings look like normal rings, containing a latewood zone with an abrupt transition from late- to earlywood. They occur mainly in Mediterranean and arid regions and are primarily triggered by short, intensive summer droughts.

Density variations of conifers are characterized by intra-annual tangential zones of thicker-walled and often smaller tracheids

in relation to previous and following zones. They occur from the tropics to the temperate zone and are almost absent in boreal and arctic zones. Intra-annual density variations are triggered by short-term climatic events such as cold or dry periods during the growing season or slight damage to crown or roots. The susceptibility varies within the conifers. Cupressaceae are more, Pinaceae are less susceptible.

Intra-annual tangential structural changes of some deciduous plants are based on rhythmic formation of different cell types such as fibers/parenchyma, fibers/vessels or parenchyma/vessels. Theses changes generally occur in fast-growing stems of trees, shrubs and herbs from the tropics to the temperate zone. Such structural changes are triggered by genetic characteristics as well as short-term climatic events.

False rings

100 µm

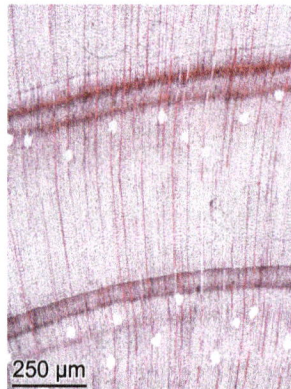

250 µm

Few density variations in conifers

250 µm

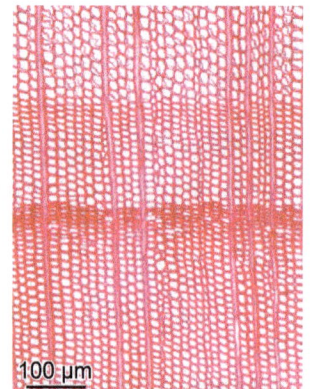

100 µm

10.21 False rings in *Pseudotsuga menziesii* growing in Arizona, USA, an area with monsoon rains. There are two rings within a growing season, with the first ring interrupted from the second by a severe summer drought.

10.22 Large rings with two density variations in the latewood of *Pinus sylvestris* growing on shallow soil in Valais, Switzerland. Late summer drought triggered irregular latewood growth.

10.23 Large rings with two density variations in the latewood of *Pinus sylvestris* growing on sand dunes in Germany. Zones of smaller and larger, and thin- and thick-walled tracheids indicate variable water availability.

10.24 Two density variations in *Cupressus sempervirens*. The more intensive variation is characterized by a few rows of small, thick-walled tracheids. The weaker variation is expressed just by smaller tracheids.

Multiple density variations

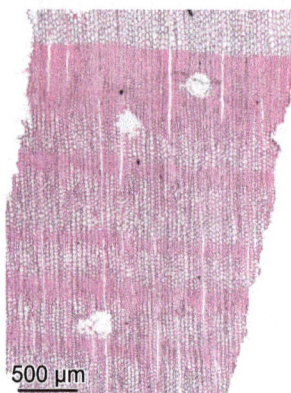

500 µm

Intra-annual structural changes

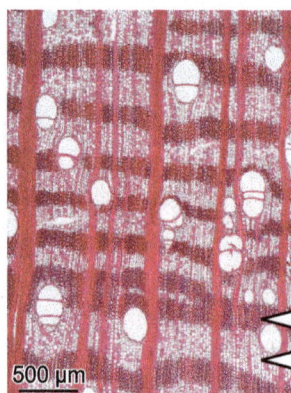

500 µm

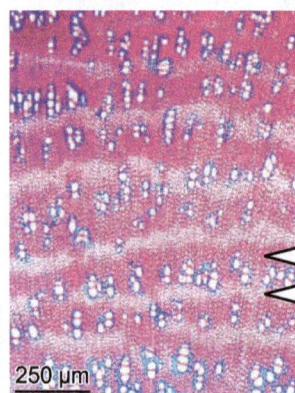

250 µm

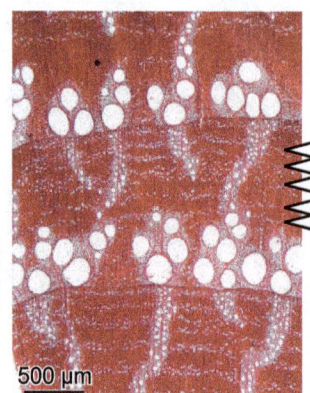

500 µm

10.25 Multiple density variations in *Pinus elliottii* in tropical Tahiti.

10.26 Tangential bands of fibers alternate with bands of parenchyma in *Ficus sycomorus*, Egypt. This is characteristic for many tropical species.

10.27 Tangential bands of thick- and thin-walled fibers alternate irregularly in a shoot of a 40 cm-tall annual herb *Lepidium campestre* in the temperate zone in Davos, Switzerland.

10.28 Tangentially arranged parenchyma cells in the latewood of the ring-porous *Quercus robur* of the temperate zone in Switzerland.

10.1.6 Tissue and fiber cracks

All cracks are the result of short-term events.

Radial cracks in living conifer trees occur just in the earlywood of individual years, or along rays over many rings. Due to insufficient water flow in summer, the thin-walled earlywood cells contract laterally and split along the weakest zone—the rays. Tangential rows of radial splits can be dendrochronologically dated. Radial cracks in dry logs are a result of anisotropic tissue contractions. Extreme shrinkage of degraded, subfossil wood produces cogwheel-like stem sections. If splits in the xylem reach the cambial zone they induce traumatic reactions. Therefore cracks can be overgrown by accelerated cell production. Radial cracks in bark are the result of enlarged stem circumferences.

Tangential cracks (ring shake) are a result of mechanical stress and occur along anatomically weak zones in various species. Genetic predisposition or extreme wind events are the most frequent causes of ring shake and resin pockets. Stems of the ring-porous chestnuts split along the earlywood vessels, those of spruces along the thin-walled earlywood tracheids. Resin pockets occur mainly in stems of conifers in windy mountainous regions. Extreme tangential splitting is characteristic for a few alpine herbs, e.g. *Saxifraga* sp. and *Androsace* sp.

Micro-cracks mainly occur in cell walls of conifer tracheids. Extremely intense, very short events of tension or pressure due to rock fall, wind storms or avalanches can break tracheids irreversibly. The stability of wood with broken tracheid walls is probably not much reduced. The abnormalities can be detected by microscopy with polarized light.

Macroscopic aspect of long radial cracks

10.29 Radial cracks in a disc of *Fraxinus excelsior* are a result of intensive tangential shrinkage.

10.30 Large radial cracks in the anaerobically degraded peripheral zone of a Neolithic, waterlogged post of *Abies alba*.

10.31 Extremely large radial cracks in the degraded, light-colored sapwood and the dark heartwood of a Neolithic, waterlogged post of *Quercus* sp.

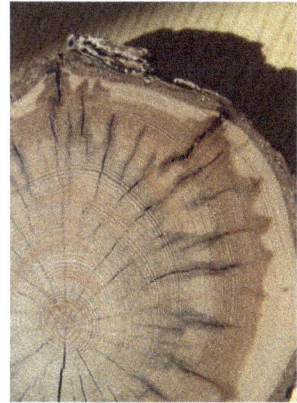

10.32 Many radial frost cracks in the heartwood of *Larix laricina* in the boreal zone. Material: Y. Begin.

Microscopic aspect of radial cracks

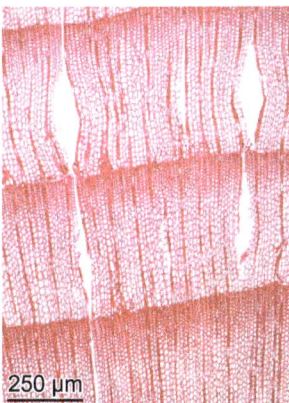

250 μm

10.33 Intra-annual radial cracks along rays in the earlywood of *Larix dahurica*.

500 μm

10.34 Frost split in *Larix sibirica*. The split opened and triggered accelerated radial growth and induced the formation of traumatic resin ducts.

Radial splits in the bark

10.35 Radial splits in the bark of *Populus nigra* are a result of stem expansion.

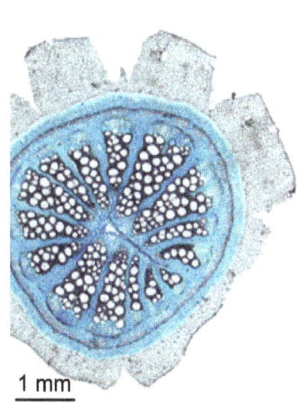

1 mm

10.36 Radial splits in the bark of *Aristolochia macrophylla* are a result of stem expansion.

Tangential cracks along weak, thin-walled zones

10.37 Tangential split in *Picea glauca*, indicating a thin-walled latewood after a volcanic event. Material: G. Jacoby.

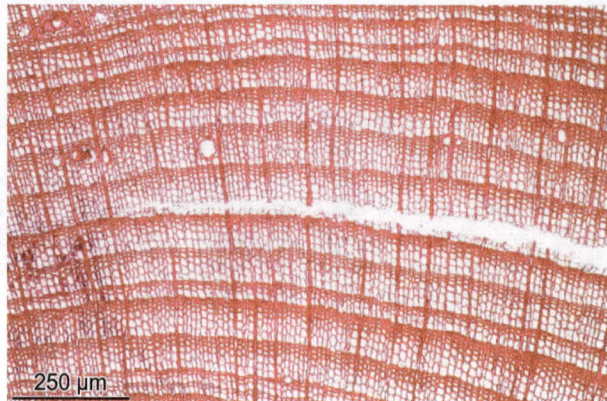

10.38 Tangential split along a thin-walled earlywood zone in *Larix dahurica*.

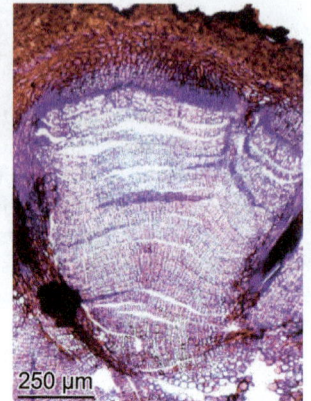

10.39 Tangential splits in a ray-less stem of the perennial herb *Saxifraga muscoides*.

Wind-induced tangential cracks

10.40 *Picea engelmannii* with one-sided branches in a windy area.

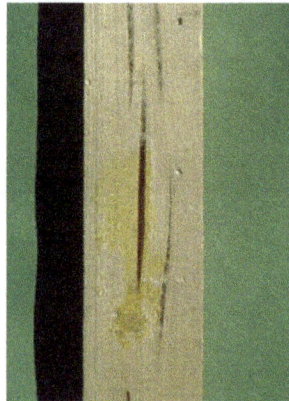

10.41 Longitudinal section trough resin pockets of *Picea abies* in an alpine region.

10.42 Macroscopic aspect of a resin pocket in *Pinus sylvestris* in an alpine region.

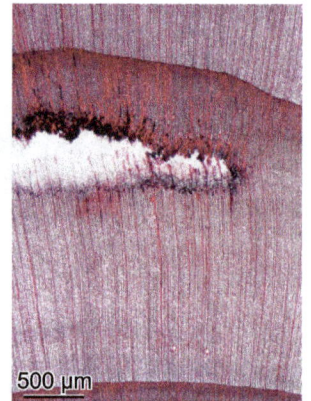

10.43 Microscopic aspect of a resin pocket in *Picea abies*. The crack borders are surrounded by callus tissues.

Shock-induced cracks in axial tracheid walls

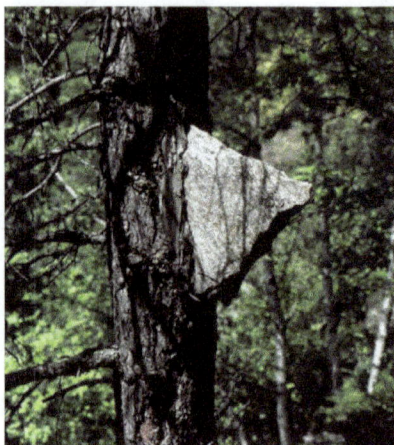

10.44 A stone hit a the stem of *Pinus sylvestris* in a rock fall.

10.45 Cracks in axial tracheid walls of *Picea abies* under polarized light.

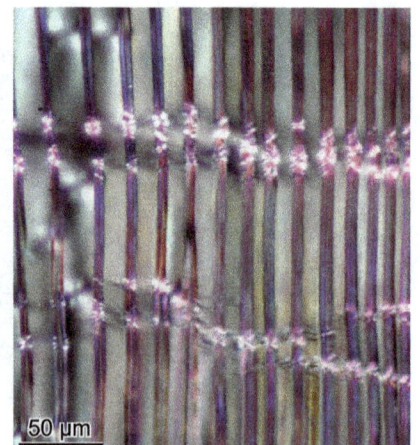

10.46 Zone of crushed tracheid walls in *Picea abies* under polarized light.

10.2 Effect of multi-annual environmental changes

10.2.1 Abrupt growth changes

Abrupt growth changes are the result of long-lasting environmental changes. A long-term reduction in growth can be the consequence of crown damage by repeated insect defoliation, environmental pollution, pruning, a change in water availability or increased competition. Sudden increased growth can be the consequence of improved light conditions, e.g. after thinning, or improved hydrological conditions. Abrupt growth changes are rarely permanent. They can in extreme cases lead to the death of individuals, however, in most cases they are reversible. Their causes can only be evaluated in relation to observations or measurements of environmental conditions.

Abrupt negative growth changes

10.47 Growth reduction in the conifer *Pinus mugo* after crown damage by a rock slide.

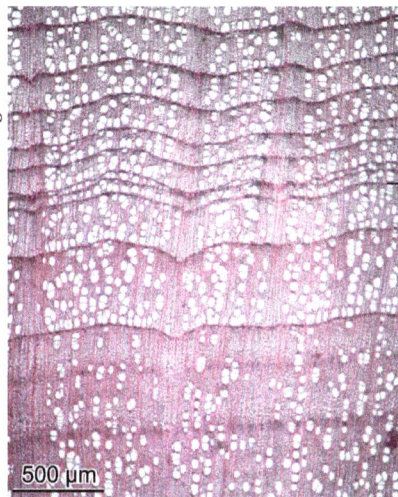

10.48 Growth reduction in the deciduous *Alnus glutinosa* after root damage during a landslide.

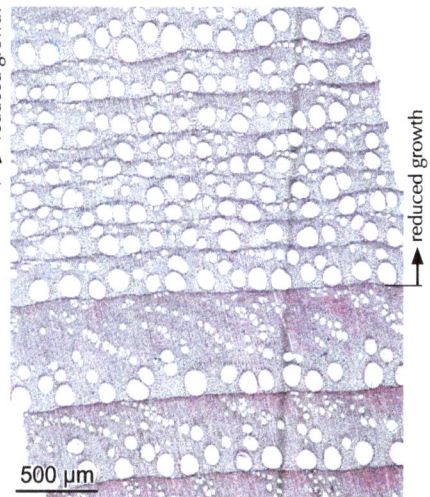

10.49 Growth reduction of the deciduous *Castanea sativa* after crown damage by the chestnut blight fungus (*Cryphonectria parasitica*).

Abrupt positive growth changes

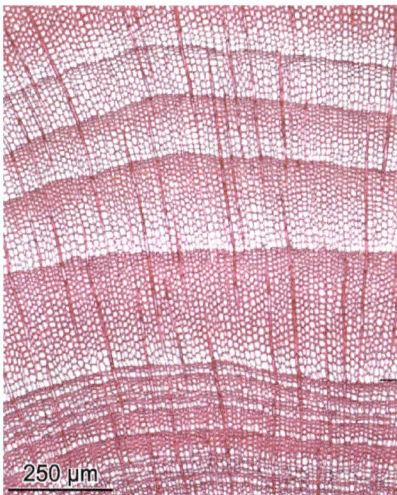

10.50 Enhanced growth in the boreal conifer *Larix dahurica*, its cause is unknown.

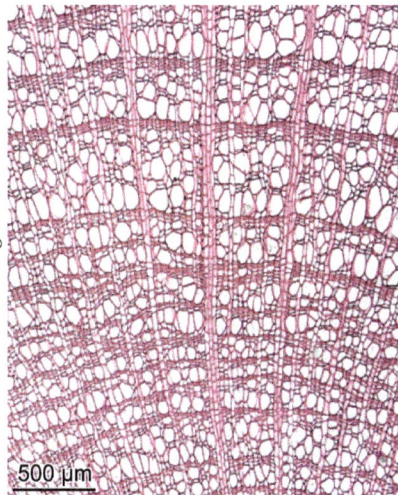

10.51 Enhanced growth in the prostrate mountain dwarf shrub *Arctostaphylos rubra*, its cause is unknown.

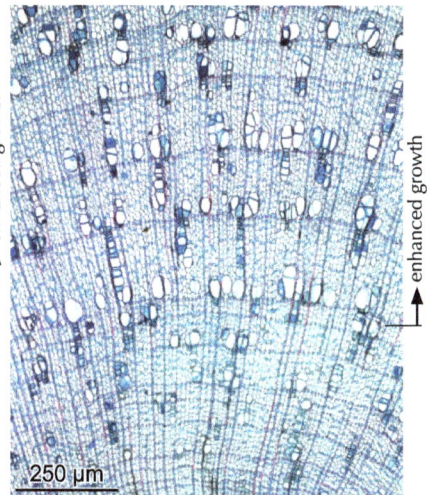

10.52 Enhanced growth in the prostrate arctic shrub *Betula nana*, its cause is unknown.

10.2.2 Structural changes

Discussed here is the anatomical structure of changes in cell type, cell size, cell-wall density and cell orientation after root exposure in conifers and deciduous trees. Structural changes are the result of positional changes within the plant body, of extensive plant destruction, or long-term environmental changes, e.g. by root exposure, stem cutting and extreme suppression of individuals. This can only be ecologically explained in the context of external observation.

Root exposures are expressed by changes of tracheid diameters, fiber dimension, fiber-wall thickness, vessel distribution patterns, tyloses, ray width and ring distinctness.

Resprouting stumps after they have been cut indicate a reorganization of tissues. After removal of the crown, conifers and deciduous plants reduce radial growth, change fiber directions and reduce ray height. Conifers reduce latewood zones and tracheid walls, deciduous trees reduce vessel frequency and diameters. These reactions are an expression of dramatic physiological changes. Instead of using their own photosynthetic resources plants after crown removal use resources of neighboring individuals through anastomosing roots.

Dying stems in beech cohorts lose their foliage. Due to reduced sap flow the proportion of water conduction to nutrient storage changes. The conducting area of vessels continuously reduces its capacity, while the amount of parenchymatous tissue (rays) increases. See also Gärtner 2003.

Macroscopic view of exposed roots

10.53 Exposed roots of *Fagus sylvatica* on a rock.

10.54 Exposed roots of *Fraxinus excelsior* at a river bank.

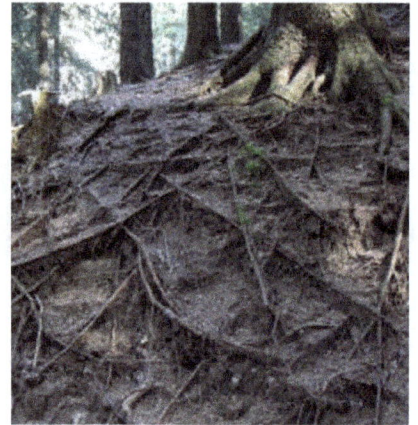

10.55 Exposed roots of *Picea abies* after an extremely intensive rainfall in the Swiss Alps.

Anatomical changes after root exposure in conifers

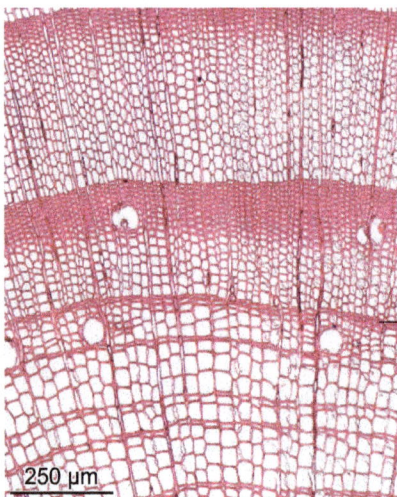

10.56 Root of *Pinus sylvestris* in an avalanche track. Only the diameters of tracheids are changing.

10.57 Root of *Juniperus sabina* on a scree field. Only the diameters of tracheids are changing. The position of the root did not change.

10.58 Root of *Pinus sylvestris* on a path. Exposure triggered changes in tracheid diameters and the formation of resin ducts. Scars in the exposed part of the root are a result of sustained injuries.

Anatomical changes after root exposure in ring-porous deciduous trees

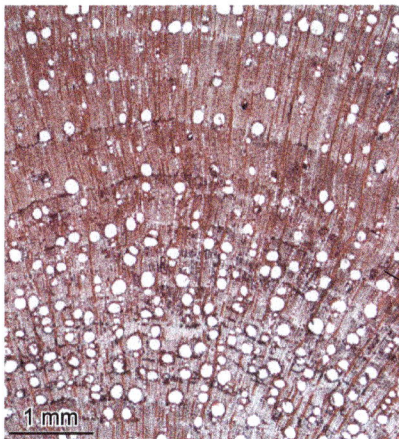

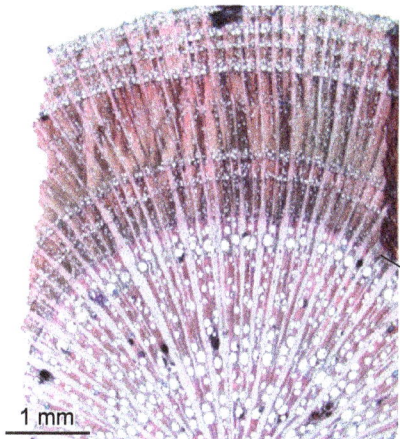

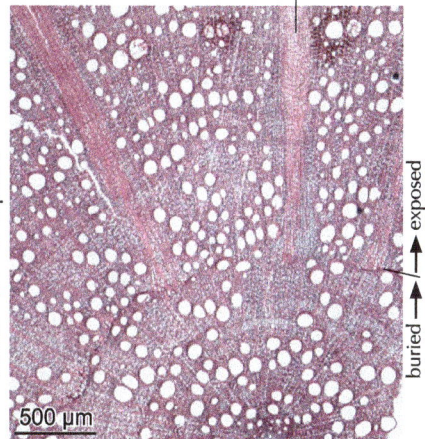

10.59 Root of *Fraxinus excelsior* on a river bank. Cell walls of fibers are thicker, vessels are less frequent, and rays are smaller after exposure.

10.60 Root of *Prunus amygdalus* on a path. Cell walls of fibers are thicker, vessels are smaller and less frequent, and rays are smaller after exposure.

10.61 Root of *Quercus petraea* on a river bank. New large rays are initiated, and rings are more distinct after root exposure.

Anatomical changes after root exposure in diffuse-porous deciduous trees

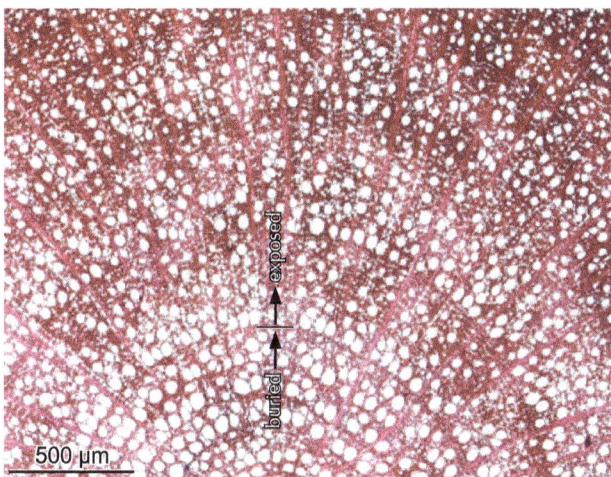

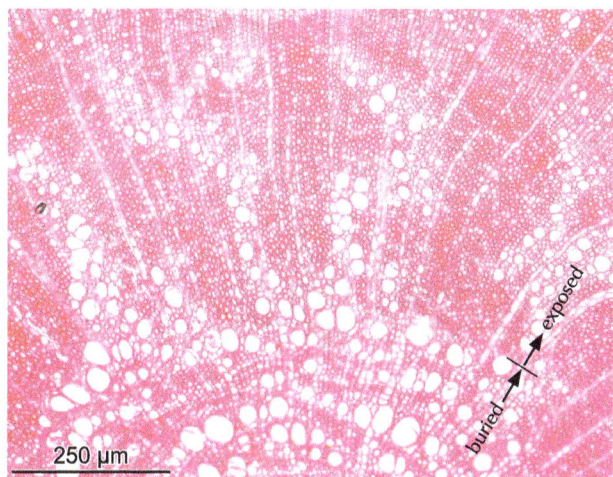

10.62 Root of *Amelanchier ovalis* on an eroded slope. Vessel diameters are continuously reduced, and fiber-wall thickness increases after root exposure.

10.63 Root of *Lycium chanar* in a river bed. Ring width increases, vessel distribution patterns distinctly change, and vessel diameter is reduced after root exposure.

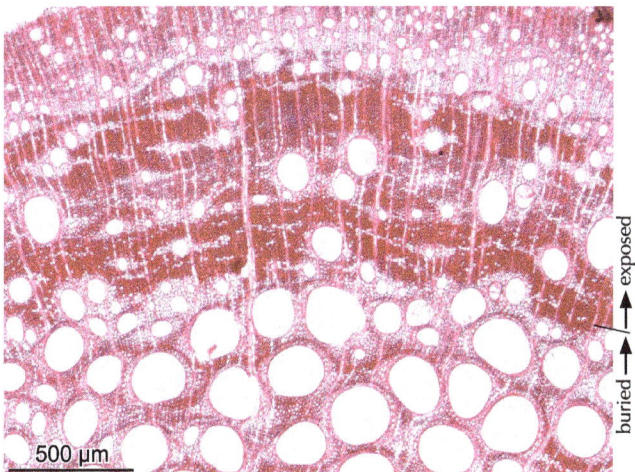

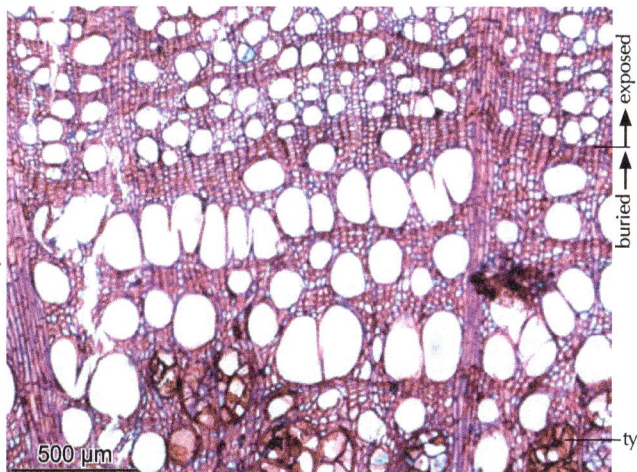

10.64 Root of *Eucalyptus* sp. on a river bank. Vessel frequency and diameter is reduced, and fiber-wall thickness increases after root exposure.

10.65 Root of *Fagus sylvatica* on an enlarged path. Tyloses are formed in the root part after exposure.

Macroscopic aspect of decapitated trees

10.66 Decaying stump of *Picea abies* with a living external part.

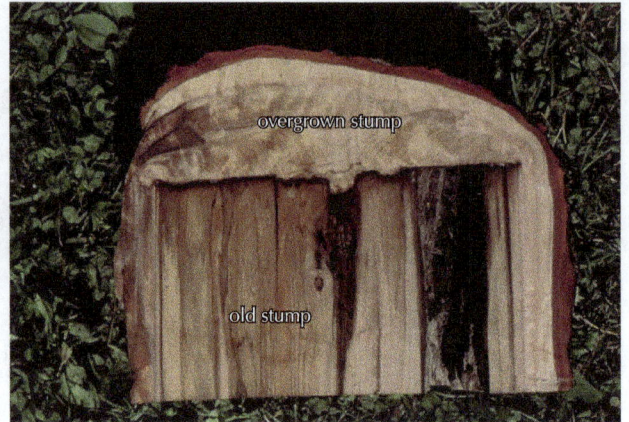

10.67 Longitudinal section of a completely overgrown stump of *Abies alba*. The tracheids are axially oriented.

10.68 Decapitated *Fagus sylvatica* with adventitious sprouts.

10.69 Intensively pollarded shrub of *Krascheninnikovia ceratoides*. The twigs are used for goat fodder in Ladakh, India.

Structural changes after decapitation in conifers

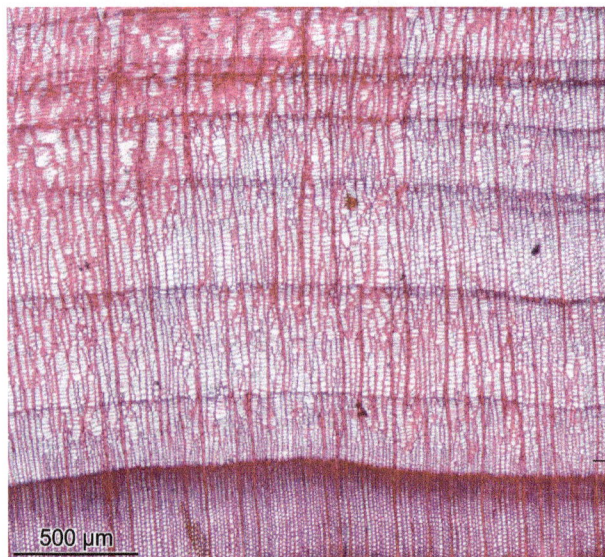

10.70 Reactions to decapitation in *Thuja occidentalis* are the reduction of radial growth, latewood formation and cell-wall thickness, and local changes in the direction of tracheids.

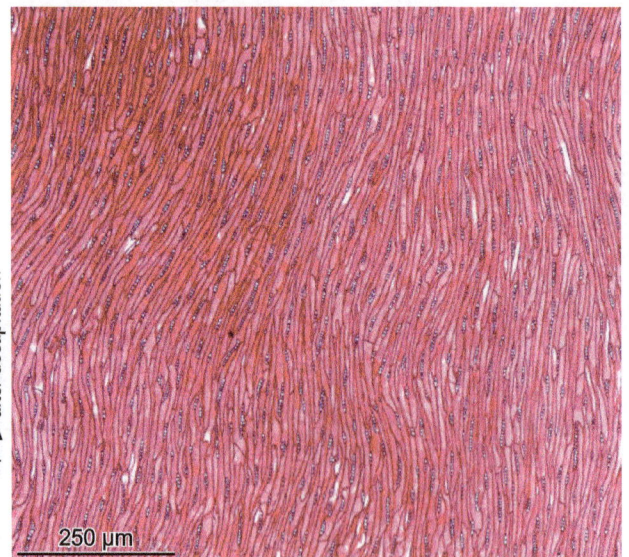

10.71 Wavy fiber direction after decapitation in *Thuja occidentalis*.

Structural changes after decapitation in conifers

10.72 Reactions to decapitation in *Picea abies* are a reduction in radial growth, the formation of traumatic resin ducts, and a change in the direction of tracheids.

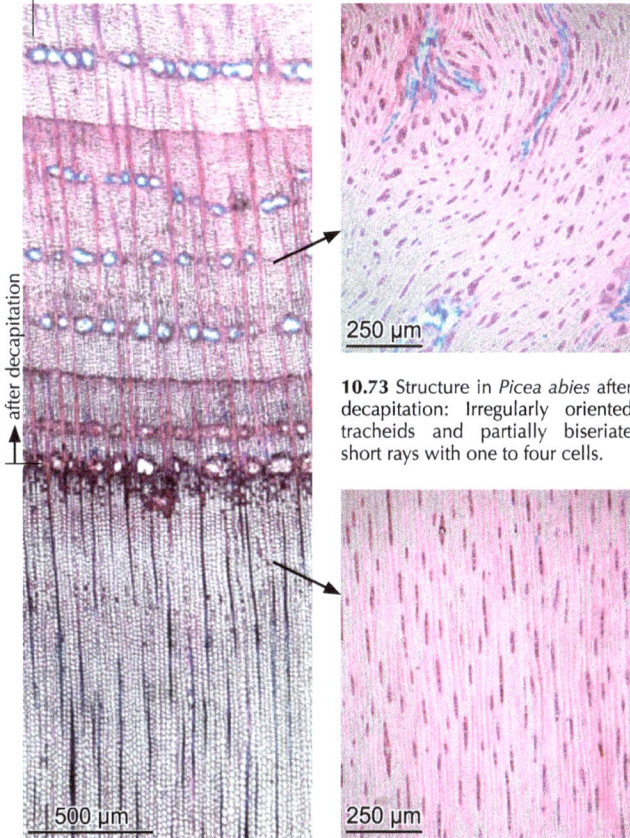

10.73 Structure in *Picea abies* after decapitation: Irregularly oriented tracheids and partially biseriate short rays with one to four cells.

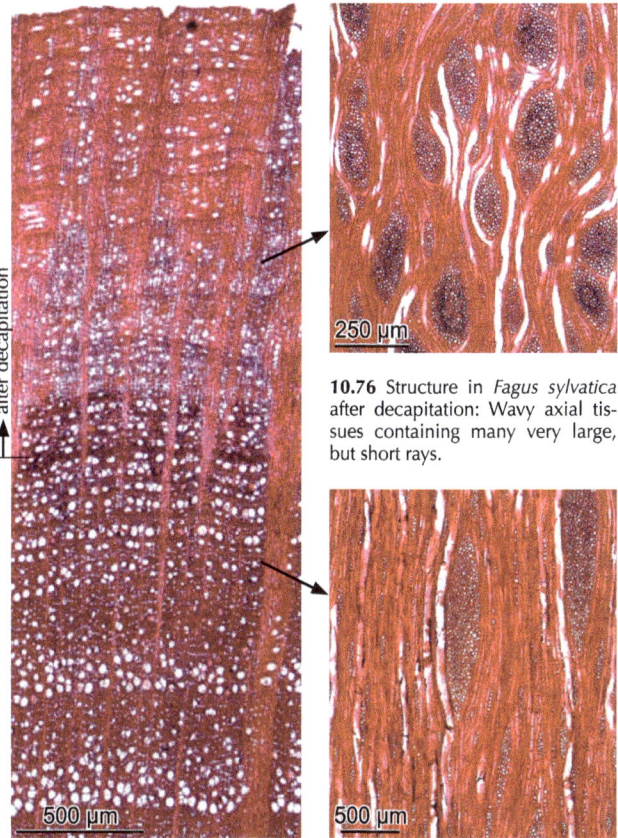

10.74 Structure in *Picea abies* before decapitation: Straight, axially oriented tracheids and exclusively uniseriate rays with four to ten cells.

Structural changes after decapitation in deciduous trees

10.75 Reactions to decapitation in *Fagus sylvatica* are a reduction in radial growth, vessel frequency and diameter, an enlargement of rays, and a change in fiber direction.

10.76 Structure in *Fagus sylvatica* after decapitation: Wavy axial tissues containing many very large, but short rays.

10.77 Structure in *Fagus sylvatica* before decapitation: Straight, axially oriented fibers and vessels, and slender small and large rays.

Structural changes in deciduous trees under extreme competition stress

10.78 Dying suppressed individuals of *Fagus sylvatica*.

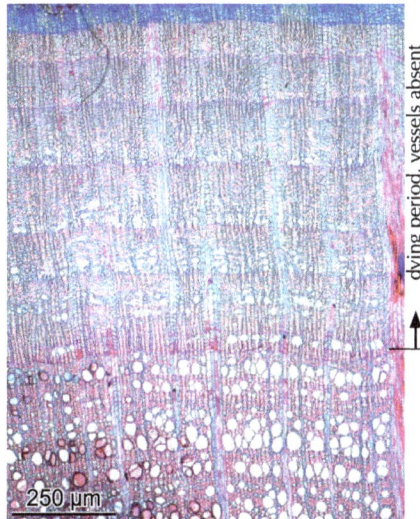

10.79 A continuous reduction in vessel frequency and diameter, and an enlargement in rays occurs after loss of foliage in *Fagus sylvatica*.

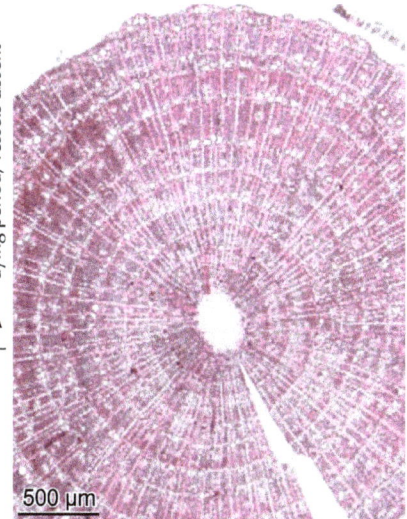

10.80 Twelve-year-old, 3 cm-tall seedling of *Fraxinus excelsior* in a meadow. Grasses compete with it for light and resources, resulting in few, small vessels, uni- to biseriate rays and thin- to thick-walled fibers.

10.3 Eccentricity and irregular stem forms

Changes in vertical position, irregularly applied tension or pressure, unfavorable growing conditions, or a genetic constitution can be expressed by eccentric growth in roots, stems and branches. Conifers generally form additional cells at the side of pressure, while deciduous plants form additional cells at the side of tension. The cambium is sensitive to short- or long-term changes. Each topological reaction is therefore reflected in a cross section.

In a cross section, circular annual rings indicate an upright, while oval rings indicate an oblique position. Growth reactions are indicators for extremely local changes within the plant. If a stem changes position the reaction is only expressed by eccentricity in this stem and not in any other plant parts.

Intra-annual dating of pressure or tension changes during the growing period is possible with some restrictions. If the cambium is not fully active or if hormonal influences are delayed, the anatomical reaction does not indicate the exact moment of the external mechanical influence. Any changes of balance during the dormant period are reflected in the next growing period, immediately in the early or in the late earlywood. Intra-annual trophic changes are most reliable in fast-growing plants because many cambial cells are at disposition for reactions.

Eccentricity only indicates that an event took place that triggered a change in inclination. The exact cause can only be reconstructed through observation of the environmental conditions.

Macroscopic aspect of trees with eccentric growth

10.81 Snow creeping causes stems of *Fagus sylvatica* to grow with a curved stem basis.

10.82 An avalanche bent these stems of *Larix decidua*. They regenerated in a vertical position.

10.83 Leaning trees of *Platanus occidentalis* with eccentric stems after several flood events.

Reaction to a single event

10.84 Leaning stem of the tree *Rhus typhina*, with an upright position of the shoot in the first year, leaning from the second year onwards. Accelerated growth occurs at the tension side.

10.85 Leaning rhizome of the perennial herb *Lythrum salicaria*, with an upright position of the shoot in the first two years, leaning from the third year onwards.

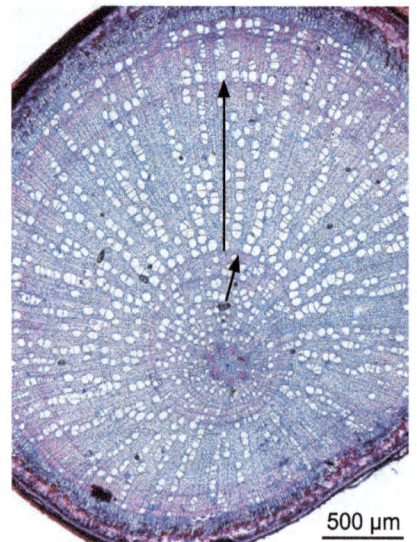

10.86 Changing inclination of a root of the shrub *Corylus avellana*. The root grew vertically in the first year and at an angle from the second year onwards.

Reaction to a multiple events

500 µm

10.87 Changing inclinations of a stem of *Picea abies* in an avalanche track. The seedling grew upright in the first few years, then got hit and inclined twice.

10.88 Changing inclination in a stem of *Pinus mugo* on an instable ground of a bog. The seedling grew upright in the first two years, after that moved twice.

Reaction to genetic constitution and mechanical events

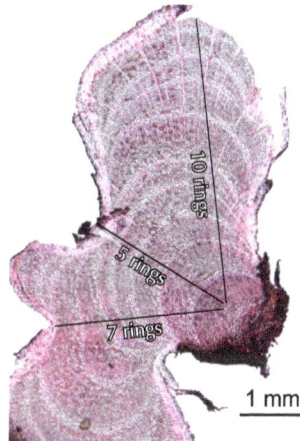

10 rings
5 rings
7 rings

1 mm

1 mm

10.89 The fluted stem of a tree of *Pyrus communis*.

10.90 Eccentric fluted rhizome of the herb *Hippocrepis comosa*. The rhizome grew toward one side for five years, then various local cambial activities formed three lobes.

10.91 Eccentric fluted stem of the dwarf shrub *Eriogonum jamesii*. The shoot grew upright in the first five years, then various local cambial activities periodically formed lobes.

Reaction to unfavorable local growing conditions

year 1150
year 1989

1 mm

10.92 Prostrate conifer *Juniperus sibirica* on a rocky dry slope in the Polar Ural. Photo: S. Shyiatov.

10.93 Very eccentric, horizontally growing stem of *Juniperus sibirica*. The cambium is only active towards the ground-facing side. Older stem sections are eroded. The ring sequence reflects cambial activity of 839 years.

10.94 Prostrate dwarf shrub *Dryas octopetala* on a wind-swept, rocky slope in the high arctic of Greenland.

10.95 Extremely eccentric, horizontally growing, 20 cm-long stem of *Dryas octopetala*. The cambium is only active towards the ground-facing side. The stem hasn't changed its position for 120 years.

10.4 Reaction wood – Reaction to mechanical stress

The formation of reaction wood is a principal survival strategy of terrestrial plants. It is responsible for the formation of directed growth, and for regeneration after gravity-related deformations.

Reaction wood is defined (after Evert 2006) as "Wood with more or less distinctive anatomical characters, formed in parts of leaning or crooked stems and branches. Compression wood occurs in conifers at the compressed parts (e.g. lower side of branches) and tension wood in angiosperms at the tension side (e.g. upper side of branches)."

The presence of intensively lignified primary walls is common for compression wood and tension wood. Each living cell reacts independently from another to mechanical stress.

Compression wood

- Occurs in all families and genera of conifers.

- Occurs at the compression side of leaning parts of trees and shrubs.

- Occurs in the xylem of stems, branches, twigs and roots.

- Occurs in tracheids, never in parenchyma.

- Secondary walls are high in lignin and low in cellulose. Tertiary walls are absent.

- Macro-fibrils are angled up to 45° towards the axis of the tracheid. The angle of spiral thickenings increases with increasing intensity of the compression wood, e.g. in *Taxus*.

- Tracheids are round in circumference, with intercellulars between the tracheids.

- Branches are stiff, with a high resistance to compression and low flexibility.

- Longitudinal shrinking is high (up to 7%).

Tension wood

- Occurs in selected families and genera of dicotyledonous plants. It does not occur in hydrophytes.

- Occurs at the tension side of leaning parts of trees, shrubs and herbs.

- Occurs in the xylem and phloem of stems, branches and twigs, never in roots.

- Occurs in fibers and fiber tracheids, never in parenchyma.

- Secondary walls contain mostly gelatinous fibers (G-layer), which are low in lignin and high in cellulose.

- Macro-fibrils are oriented parallel to the axis of the fiber.

- Fibers are angular in circumference. Intercellulars are absent.

- Branches are flexible, with a high resistance to tension and high flexibility.

- Longitudinal shrinking is low (up to 1%).

See also Gardiner *et al*. 2014, Ghislain & Clair 2017 and Onaka 1949.

Occurrence of compression wood and tension wood

10.96 Branch and twigs of the conifer *Picea abies*. Compression wood occurs at the underside of branches.

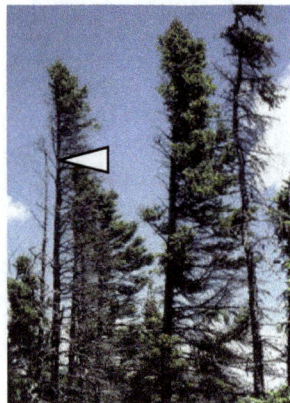

10.97 Wind-exposed trees of *Picea engelmannii*. Compression wood occurs at the lee side of the stems.

10.98 Small tree of the dicotyledonous *Euphorbia balsamifera*. Tension wood occurs at the upper side of the branches.

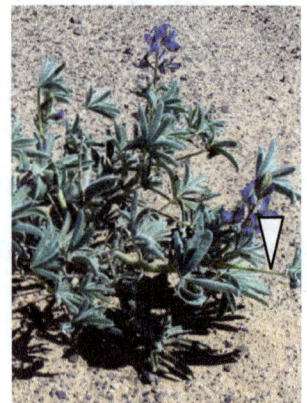

10.99 Dicotyledonous herb *Lupinus* sp. Tension wood occurs at the upper side of the branches.

10.4.1 Compression wood in conifers

Compression wood appears on polished discs as dark brown zones, and in stained slides as intensively stained layers. Directed reactions to gravity influences are obvious.

Macroscopic aspect

10.100 Polished disc of *Picea abies*. The dark zones represent compression wood. Its position indicates the influence of gravity (arrows).

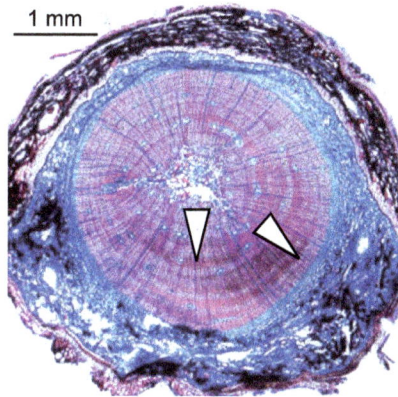

10.101 Compression wood in a twig of *Pinus sylvestris*.

Microscopic aspect

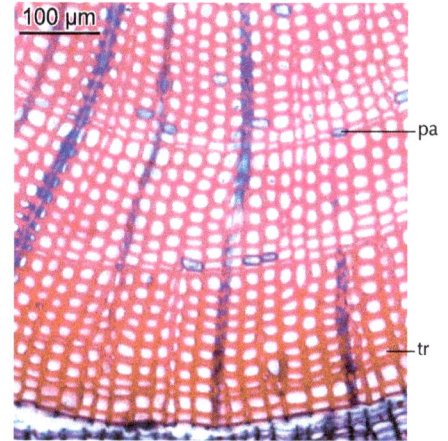

10.102 Compression wood in a twig of *Taxus baccata*. Only tracheids form compression wood. Axial parenchyma cells and rays are not responsive.

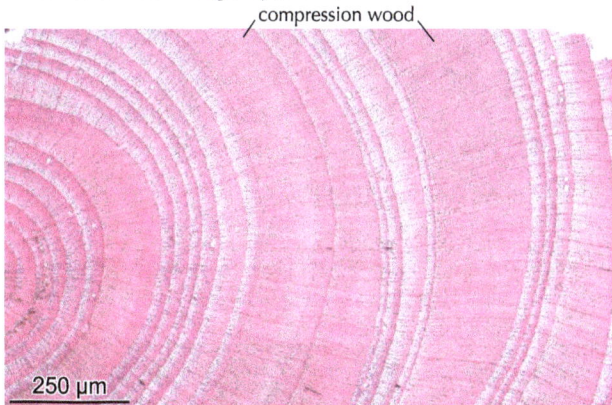

10.103 Thick-walled round latewood tracheids with intercellulars in a leaning small *Picea abies* tree.

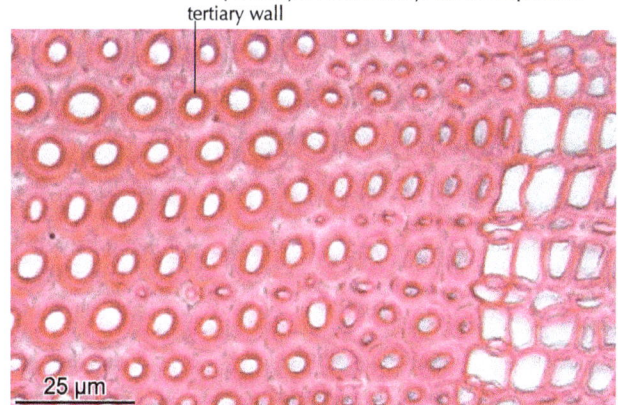

10.104 Round, thick-walled latewood tracheids with large, distinct tertiary walls in *Pinus mugo*.

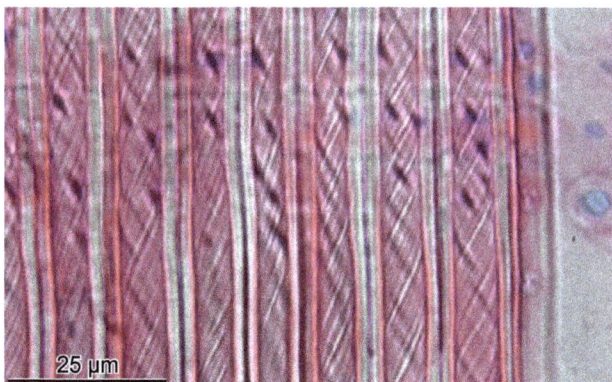

10.105 Longitudinal spiral-like cracks in latewood tracheids walls indicate macro-fibrils in the compression wood of *Picea abies*.

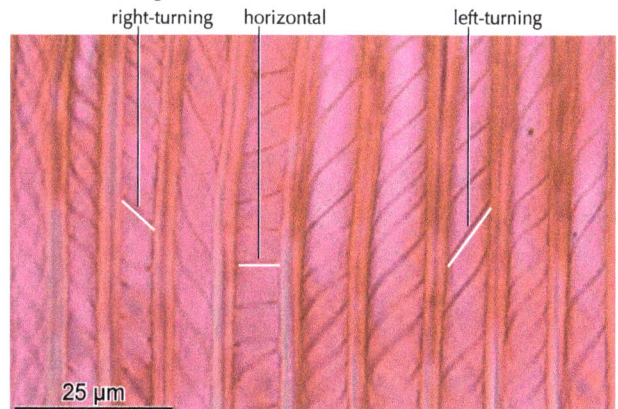

10.106 Changing direction in the angle of helical thickenings in *Taxus baccata* indicates various gravitational influences. Some helical thickenings in tracheids are left-turning, and some are right-turning.

10.4.2 Tension wood in angiosperms

Tension wood is difficult to recognize on polished wood discs. Macroscopically it appears on planed longitudinal sections as a felt-textured surface. The chemical composition and the position of gelatinous fibers within plants differ. Tension wood primarily occurs on the tension side in the xylem of angiosperm trees, shrubs and herbs, but also in the cortex and the phloem of angiosperm herbs (e.g. in *Euphorbia* sp.), shrubs (e.g. *Daphne* sp.) and trees (e.g. *Broussonetia papyrifera*) and in gymnosperms (e.g. *Gnetum gnemon*). The presence of lignified primary walls is common for gelatinous fibers in the xylem. The anatomical and chemical composition of tension wood varies, which is shown by differences in acceptance of Astrablue/Safranin stain. The function of gelatinous fibers probably varies between different types. Their occurrence in species with extremely different positions within the phylogenetic tree leads to the assumption that gelatinous fibers, at least those in the cortex, are of polyphyletic origin.

Macroscopic aspect

Microscopic aspect – Location within the xylem and bark

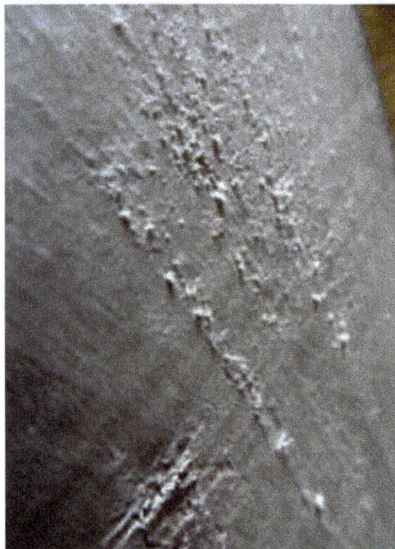

10.107 Planed board of a *Populus* sp. The felt-like surface is characteristic for tension wood. Photo: A. Crivellaro.

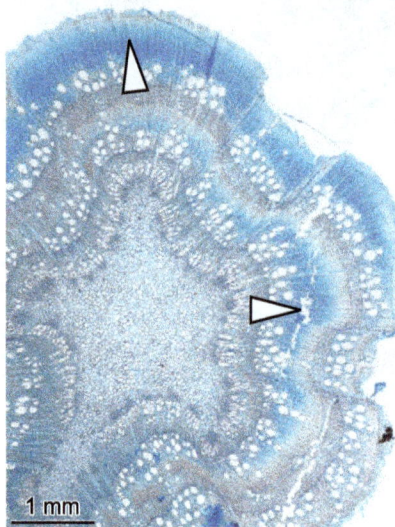

10.108 Twig with tension wood (blue) in the latewood of the second and third ring in a cross section of *Quercus robur*.

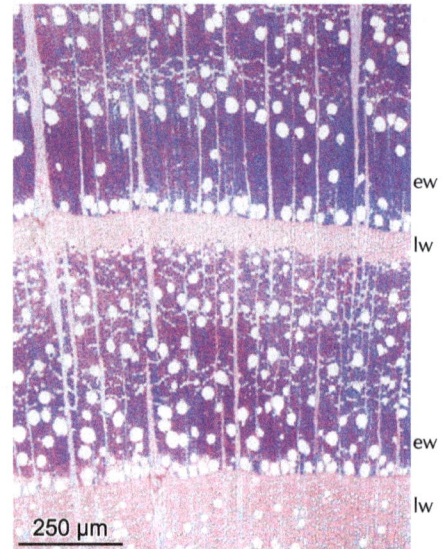

10.109 Stem of *Fagus sylvatica* with tension wood (violet) in the earlywood.

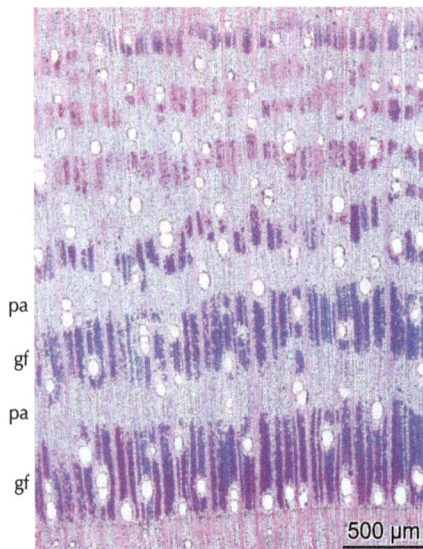

10.110 Stem of *Ficus carica* with tangential bands of different intensities of tension wood (violet) in the earlywood.

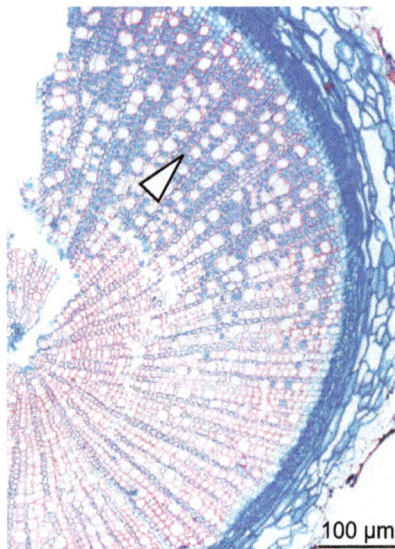

10.111 Shoot of *Linum bienne* with tension wood (blue) in the whole ring of the annual plant.

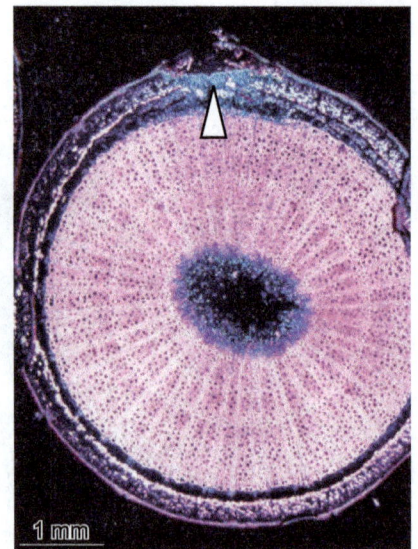

10.112 Tension wood in the bark of the Gnetaceae *Gnetum gnemon*, polarized light.

Gelatinous fibers in the xylem

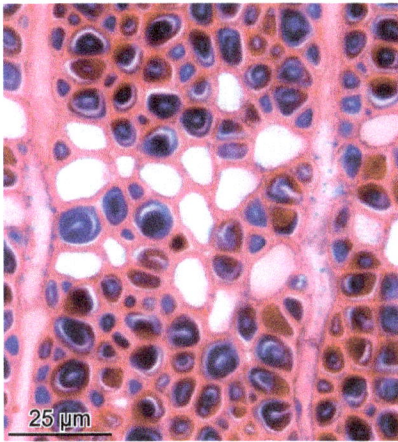

10.113 Dark-blue- to red-stained secondary walls of the tree *Fagus sylvatica*, Fagaceae.

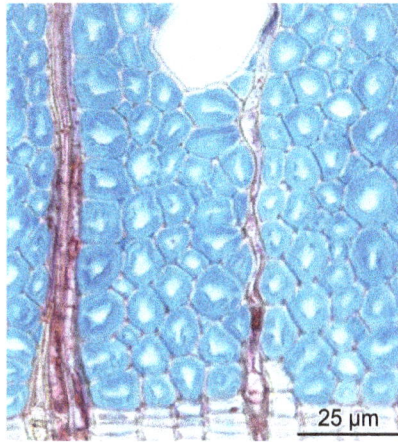

10.114 Light-blue-stained secondary walls of the tree *Betula pendula*, Betulaceae.

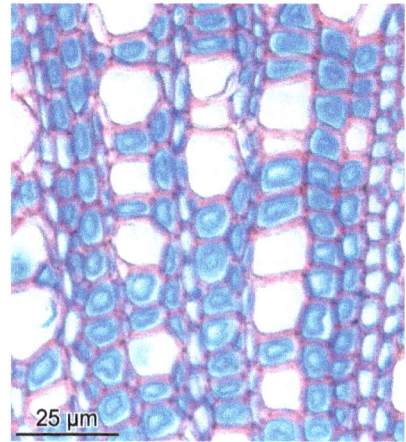

10.115 Light-blue-stained secondary walls of the annual herb *Linum bienne*, Linaceae.

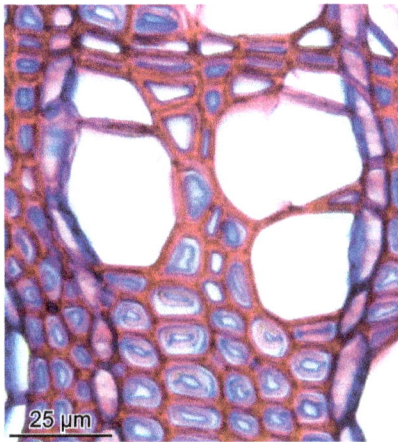

10.116 Pink-stained secondary walls and blue-stained tertiary walls in the alpine prostrate dwarf shrub *Salix retusa*, Salicaceae.

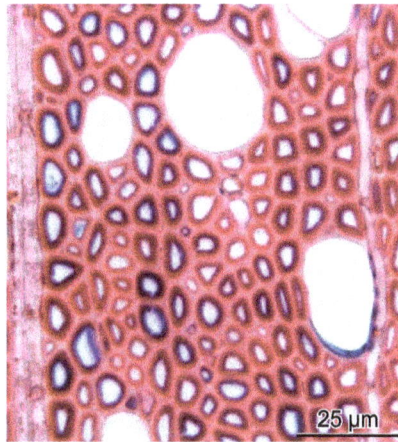

10.117 Red- to dark-blue-stained secondary and tertiary walls in *Sorbus aucuparia*, Rosaceae. Tension wood is hardly expressed.

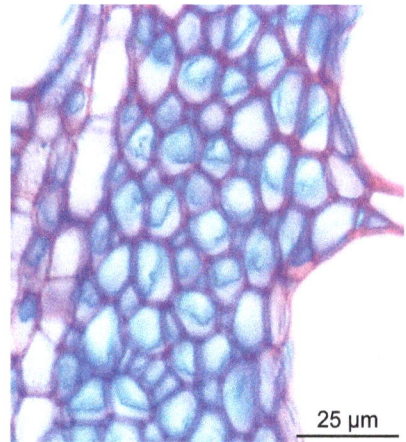

10.118 Doubtful tension wood. The fibers with thin blue secondary or tertiary walls are located at the tension side of the annual herb *Cannabis sativa*, Cannabaceae.

Gelatinous fibers in the cortex

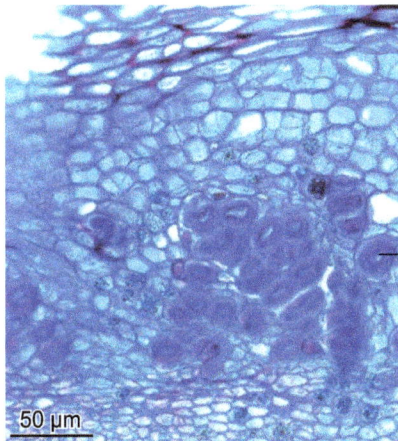

10.119 Dark-blue-stained gelatinous fibers in *Urtica dioica*, Urticaceae. These fibers were used for textiles.

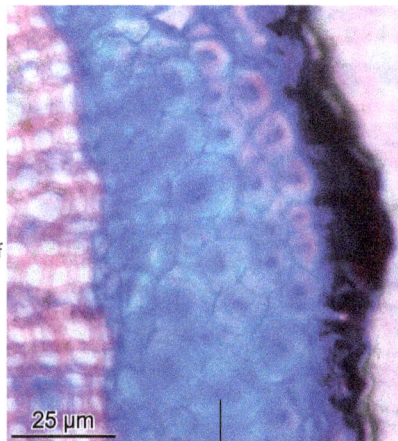

10.120 Blue-stained gelatinous fibers in *Linum usitatissimum*, Linaceae. These fibers are used for textiles (linen).

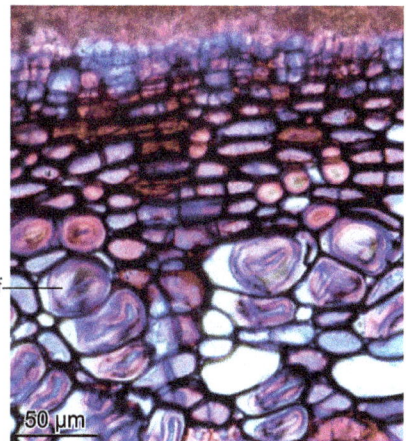

10.121 Pink-stained gelatinous fibers in *Gnetum gnemon*, Gnetaceae. These fibers are used for strings of musical instruments.

10.5 Cell collapse and lateral ray compression

Cell collapse occurs in the xylem, the phloem, the cortex and the phellem. As soon as the turgor in thin-walled cells exceptionally decreases, the negative pressure makes cells collapse.

Cell collapse in the xylem is the first step of a chain reaction. Direct mechanical damage, e.g. by hail stones or woodpeckers, primarily causes cell damage which induces callus formation. Physiological imbalances between transpiration and available water resources can cause negative pressure in water conductive areas. Therefore thin-walled, unlignified vessels, tracheids or fibers can collapse. This occurs e.g. after browsing, pollarding, heavy frost damage in the leaf area, or extreme summer droughts.

Cell collapse in the phloem is a normal, ontogenetic phenomenon for many species. As soon as sieve elements are no longer turgescent, they collapse. The occurrence of collapsed sieve elements is an indication for a non-conductive phloem. Periodic collapse of tangential bands of sieve tubes leads to a radial contraction and, in consequence, to bent rays. Cork cells in the phellem mostly collapse soon after their formation.

The course of a ray in the xylem is influenced by the pressure in vessels. Vessel development normally has the highest priority within the xylem formation. Due to high pressure in vessels, adjoining rays may change their course. They get compressed, but do not collapse.

Subfossil and fossil wood is often deformed, e.g. by heavy ice or sediment loads. The weakest components of the xylem—mostly earlywood tracheids—collapse. Anaerobically degraded wet wood from archaeological sites shrinks dramatically, which is microscopically expressed by deformed, compressed cells.

Macroscopic phenomena causing cell collapse

10.122 Woodpecker marks on a *Salix* sp.

10.123 Browsed *Fraxinus* sp.

10.124 Frost damage on the dwarf shrub *Arctostaphylos uva-ursi*.

10.125 Wounded bark of an old *Salix* sp.

Cell collapse in the xylem

10.126 Collapsed vessels in the alpine *Salix glaucosericea* after a heavy late frost.

10.127 Collapsed tracheids and bent rays in *Larix sibirica* after a heavy late frost.

Cell collapse and cell deformation in the phloem

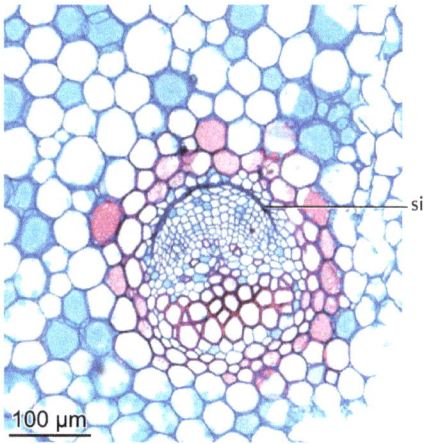

10.128 Collapsed and compressed sieve elements at the external part of a vascular bundle in the petiole of *Cycas revoluta*.

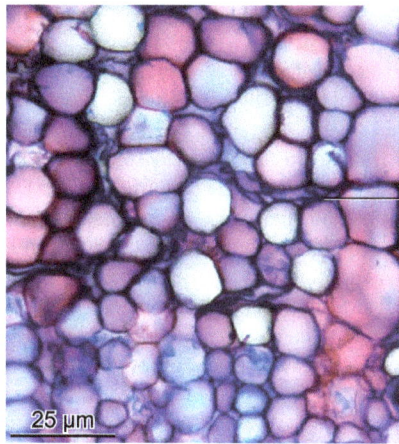

10.129 Collapsed sieve elements between parenchyma cells in the phloem of the dicotyledonous tree *Schinus molle*.

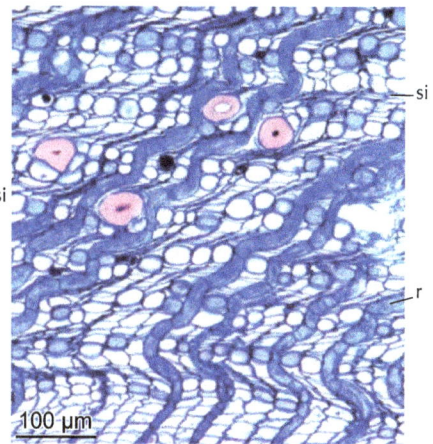

10.130 Compressed and deformed tangentially arranged sieve elements and bent rays in the conifer *Larix decidua*.

Cork cell collapse in the phellem

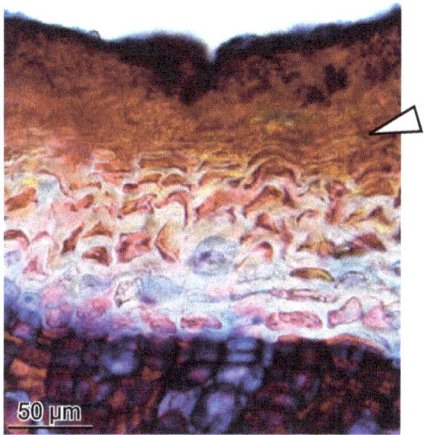

10.131 Collapsed dead cork cells in the phellem of *Fagus sylvatica*.

Lateral compression of rays

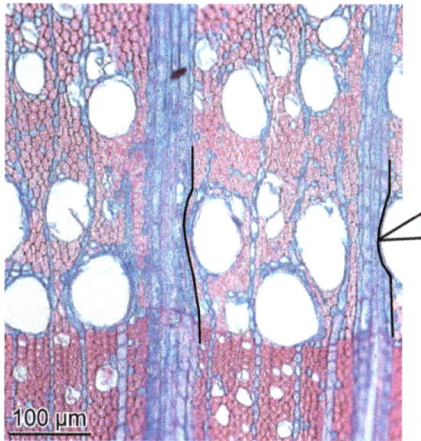

10.132 Living rays, bent in the area of large vessel in *Rosa elliptica*.

Macroscopic aspect of subfossil wood

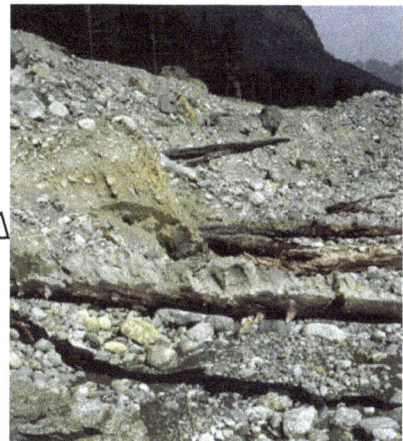

10.133 Subfossil stems in Quaternary river sediments of the Alps.

Microscopic aspect of subfossil and fossil wood

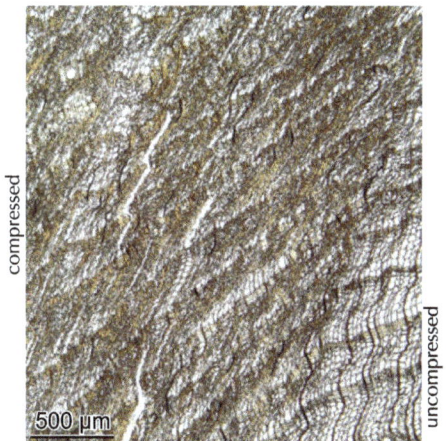

10.134 Compressed xylem of a stem of *Larix decidua* in a glacier moraine.

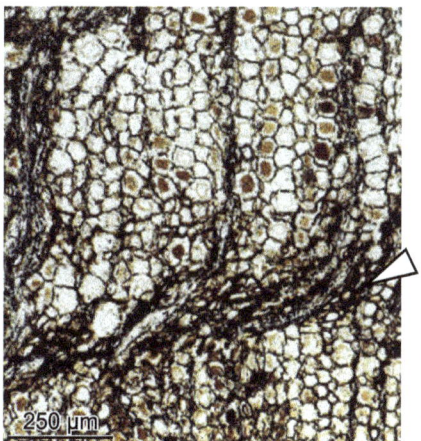

10.135 Compressed tracheids in the Carboniferous petrified wood of *Dadoxylon* sp.

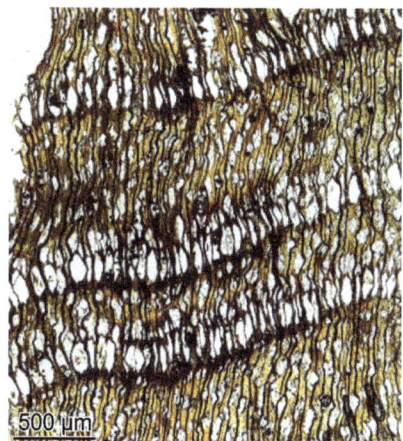

10.136 Laterally compressed xylem in the Neolithic, anaerobically decomposed and dehydrated wood of *Fraxinus excelsior*.

10.6 Cambial wounding – Callus formation, overgrowing of wounds

All plants with secondary growth have been confronted with cambial wounding by mechanical, environmental, biological or pathogenic causes for more than 300 million years. Repair mechanisms have therefore been developed to isolate the living tissues from destruents and pathogens. Wounds can be caused mechanically by environmental factors, such as hail, lightening, storms, fire, rock fall, floods, avalanches, ice plates, pruning or decapitation, and—human-induced—by grafting. Many wounds are biologically induced by animals, such as insects, birds, herbivores, rodents, and many others. Pathogenic infestations can also cause cambial reactions, e.g. mistletoes, witches' brooms, fungi, or cancerous agents.

If meristematic cells are affected, a reaction chain starts. The first anatomically visible step is the formation of a chemical and mechanical protective zone, called barrier zone. Chemical boundaries compartmentalize parts of stems adapted to their anatomical structure (Shigo 1989; see also Chapter 12.4). All cell elements can be plugged with phenols, and vessels with tyloses. Protective reactions are weak along the stem axis, but intense in radial and tangential directions. In a second step, callus tissue formed. Living meristematic and parenchymatic cells produce undifferentiated cells. The further wound occlusion process is characterized by accelerated growth, the transformation of undifferentiated callus cells to original xylem and phloem cells, and the reorganization of the cell structure corresponding to the stem axis.

Monocotyledonous plants are not able to close wounds. The excretion of phenolic slime, and the formation of tyloses and suberin layers around wounds protect the culms of bamboo against decay (Liese & Köhl 2015).

Macroscopic aspect of scars

10.137 Fire scars on *Castanea sativa*.

10.138 Open scar on a stem of *Eucalyptus capillosa*.

10.139 Surviving callus after fire on a stem of *Robinia pseudoacacia*.

Barrier zones

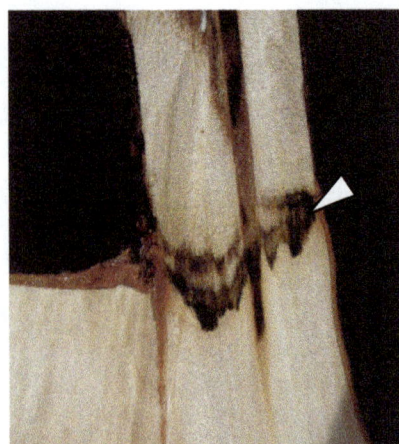

10.140 Barrier zone at the base of a dead branch of *Acer pseudoplatanus*.

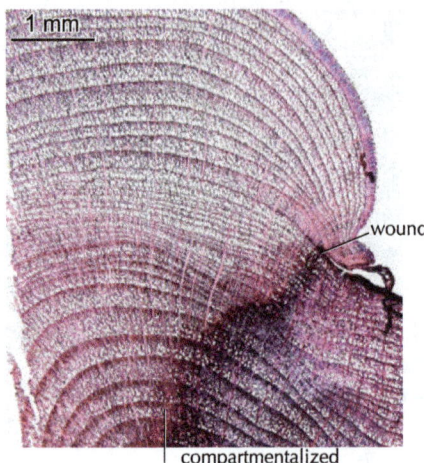

10.141 Compartmentalized zone of a mechanical wound in the dwarf shrub *Arctostaphylos uva-ursi*.

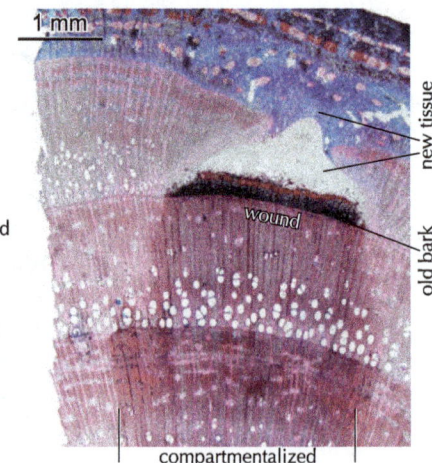

10.142 Compartmentalized and overgrown zone of a mechanical wound in the tree *Fraxinus excelsior*.

Callus formation and open scars

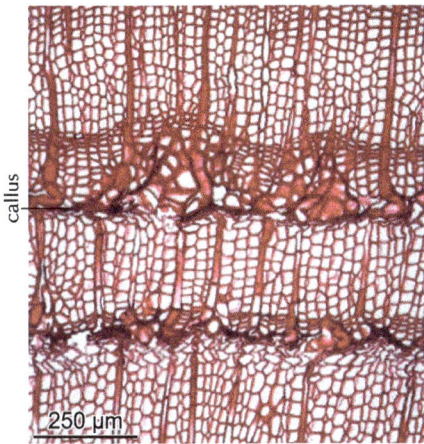

10.143 Frost rings with callus cells in *Larix decidua*.

10.144 Frost crack with lateral callus cells in *Larix decidua*.

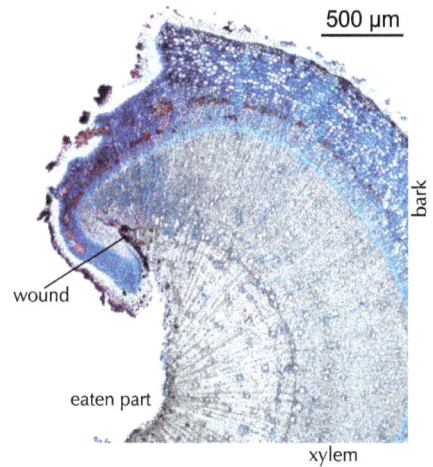

10.145 Scar from a rodent on a shoot of *Fraxinus excelsior*.

Open scar

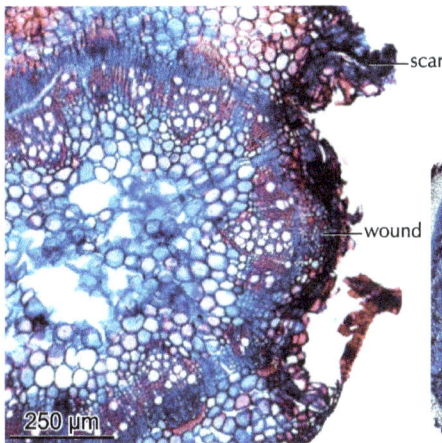

10.146 Scar on an annual stem of the 2 cm-tall herb *Knorringia pamirica*.

Overgrown scars

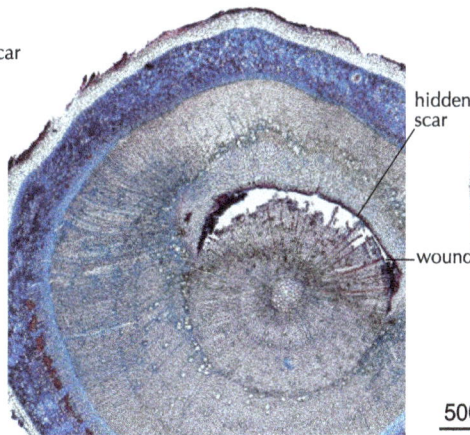

10.147 Completely healed wound with a hidden scar on a 10 cm-tall seedling of *Fraxinus excelsior*.

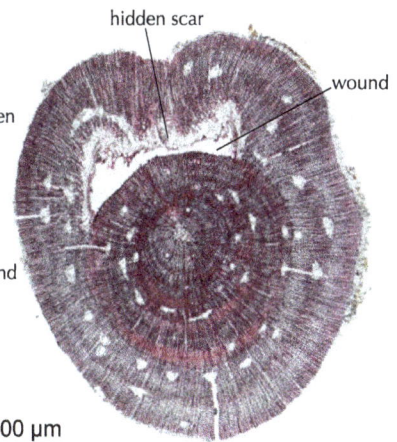

10.148 Completely healed hail wound with a hidden scar on a small branch of *Pinus mugo*.

Reorientation of the axial fiber structure in an overgrowing scar

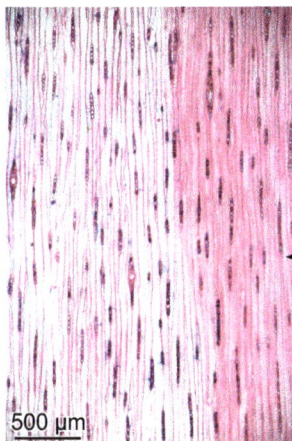

10.149 Axial parallel fiber and ray structure on the inside of the scar.

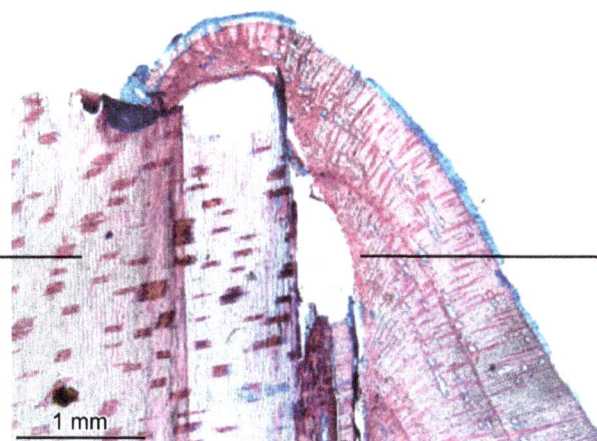

10.150 Scar on a broken stem of *Picea abies*.

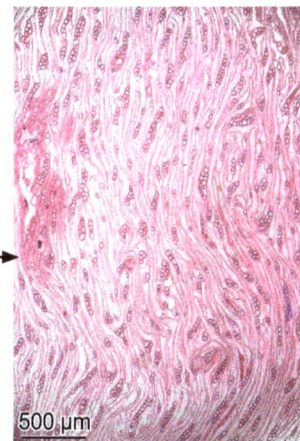

10.151 Chaotic fiber and ray structure on the outside of the scar.

10.7 Prevention of wounds

Ontogentic processes in plants have the potential to create wounds. Obvious are scars created by dropping buds, leaves, twigs, fruits and rhizoids. Plants prevent wounds by the formation of periderms, especially phellem (cork), before the potential wound is exposed to destruents (see also Chapter 6.1.3).

Stem expansions created by radial growth cause tangential tension in the bark. Before bark cracks, living parenchymatic cells react with accelerated cell-wall growth and cell division, the phloem and cortex therefore dilate. Parallel to the expansion, phellem layers seal the endangered zone with cork (see also Chapter 6.2.1). Many small plants never form complete stems, they keep the form of vascular bundles for many years, however, they are linked by inter-fascicular cambia. Stem or root segregation is a widespread phenomenon in many taxonomic

units, and in many biomes. A few examples may illustrate this here. Plant stems reorganize themselves by forming new phloem or periderm belts around a part of the xylem. Stems of many alpine cushion plants segregate near the soil surface. In consequence, living parenchyma cells change their physiological mode and form phloem, and partially phellem, all around the partial stems. The cactus *Carnegia gigantea* in American deserts has a full stem at the base, but a circle of single stems inside the cortex. A few species within the family of Primulaceae form little stems within a cortex without cambium.

Pathogenic organisms are able to penetrate meristems. Hosts and parasites live in a symbiosis. Hosts react to parasites, e.g. mistletoes, with accelerated growth without obvious structural disturbances.

Leaf scars

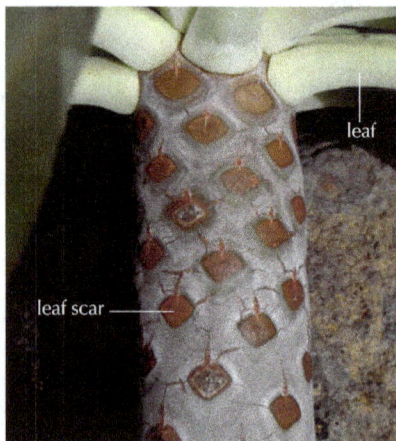

10.152 Leaf scars on a stem of the succulent *Aeonium arboretum*. Brown spots represent cork layers.

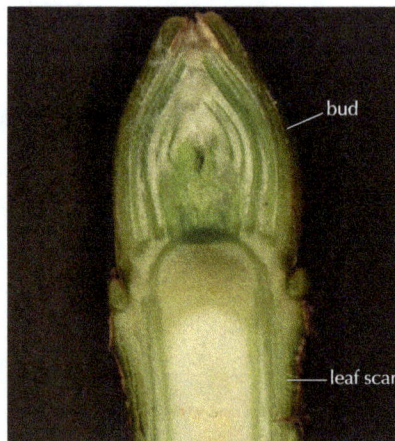

10.153 Macroscopic aspect of leaf scars on a long shoot of a twig of *Acer pseudoplatanus*.

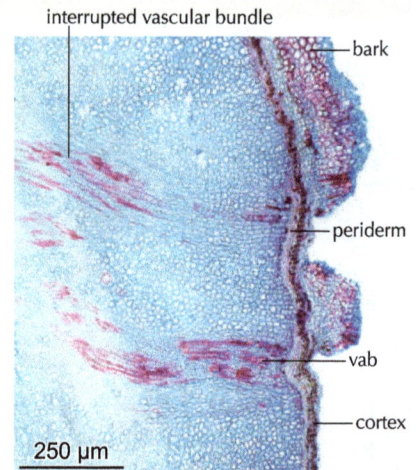

10.154 Microscopic aspect of leaf scars on a shoot of *Castanea sativa*.

Twig abscission

10.155 Macroscopic aspect of the scar of a shed twig in *Quercus robur*.

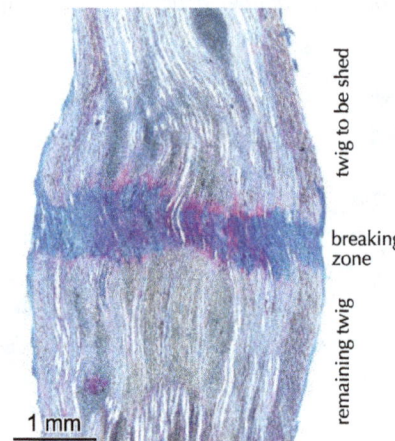

10.156 Longitudinal section trough a breaking zone of a twig of *Quercus robur*.

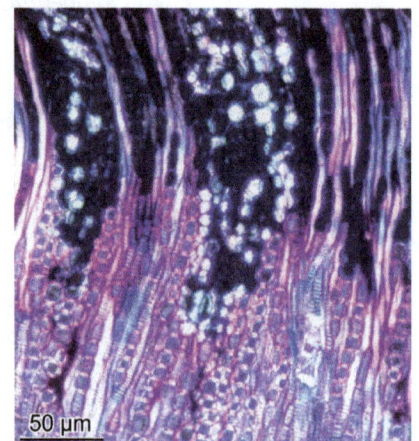

10.157 Microscopic detail of the breaking zone of a twig of *Quercus robur*, ploarized light. Crystal druses are characteristic.

Fruit abscission

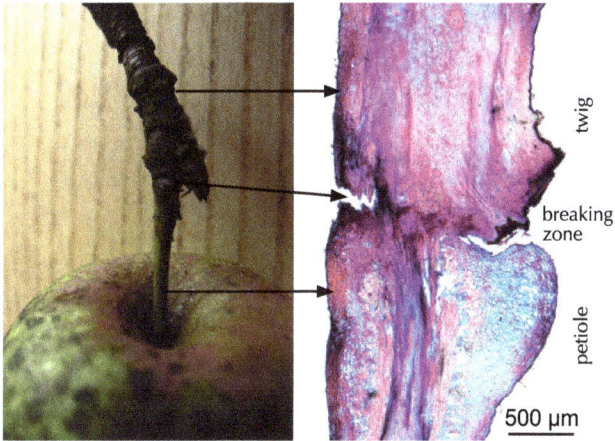

10.158 Apple (*Malus sylvestris*) with petiole on a twig before shedding.

10.159 Microscopic aspect of the abscission zone between petiole and twig in *Malus sylvestris*.

Rhizoid abscission

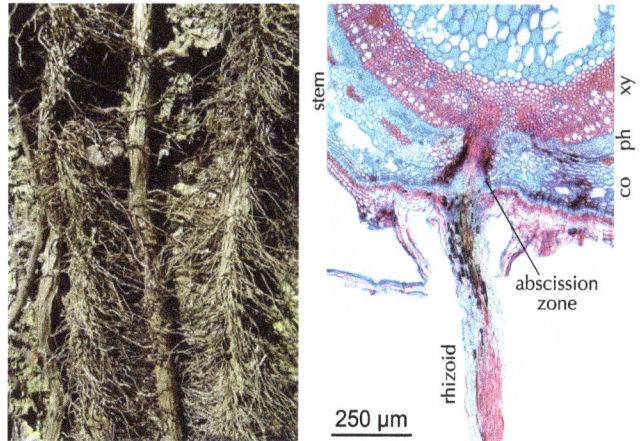

10.160 Dead rhizoids on the stem of the climber *Hedera helix*.

10.161 Interrupted connection between the stem and a rhizoid in *Hedera helix*.

Stem thickening and prevention of cracks in living stem parts

10.162 Longitudinal crack in the bark of *Betula pendula*.

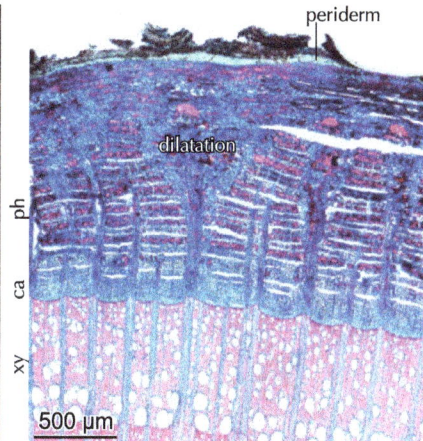

10.163 Phloem enlargement by dilation and formation of phellem in *Rosa elliptica*.

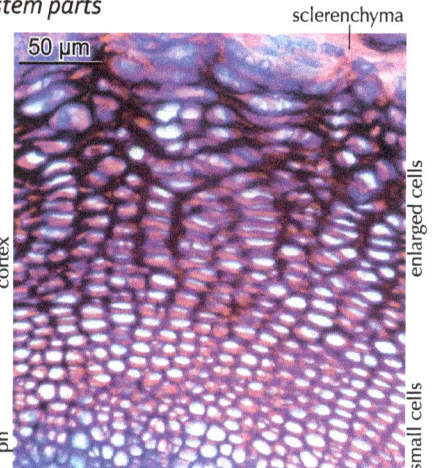

10.164 Cortex enlargement by cell wall growth, and reinforcement by sclerenchymatic tissues in the alpine herb *Androsace villosa*.

Initial stages of stem segregation in herbs

10.165 Alpine herb *Antennaria dioica*.

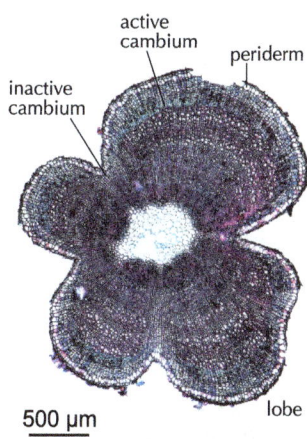

10.166 Formation of lobes is the first stage of stem segregation, *Antennaria dioica*.

10.167 Alpine herb *Arnica angustifolia*.

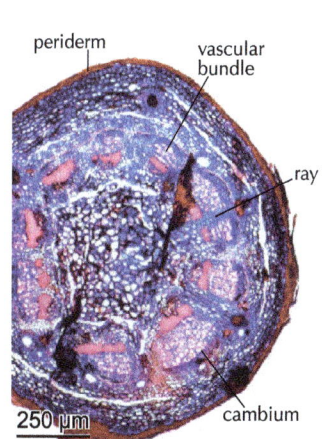

10.168 Vascular bundles in the rhizome of *Arnica angustifolia* represent incompletely segregated stems.

Segregated stems of herbs

10.169 Alpine herb *Potentilla nitida*.

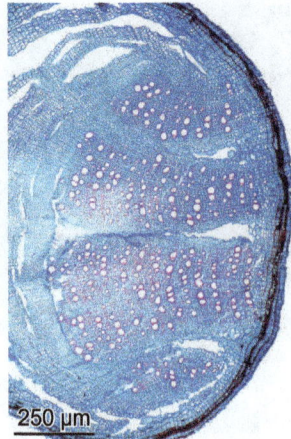

10.170 Microscopic aspect of a segregated rhizome of *Potentilla nitida*, completely surrounded by a periderm.

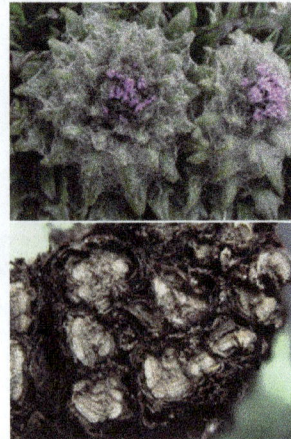

10.171 Top: Alpine cushion plant *Saussurea hypsipeta*. Bottom: Segregated stems of *Saussurea glanduligera*. Material: J. Dolezal.

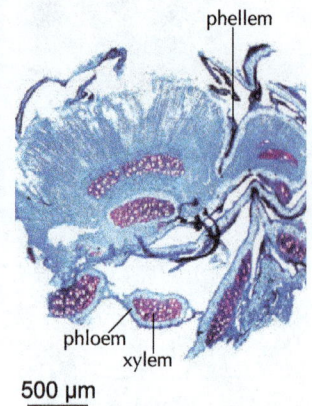

10.172 Microscopic aspect of a segregated stem of *Saussurea andryaloides*. Material: J. Dolezal.

Segregated stem of a cactus

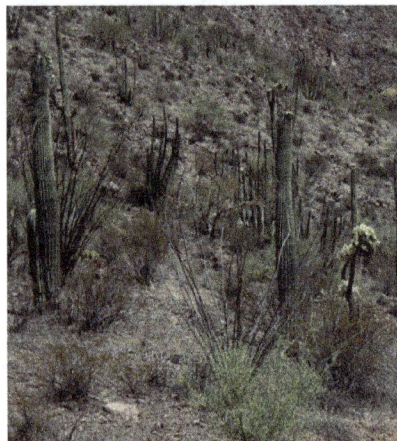

10.173 *Carnegia gigantea* in a desert in northern Mexico.

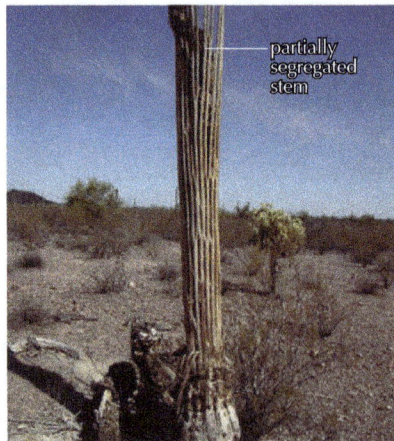

10.174 Macroscopic aspect of a segregated stem of *Carnegia gigantea*. The cortex is rotten.

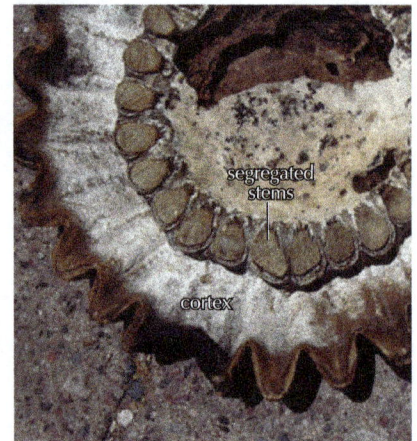

10.175 Macroscopic aspect of a cross section of a segregated stem of *Carnegia gigantea*.

Suppression of a rejection reaction between host and parasite

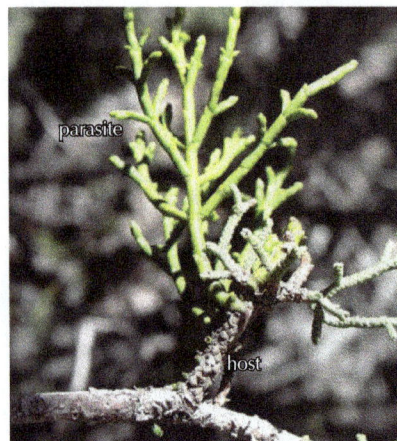

10.176 Semiparasite *Phoradendron* sp. on the conifer *Juniperus* sp.

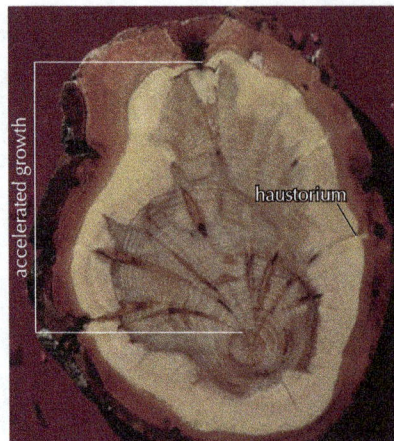

10.177 Macroscopic aspect of mistletoe haustoria (*Viscum album*) in a stem of *Abies alba*.

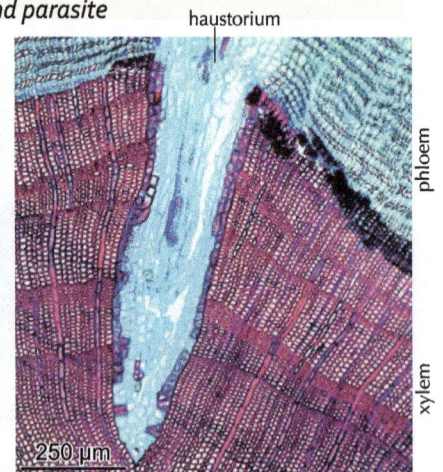

10.178 Microscopic aspect of a haustorium of *Phoradendron* sp. in the stem of *Juniperus* sp. Callus tissues are absent.

10.8 Resin and gum ducts

Resin ducts are part of the defense system of some plants (see also Chapter 4.12), and secrete resin and gums in the living parts of roots, stems, twigs and needles. Resin canals are axially and radially linked and form a network. The occurrence and distribution of resin ducts in conifers is primarily related to taxonomy. In some genera, e.g. *Pinus* and *Picea*, resin ducts occur in the bark and the xylem. In other species, resin ducts occur only in the bark, and in some they are completely absent.

The frequency of resin ducts in conifers is related to stress, and increases with higher stress levels. If stress occurs mechanically in the cambial region, e.g. after woodpecker attacks, or through intensive fungal infections or insect infestations, tangential rows of ducts in the xylem and phloem make intra-annual dating of the event possible. If stresses occur in the crown, e.g. by defoliators, the frequency of resin ducts is increased, but the diffuse distribution of ducts make intra-annual dating impossible. Tangential rows of resin ducts after wounding indicate sporadic long-term reactions.

Conifers with resin ducts

10.179 *Pinus canariensis* with many resin ducts.

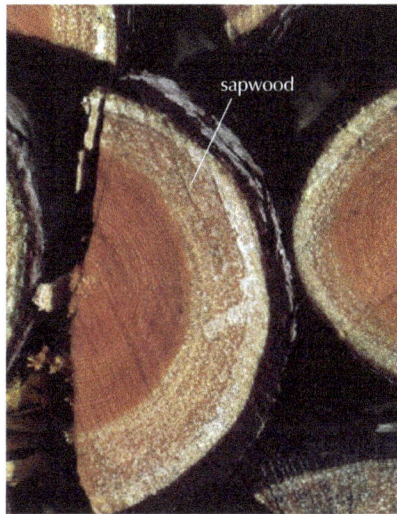

10.180 Sapwood of *Pinus sylvestris* with living resin ducts.

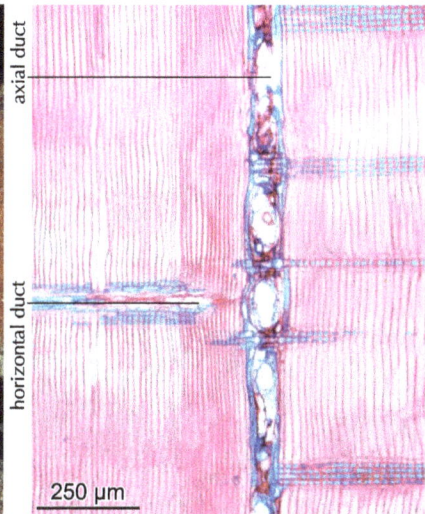

10.181 Longitudinal section of the intraxylary network of resin ducts in *Pinus sylvestris*.

Taxonomic significance of resin ducts

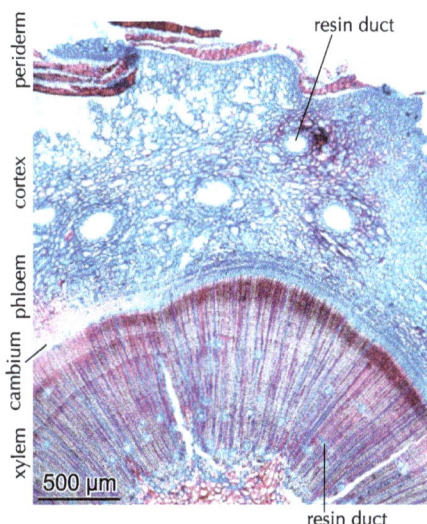

10.182 *Pinus ponderosa* with resin ducts in the xylem and the cortex.

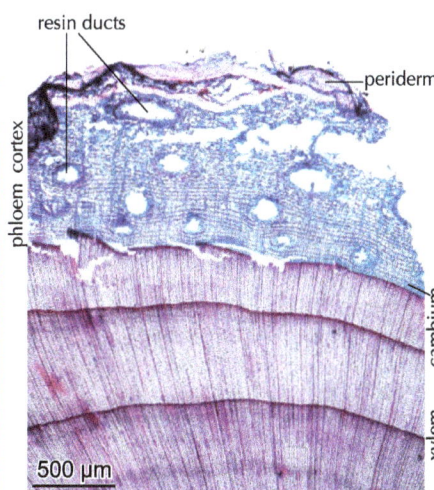

10.183 *Microbiota decussata* with resin ducts in the phloem and the cortex, but no ducts in the xylem.

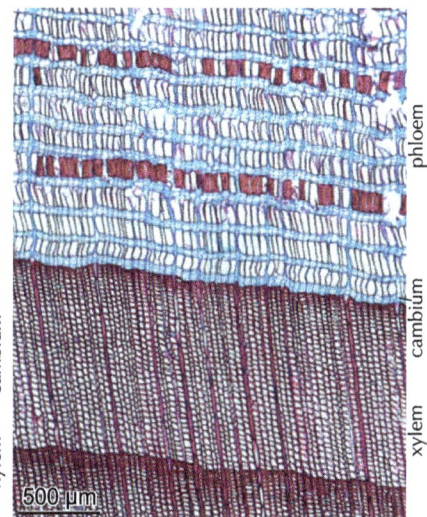

10.184 *Pilgerodendron uviferum* without resin ducts.

Long-term reaction to crown and stem damage

10.185 Spruce bud worm (*Choristoneura fumiferana*) infestation on *Picea engelmannii*.

10.186 The frequency of resin ducts is extremely high for *Pseudotsuga menziesii*.

10.187 Windbreak on *Picea abies*.

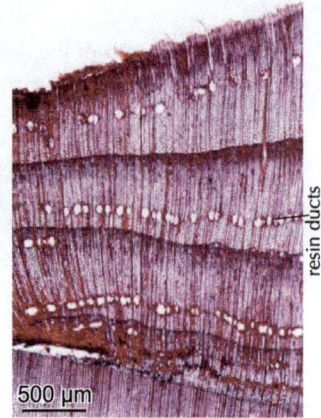

10.188 Scar on *Picea abies* with periodic formation of traumatic resin ducts.

Short-term reaction to woodpecker attacks

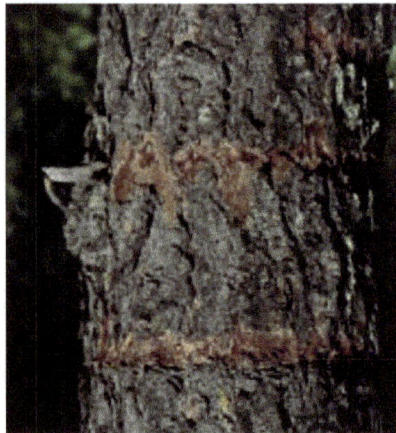

10.189 Woodpecker scars on a stem of *Pinus mugo*.

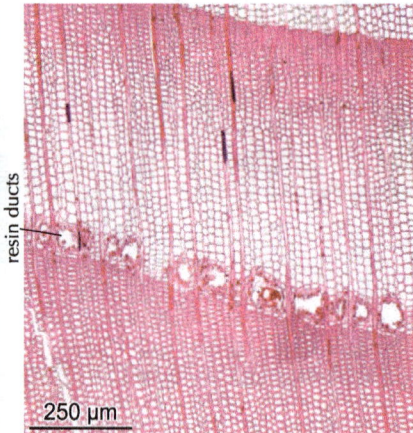

10.190 Traumatic resin ducts in the xylem of *Abies alba*. The damage occurred in the dormant period or just at the beginning of the growing season. Firs do not normally form resin ducts in the xylem.

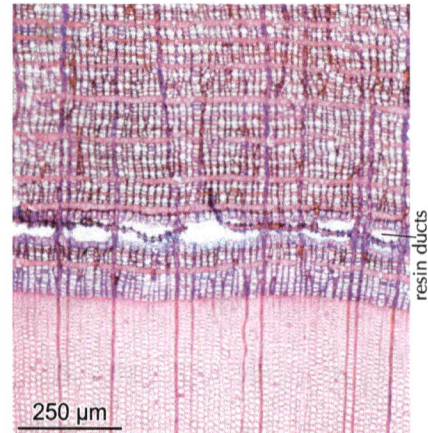

10.191 Traumatic resin ducts in the phloem of *Juniperus oxycedrus*, caused by a bird attack.

Reactions to fungal attacks   Unknown cause

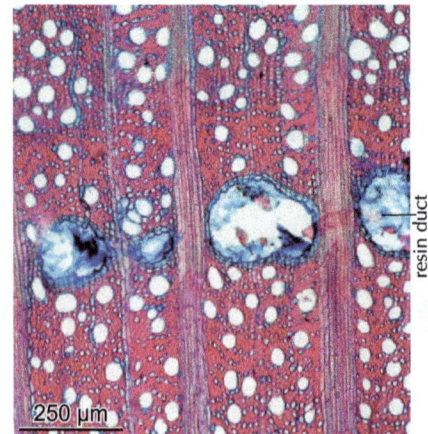

10.192 Scar caused by the fungus *Monilia* sp. on a twig of *Prunus armeniaca*. The fungus attacks the cambium, initiating scars and dieback of twigs.

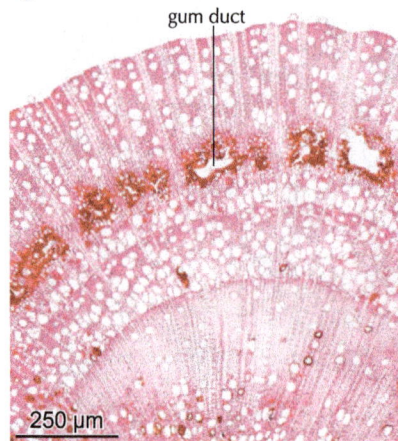

10.193 Tangential, traumatic gum ducts in the xylem of *Prunus cerasus*. The fungal attack occurred in early summer, just after earlywood formation.

10.194 An unknown attack triggered traumatic ducts in the xylem of *Elaeagnus pungens*.

Plants, Algae and Fungi: A Symbiotic Association

11.1 Mycorrhizae – Coexistence of vascular plants and fungi

The symbiotic relationship between fungi and the roots of vascular plants is called mycorrhiza. Hyphae transport inorganic nutrients—mainly carbohydrates—from the soil to the plants, while the fungi profit from organic substances provided by the vascular plants. Mycorrhizae occur in most of the terrestrial plants.

Ectomycorrhizae
Fine roots are coated by a mycelium of numerous fungus species. The hyphae penetrate the intercellulars between the endodermis (*syn*. rhizodermis) cells and form the Hartig net. The hyphae functionally replace the fine roots. Ectomycorrhiza is most abundant in conifers (e.g. in *Picea*, *Abies* and *Pinus)*, and in deciduous trees (e.g. *Fagus* and *Quercus)* of the temperate zones in the Northern Hemisphere.

Endomycorrhizae
Hyphae within the cortex cells of plants characterize endomycorrhiza. The most abundant endomycorrhiza is arbuscular mycorrhiza (vesicular-arbuscular mycorrhiza, or VAM). It is found in the majority of terrestrial plants. Anatomically characteristic are globular (vesicle) or irregular, tree-like (arbuscular) terminal ends of hyphae within the living cortex cells of roots.

Mycorrhizal symbiosis is essential for orchids. Since orchid seeds don't have their own nutrient reserves, they are relying on symbiotic fungi (basidiomycetes) for successful germination. Fungi also provide organic and inorganic nutrients to orchids without photosynthetic capacity, e.g. *Neottia nidus-avis*. In this case the vascular plant acts as a parasite of the fungus (holoparasite).

Ectomycorrhiza

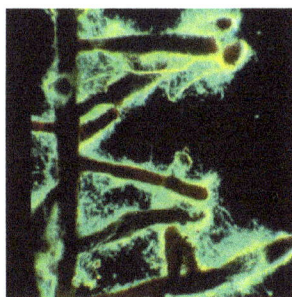

11.1 Ectomycorrhiza on *Picea* fine roots. Slide: S. Egli.

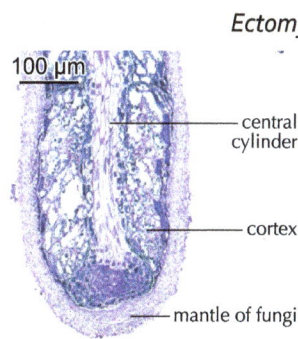

11.2 Ectomycorrhiza on a fine root of *Picea abies*. Section stained with Lugol's iodine, hyphae appear purple. Slide: S. Egli.

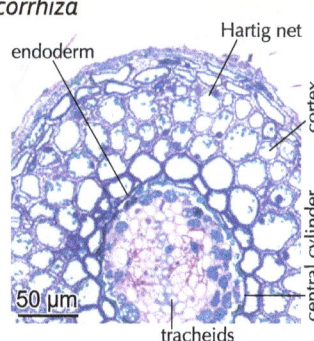

11.3 Ectomycorrhiza on a fine root of *Picea abies*. Hyphae surround the root and penetrate the rhizodermis. Slide: S. Egli.

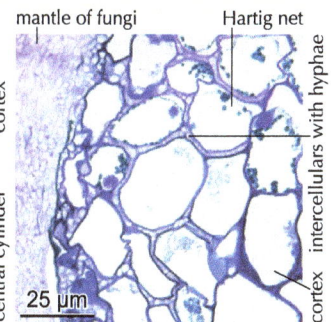

11.4 Ectomycorrhiza in intercellulars of the rhizodermis of a fine root of *Picea abies*. Slide: S. Egli.

Endomycorrhiza

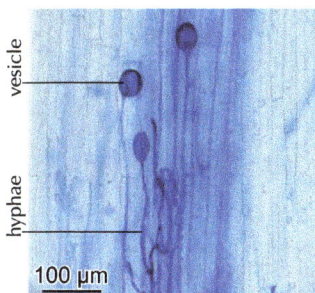

11.5 Endomycrorrhiza with globular vesicles in the roots of *Allium porrum*. Slide: S. Egli.

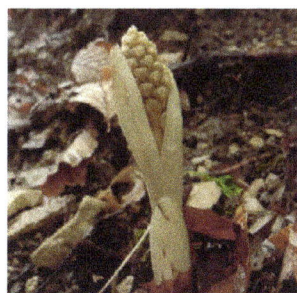

11.6 *Neottia nidus-avis*, a holoparasite without chlorophyll. Photo: A. Moehl.

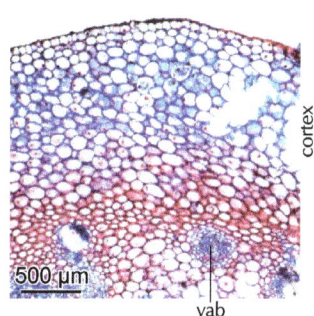

11.7 Cross section of *Neottia nidus-avis* with vascular bundles in the center and a parenchymatous cortex.

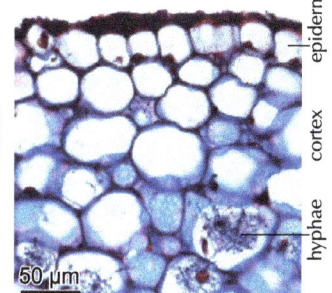

11.8 Endomycorrhiza in living cortex cells of *Neottia nidus-avis*. Groups of hyphae surround cell nuclei.

11.2 Lichens – Coexistence of algae and fungi

Hyphae of fungi and algae live together in a symbiotic association and form a morphological and physiological unit. Partners are principally ascomycetes and cyanobacteria. More than 20,000 species occur from the tropics to the arctic, and from extremely dry to aquatic sites. Some thousand species form leaf-like, not self-supporting thalli, some form small, self-supporting and upright, horizontal or hanging stems. Fungi build the corpus and are responsible for water uptake, while the photosynthetic algae provide organic nutritive substances. The periphery is light-permeable and extremely hydrophilic. This allows short-term photosynthetic reactions of the algae by moistening the surface. Lichens produce species-specific acids, which partially crystallize. The arrangement of hyphae is related to growth forms. Upright types form a dense external tube, hanging types form a dense central cable-like strand, and leaf-like forms do not have stabilizing elements. The following images describe the principal lichen structure and some anatomical variations of different growth forms.

Macroscopic aspect of lichen growth forms

11.9 Upright *Cladonia* sp. with 4 cm-tall stems with reproductive organs.

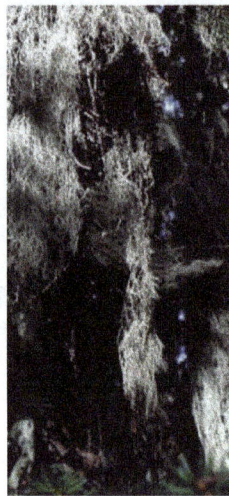

11.10 Hanging *Usnea barbata* with up to 20 cm-long strands.

11.11 Leaf-like, not self-supporting *Hypogymnia physodes* with 3 cm-long, flat thalli.

Principal structure of lichens

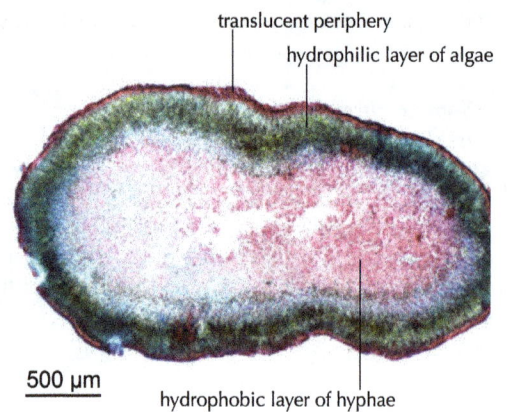

translucent periphery
hydrophilic layer of algae
500 µm
hydrophobic layer of hyphae

11.12 Structure of *Roccella* sp. Characteristic are the translucent periphery, the green hydrophilic layer of hyphae with algae, and the central hydrophobic layer of hyphae. Material: C. Scheidegger.

Structure of different lichen growth forms

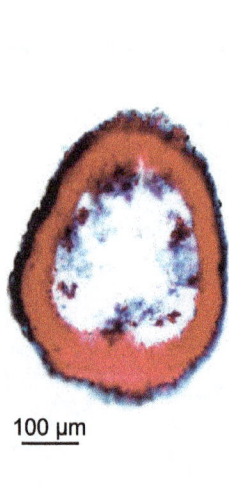

100 µm

11.13 Upright *Alectoria nigricans* with a dens tube-like stem.

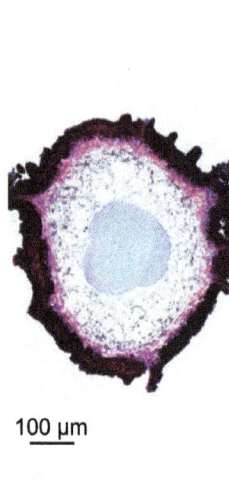

100 µm

11.14 Hanging *Usnea hirta* with a dense central strand.

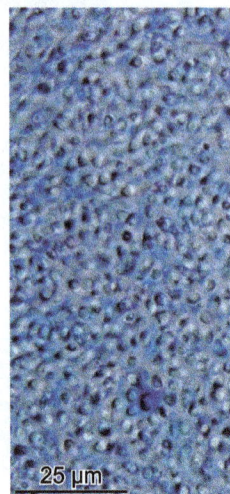

25 µm

11.15 Thick-walled hyphae in the central strand of *Usnea hirta*.

250 µm

11.16 Leaf-like *Hypogymnia physodes* without dense hyphae zones.

Crystals

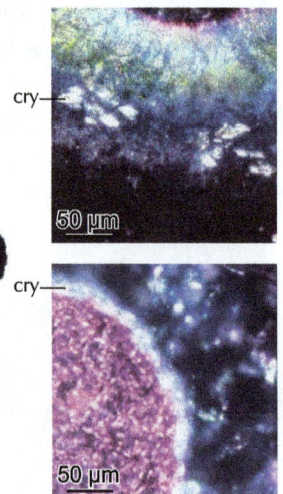

cry
50 µm

cry
50 µm

11.17 Top: Large crystals in *Roccella* sp. Bottom: Small crystals around the central strand of *Usnea barbata*, polarized light.

12

Abiotic and Biological Decomposition of Wood

The „system Earth" is based on a closed carbon cycle. In one half organisms store the carbon in organic substances. This is partially described for the "system plant" in the previous chapters. In the other half, different organisms disaggregate the organic substances into inorganic substances. Described here briefly are traces of processes and organisms which fragment wood into smaller particles. Wood can be decayed by photochemical processes, bacteria, archaea, fungi, insects and vertebrates.

12.1 Abiotic decomposition

Ultraviolet light and acid rain determine the aspect of old wooden houses on sunny slopes. The originally light-colored lignin turns into a dark brown through photodegradation. Cell walls in dense latewood are broken and stain dark red with Safranin. Since latewood cells decay slower than earlywood cells, mechanical erosion by wind and rain creates a wavy density profile on the surface. Cells on the surface of silvery shining wooden boards are delignified. See also Evans *et al.* 1996 and George *et al.* 2005.

Macroscopic and microscopic aspect of abiotic wood decomposition

12.1 Five-hundred-year-old wooden alpine house. The aged wood appears brown.

12.2 Logs of *Larix decidua* in an alpine house. Ultraviolet light stained the lignin brown and mechanical weather conditions modeled the surfaces.

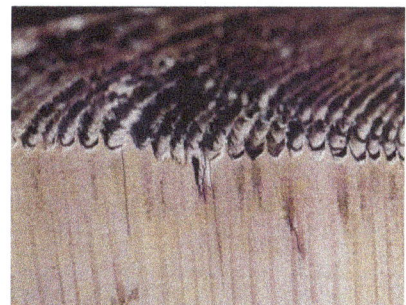

12.3 A natural density profile in the wood of *Larix decidua* due to ultraviolet insolation and weathering.

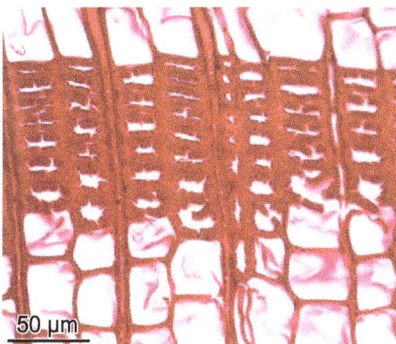

12.4 Broken secondary walls in the latewood of a beam of *Larix decidua* due to ultraviolet radiation. Slide stained with Safranin.

12.5 Wooden shingles of *Picea abies* on a roof appear gray due to the impact of acid rain.

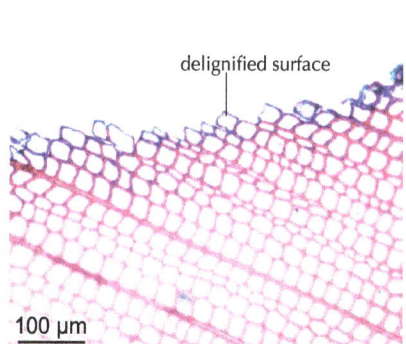

delignified surface

12.6 Delignified surface of a rain-exposed beam of *Picea abies*. The exposed secondary walls of tracheids appear blue when stained with Astrablue/Safranin.

12.2 Anaerobic decay – Absence of oxygen

Subfossil, non-petrified wood is well known in dendrochronology from riverbeds, bogs and clay pits. Subfossil wood is preserved because shortly after death it was covered by loamy sediments. Due to a lack of oxygen, only some bacteria and archaea are able to decompose the wood. Different stages of decomposition are expressed by discoloration, the loss of weight and shrinkage.

The alterations by microscopic anaerobic degradation are demonstrated exemplarily on Astrablue/Safranin-stained slides taken from subfossil late glacial pine stumps (*Pinus* cf. *sylvestris*) of a

clay pit in Zurich, Switzerland, oak stems from riverbeds of large European rivers, and posts from a prehistoric lake dwelling settlement in Switzerland.

In the first stage, decay in conifers occurs mosaic-like, cell by cell. In the next stage, the cellulose structure of secondary walls is broken up. In the last stage, the secondary walls contract. During all stages, the primary walls remain largely chemically untouched and keep their original form. Therefore the general wood structure is preserved, and subfossil and fossil wood can be taxonomically identified.

Macroscopic and microscopic aspect of pine stumps decayed under anaerobic conditions

12.7 Late glacial stump of *Pinus sylvestris* in-situ deposited in grey loam, Zurich, Switzerland, 13,000 years BP. Photo: U. Büngen.

12.8 Cross section of a late glacial stump of *Pinus sylvestris*. The dark center is preserved by resin, and the light, contracted periphery is heavily anaerobically decayed.

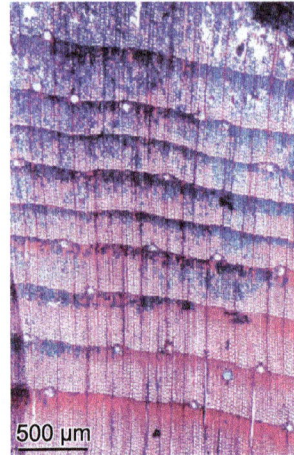

12.9 Cross section overview of a late glacial stump of *Pinus sylvestris*. Transition between areas preserved by resin (red) and areas in decay (blue).

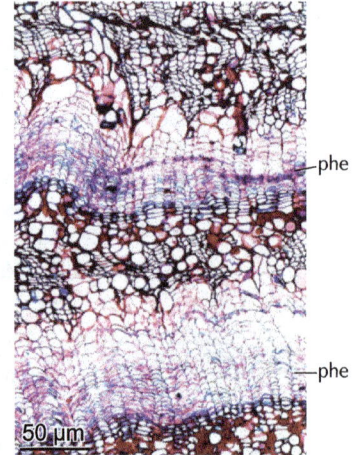

12.10 Bark structures at the periphery of a heavily decomposed stump of *Pinus sylvestris*. The anatomy of the bark is perfectly preserved.

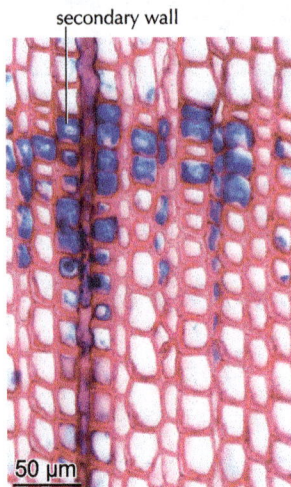

12.11 Cross section of the transition zone between the well-preserved and the decayed zone in a stump of *Pinus sylvestris*. A few latewood tracheids have delignified secondary walls (blue).

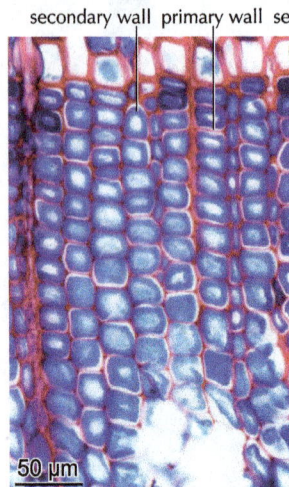

12.12 Cross section of the latewood in a slightly decayed zone of a stump of *Pinus sylvestris*. All secondary walls are delignified (blue). The primary walls of tracheids and rays are not delignified (red).

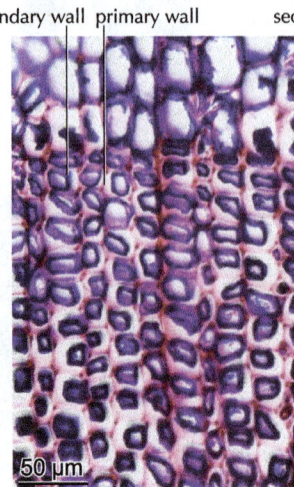

12.13 Cross section of the latewood in a heavily decayed zone of a stump of *Pinus sylvestris*. The secondary walls are delignified and contracted (dark blue). The primary walls are still lignified (red).

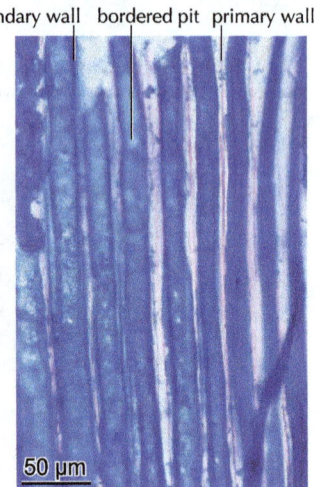

12.14 Radial section of the latewood in a slightly decayed zone of a stump of *Pinus sylvestris*. Pits in the secondary walls are largely decomposed.

Macroscopic aspect of stems decayed under anaerobic conditions

12.15 Holocene stems of *Quercus* sp. in sediments of a Central European river bed. Photo: W. Tegel.

12.16 Holocene stem of *Pinus sylvestris* in a lake in the boreal zone of northern Scandinavia. Photo: T. Bartholin.

12.17 Conifer posts in a Neolithic lake dwelling settlement in Northern Italy (Fiave). Photo: W. Schoch.

Macroscopic and microscopic aspects of deciduous tree stems decayed under anaerobic conditions

12.18 Wet cross section of a Holocene *Quercus* stem. The sapwood is light-colored and the heartwood appears black. Different stages of degradation are expressed by different degrees of shrinkage (see Chapter 13.5). Photo: W. Tegel.

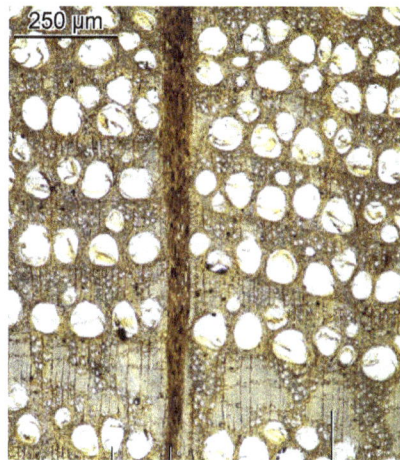

12.19 Microscopic unstained cross section of the black heartwood zone of a subfossil *Quercus* sp. All anatomical features including tyloses are preserved.

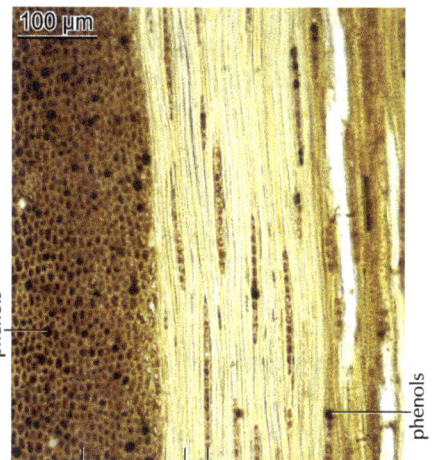

12.20 Microscopic tangential section of a black heartwood zone of *Quercus* sp. All parenchyma cells are filled with dark substances (phenols), giving the heartwood its macroscopic black appearance.

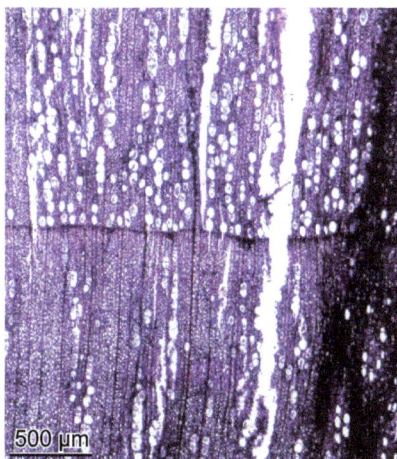

12.21 Microscopic Astrablue/Safranin-stained cross section of a Neolithic post of *Alnus* sp. Despite intensive degradation, species-specific features are well preserved.

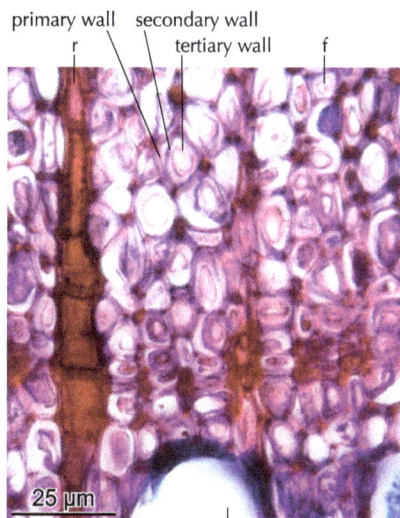

12.22 Cell-wall degradation in a very soft, decayed *Alnus* stem. Primary walls determine the species-specific structure. Secondary walls are almost completely degraded. Tertiary walls are preserved, but separated from the primary walls.

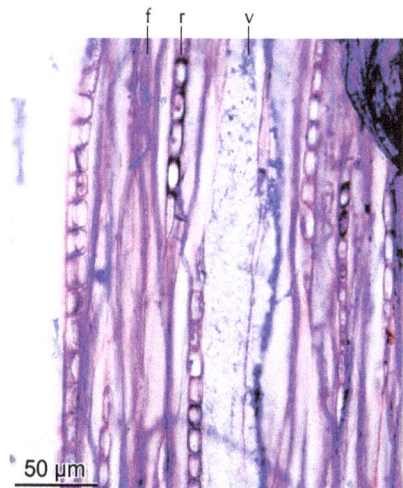

12.23 Different levels of cell-wall degradation in a very soft, decayed *Alnus* stem. Rays consist of parenchyma cells and are well preserved. Cell-wall structures of axial elements (fibers and vessels) are decomposed.

12.3 Aerobic decay – Wood-decaying fungi

Hundreds of fungus species decay wood. Fungi can be differentiated by their fruiting bodies, while differentiation by hyphae is very limited. Some fungi grow as parasites on living trees, but most decompose dead wood. Some fungi attack carbohydrates, while others attack lignin or suberin (lipid polymers). Fungi grow from cell to cell through pit apertures or enzymatically dissolve cell walls. They cause discoloration as blue stain or randomly green wood, or decay patterns as brown rot, soft rot and white rot. Some decay patterns are shown here exemplarily. Their natural variability is much more diverse.

Blue stain fungi produce radial blue stripes in the sapwood of freshly felled logs, mainly in *Pinus* and *Larix*. The blue stripes are just an optical effect due to light refraction. Fungi, mainly ascomycetes, do not degrade cell walls nor do they reduce the stability of the wood. Hyphae absorb sugars, starch proteins and lipids in parenchyma cells. Thick, brown hyphae primarily follow rays and grow from cell to cell through pit apertures. Cell composition as well as cell-wall structures are perfectly preserved.

Brown rot fungi, mainly of the family of Polyporaceae, decay living and dead wood. Characteristic is the cubiform brown decay pattern and the dramatic loss of weight and bending strength. Enzymes decay cellulose and hemicellulose. The cellular composition of decayed wood does not change. Anatomical details of rays remain, however, those in tracheids disappear or are indistinct. Since the cellulosic structure is decomposed, wood structure disappears in polarized light. Decomposition artifacts limit an anatomical species determination.

Soft rot fungi mainly decay construction wood of conifers and deciduous trees. Characteristic are cubiform and fibrous decay patterns on logs and boards in humid conditions. It is mainly ascomycetes that primarily decay cellulose and hemicellulose and—to a small amount—also lignin.

White rot fungi produce simultaneous rot and successive white rot. Both types occur mainly on deciduous trees but also on conifers in the forest. Decayed wood is spongy and often appears whitish. Characteristic for **simultaneous white rot** are irregular dark lines which consist of concentrations of dark-colored hyphae. Affected wood loses a lot of weight. Basidiomycetes and ascomycetes first attack cellulose and lignin, and later also hemicellulose. In **successive white rot**, basidiomycetes and ascomycetes first attack lignin and hemicellulose, and later also cellulose. Characteristic for damages from the fungus *Ganoderma lipsiense* are enlarged decay zones.

Macroscopic aspect of major discolorations and rottenness caused by fungi

12.24 Blue stain in *Pinus sylvestris*. Characteristic are radial dark stripes in the sapwood of pines and larches.

12.25 Green rot in *Fagus sylvatica*.

12.26 Concentrated hyphae (mycelium) of *Armillaria* sp. in the cambial zone of a tree.

12.27 Brown rot in a stem of *Picea abies*. Characteristic are cubiform, dark brown areas in conifers.

12.28 Red rot fungus (*Fomes annosus*) in a root and a stem of *Picea abies*.

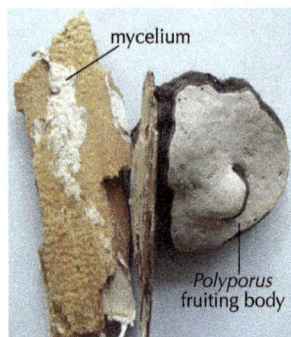

12.29 Soft rot caused by a species of *Polyporus*. Characteristic is soft, fibrous wood in moist conditions.

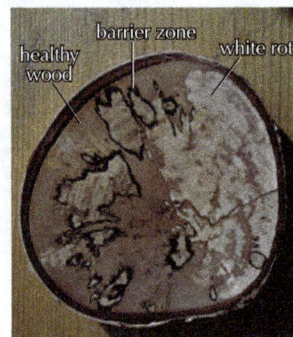

12.30 Simultaneous white rot with bleached parts and dark demarcation lines in *Betula pendula*.

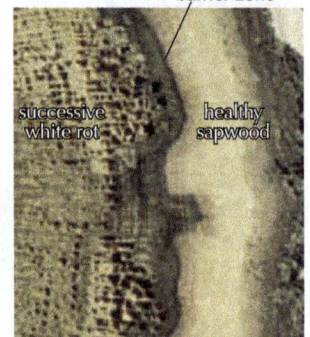

12.31 Successive white rot and a dark demarcation zone in *Quercus* sp. The little hollows indicate the decay.

Microscopic aspect of fungal hyphae in wood

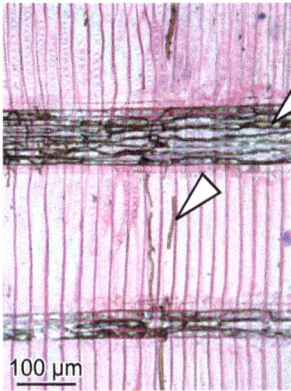

12.32 Radial section of a blue-stained *Pinus sylvestris* with thick brown hyphae in rays and tracheids. The hyphae do not stain with Astra-blue/Safranin.

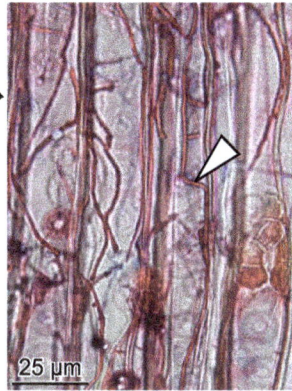

12.33 Radial section of a brown rotten *Pinus sylvestris* with thick, red-stained, non-septate hyphae in tracheids.

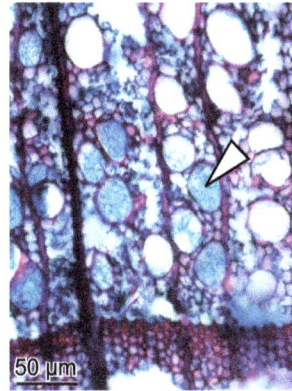

12.34 Cross section of white rotten *Fagus sylvatica* with thin, blue-stained hyphae.

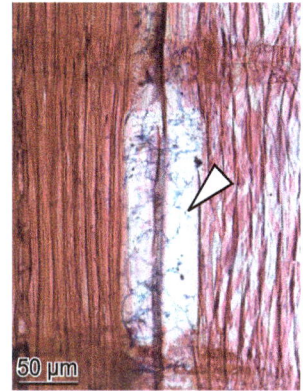

12.35 Radial section of a green rotten *Betula pendula* with thin, blue-stained hyphae in a vessel.

Hyphae growing through cell walls

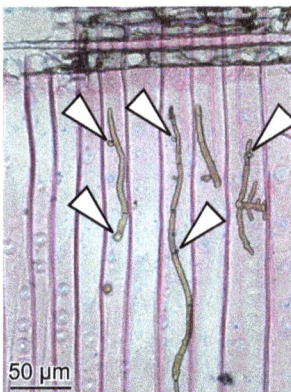

12.36 Septate hyphae growing through pit openings in *Pinus sylvestris*.

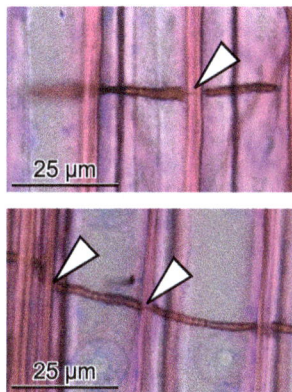

12.37 Hyphae enzymatically dissolve cell walls in *Pinus sylvestris*.

Brown rot decay

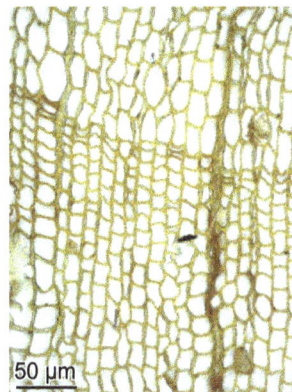

12.38 Preserved cell wall structure in *Pinus mugo* (unstained slide).

12.39 Largely decomposed secondary cell-wall structures in tracheids and preserved cell-wall structures in rays of *Pinus mugo*.

Soft rot decay

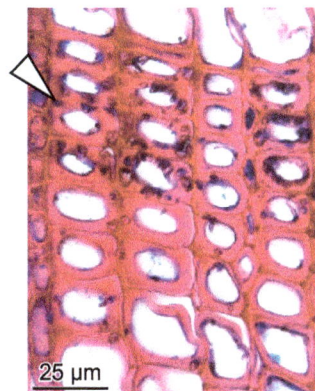

12.40 Hyphae within the walls of latewood tracheids in *Pinus sylvestris*.

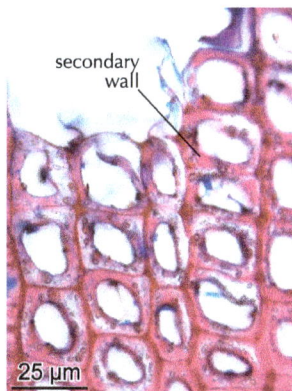

12.41 Advanced decay in *Pinus sylvestris*. Hyphae decomposed and delignified secondary walls. Primary walls are still lignified. Tertiary walls are structurally preserved but slightly delignified.

Simultaneous white rot decay

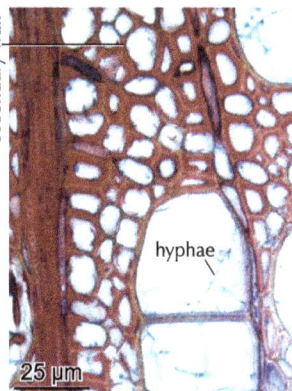

12.42 White rot decay in *Betula pendula*. Secondary walls of most fibers are broken and dissolved.

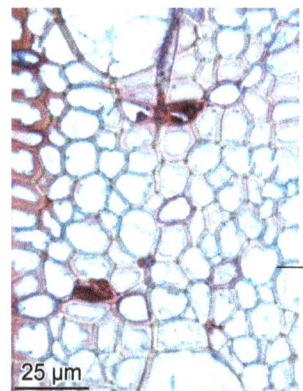

12.43 Advanced decay in *Betula pendula*. Some secondary walls are broken, some are delignfied, and some are completely gone. Primary walls of decomposed cells are delignified.

Successive white rot decay

delignified   delignified   delignified

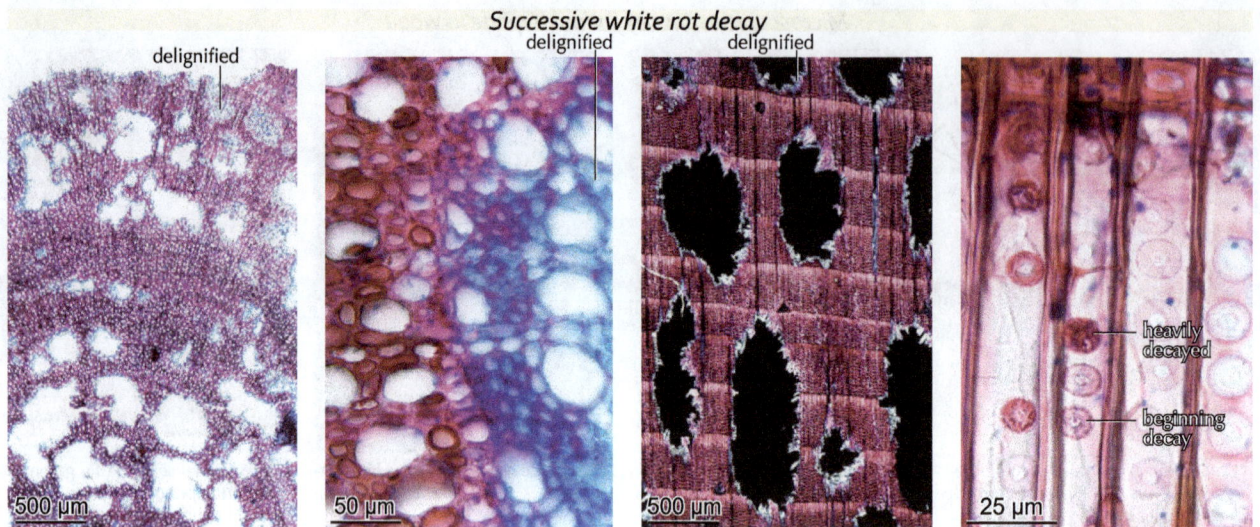

12.44 Holes in a stem of the dwarf shrub *Calluna vulgaris*.

12.45 Selective degradation of cells around a hole in a stem of *Calluna vulgaris*.

12.46 Holes in a stem of the conifer *Pinus mugo*, polarized light. Tracheids around the holes are delignified.

12.47 Various degrees of decay of pits around a hole in a stem of *Pinus mugo*.

12.4 Compartmentalization – The natural limit to fungal growth

Shigo 1989 described the CODIT concept (COmpartmentalization of Decay In Trees). It states that hyphae of fungi cannot grow unlimited because stems form radial tangential and axial fungicide barrier zones. Hyphae initiate phenolic excretion of living parenchyma cells (rays and axial parenchyma). Four walls limit the expansion of hyphae.

The CODIT concept is not a means of healing stems, but it can compartmentalize damages.

- **Wall 1** prevents expansion in axial direction. This is the weakest barrier in the system.
- **Wall 2** moderately prevents stem-inward expansion.
- **Wall 3** moderately prevents lateral expansion.
- **Wall 4** prevents expansion towards newly formed cells after injury. This is the strongest zone. Suberin layers in the new cells on the side facing the injury prevent growth of hyphae.

Macroscopic and microscopic aspect of compartmentalization

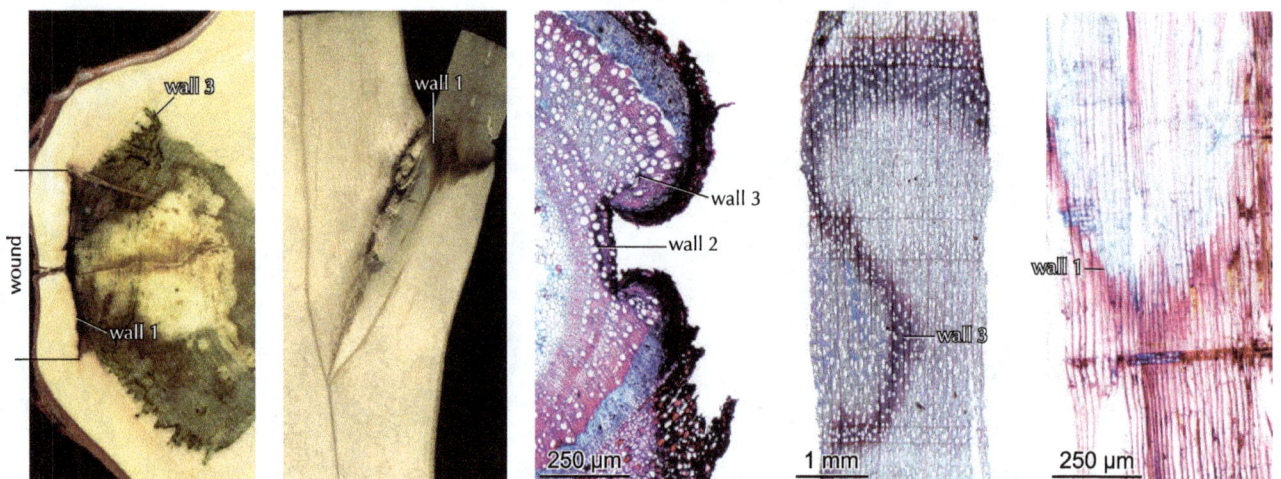

12.48 Cross section of a compartmentalized wound in *Acer* sp.

12.49 Longitudinal section of a compartmentalized dead branch of *Acer* sp.

12.50 Cross section of a compartmentalized wound of a rhizome of the herb *Mentha* sp.

12.51 Cross section of a barrier zone in *Betula pendula*.

12.52 Radial section of a barrier zone in *Pinus mugo*.

12.5 Decay by xylobiontic insects

Thousands of insect species are part of the recycling process of wood and bark. Some specialize in feeding on living plants, many prefer dead logs and timber, and a large group is responsible for the decomposition of rotten wood. Xylobionts evolved in many taxonomic units, e.g. in beetles, termites, wasps, bees, miner flies, ants and woodlice. Each insect species prefers specific wood conditions. All decomposers leave feeding traces in the wood or in the bark. Beetles destroy the cambium, the peripheral parts of the living xylem, and the most active part of the phloem. Affected trees can therefore die. Larvae of miner flies (Agromyzidae) feed up and down in the cambial zone where they consume nutrients. Since they do not destroy all meristematic cells, these galleries can be closed by callus cells. These scars are called pith flecks. Agromyzid-affected stems

don't die. Longhorn beetles (Cerambycidae), wood borers (Anobiidae) and carpenter ants (*Camponotus* sp.) feed on dense, dry, dead wood, where they form galleries of various forms. The remaining burrows are mostly filled with coprolites. Termites have the most powerful mandibles, and destruct extremely dry wood. Various insect larvae, e.g. of goat moth (Cossidae) and the Asian longhorn beetle, form large scale galleries, and can often kill the trees. Wasps peel externally eroded plants, and use the wood fibers for the construction of their nests. Wasp delignify the fibers chemically so the fibers become flexible. Woodlice (Oniscidea) are the last members in the wood decay chain, they live in moist mull.

For more information see e.g. Wermelinger 2017.

Galleries of bark beetles—cambium miners which can kill the host trees

12.53 The bark beetle *Ips typographus*. The beetle lives in the cambial zone of *Picea abies* in Europe. Photo: B. Wermelinger.

12.54 Galleries of the European spruce bark beetle *Ips typographus* in the bark of *Picea abies*.

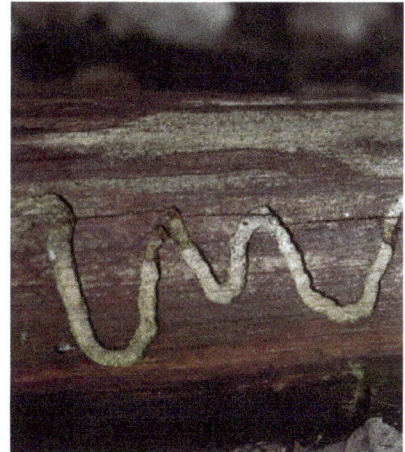

12.55 Beetle galleries in the wood of a post.

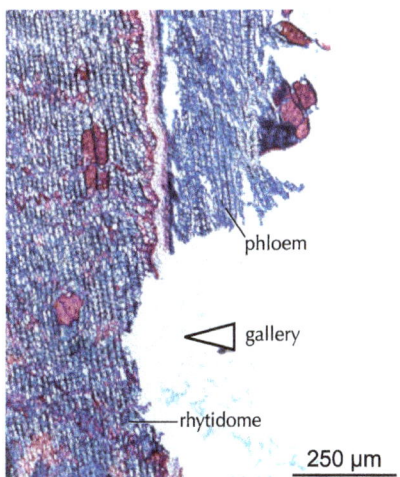

phloem

gallery

rhytidome

250 µm

12.56 Gallery of a bark beetle in the bark of *Picea abies*. Affected is the whole living phloem as well as the dead rhytidome.

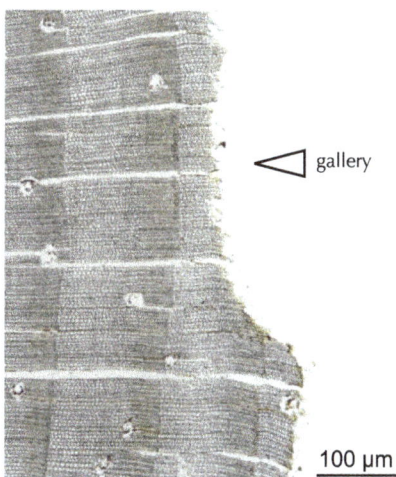

gallery

100 µm

12.57 Bark beetle gallery in the wood of *Picea abies*. Affected are the two last rings.

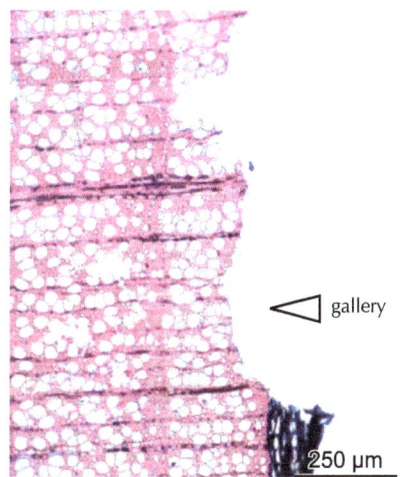

gallery

250 µm

12.58 Bark beetle gallery in the wood of *Fagus sylvatica*. Affected is just the last ring.

Galleries of cambium miner flies (Agromyzidae)—which do not kill the tree

12.59 Peripheral xylem zone of *Salix* sp. with galleries. Larvae of miner flies feed in the soft cambial zone before they leave the tree trough the bark.

12.60 Pith flecks in all rings of a young *Salix* tree.

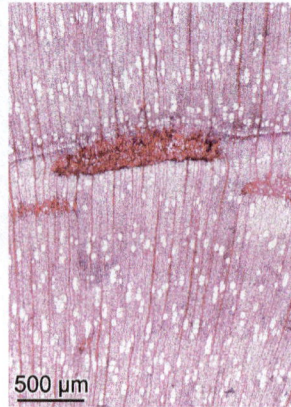

12.61 Pith flecks in the latewood of *Betula pendula*. The flecks are filled with callus tissue.

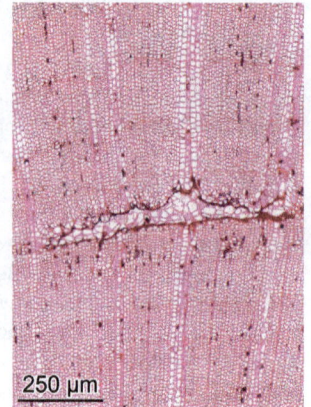

12.62 Pith flecks in the earlywood wood of *Tasmannia xerophila*. The flecks are surrounded by collapsed cells and filled with callus tissue.

Sapwood destruents

12.63 Heavily decayed sapwood of a beam of *Quercus* sp.

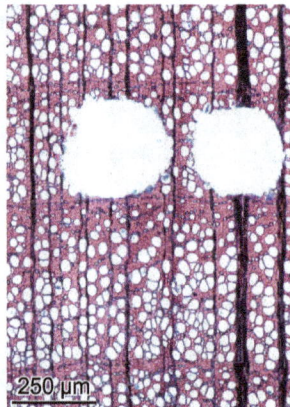

12.64 Traces of a beetle in a dry beam of *Fagus sylvatica*.

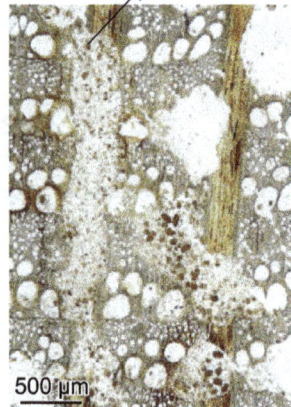

coprolite

12.65 The irregularly formed galleries are filled with frass, composed of coprolites and fine wood particles. The insect eats rays, fibers and vessels.

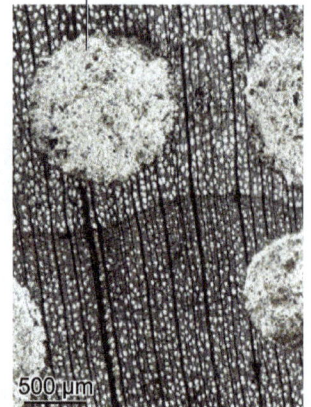

frass

12.66 The round galleries are filled with frass (coprolites and fine wood particles).

Sapwood and heartwood destruents

12.67 Longhorn beetle (*Hylotrupes bajulus*, Cerambycidae). Photo: B. Wermelinger.

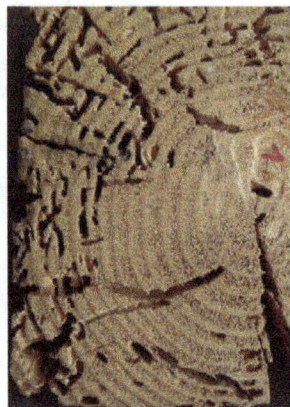

12.68 Galleries of larvae of longhorn beetles occur mainly in the earlywood.

12.69 The Asian longhorn beetle (*Anoplophora glabripenni*) lives in the wood of various deciduous trees. Photo: D. Hölling.

12.70 The larvae of the goat moth *Cossus cossus* feed in tree trunks, here in a stem of *Salix* sp.

Sapwood and heartwood destruents—Anobiidae wood borers

12.71 Exit holes of wood borer larvae in a wooden tool made of *Fagus sylvatica* wood.

12.72 Irregular internal galleries of wood borer larvae in wood of *Fagus sylvatica*.

12.73 Round wood borer galleries, polarized light. The insect larvae prefer to feed on the soft earlywood.

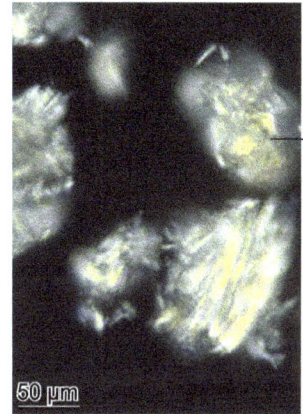

12.74 Frass with wood dust in coprolites, polarized light. Particles have a length of 10–20 µm.

Sapwood and heartwood destruents—carpenter ants

Mull consumers—woodlice

12.75 Carpenter ant *Camponotus* sp. Photo: B. Wermelinger.

12.76 Galleries of carpenter ants.

12.77 Woodlouse in a rotten stem of *Fagus sylvatica*.

Fiber collectors—wasps

12.78 Common wasp *Vespula vulgaris*. Photo: B. Wermelinger.

12.79 Nest of *Vespula vulgaris*. Photo: B. Wermelinger.

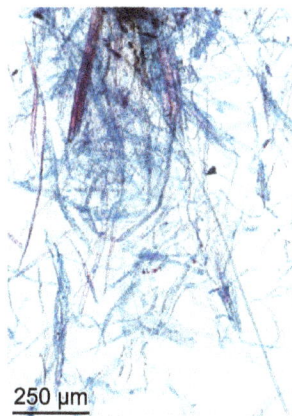

12.80 Mostly delignified, large fibers of many different plants in a nest of *Vespula vulgaris*. Ray cells and vessels are absent.

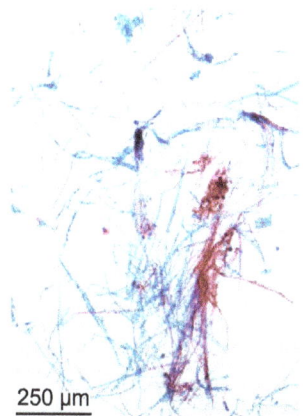

12.81 Mostly delignified, small fibers of an uncovered nest of *Vespula vulgaris* in a wet meadow.

Wood: Fossilization, Permineralization, Carbonization and Conservation

13.1 Fossilization

When remains or traces of organisms are preserved in sediments they are called fossils. The process of fossilization (syn. petrification or permineralization) begins when organisms, e.g. stems, are buried and remain in anaerobic conditions. Most petrified stems are therefore in a horizontal position. Many fossil stems were probably driftwood that was buried by fluvial sediments. Stems in upright position were more likely embedded in volcanic ashes. When supersaturated groundwater penetrates the wood, the minerals precipitate within cellular spaces and crystalize. Most frequent are calcium carbonate ($CaCO_3$) or silicate minerals (SiO_2). Fossilization occurs mainly in association with marine or volcanic hydrothermal water. Characteristic for fossils is the conservation of microscopic structures that can be observed on polished disks or on micro-sections. Observations on fossilized wood leads to four principal findings:

◦ Taxonomic classifications are still possible on well-preserved fossils.

◦ Assemblages of fossils (petrified wood and pollen) in stratigraphically uniform layers allow a limited reconstruction of past vegetation types.

◦ Conclusions about the seasonality of past climates are possible on the basis of the presence or absence of growth rings. Banded agates can resemble growth rings in petrified wood, but they represent periodic mineral deposits, e.g. in stalactites.

◦ Diagenetic processes (alteration during fossilization) can be reconstructed on the basis of the preservation of tissue and cell-wall structure, and the crystalline composition of minerals. Hyphae, barrier zones and coprolites indicate a decay under aerobic conditions, while cell-wall degradation, cell deformation and root inclusions indicate decay under anaerobic conditions before petrification. Different colors on polished discs are related to different minerals, e.g. red is related to iron, blue and transparent to silicate, green to copper and pink to manganese. Mineral composition can be quantified on a microscopic section under polarized light.

See also Taylor *et al*. 2009 and Selmeier 2015.

13.1 Stem disc of a Triassic *Araucarioxylon* stem from a petrified forest in Arizona, USA.

Geological context of petrified wood

13.2 Eroded petrified stems from fluviatile sediments of the Upper Triassic, approx. 200 million years ago, Petrified Forest, Arizona, USA. Photo: V. Markgraf.

13.3 Broken and dislocated petrified stems from fluviatile sediments. Upper Triassic, approx. 200 million years ago, Petrified Forest, Arizona, USA. Photo: V. Markgraf.

13.4 Reconstruction of the *Araucaria* forest at the Chinle Formation (Trias). Display at the Visitor Center of the Petrified Forest.

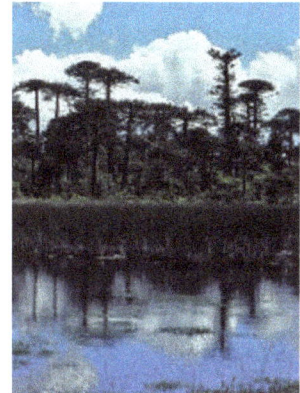

13.5 Modern *Araucaria araucana* forest in Patagonia, Chile.

Macroscopic aspect of polished disks

13.6 *Quercoxylon* sp. with distinct annual rings. Miocene, approx. 20 million years ago, Great Basin, Idaho, USA.

13.7 Conifer with indistinct annual rings. Miocene, approx. 20 million years ago, Great Basin, Idaho, USA.

13.8 *Palmoxylon* sp., of unknown age and locality, with distinct vascular bundles.

13.9 Petrified liana of unknown age and locality.

Taxonomic classification of petrified wood

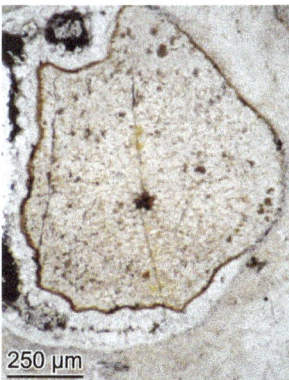

250 µm

13.10 *Rhynia gwynne-vaughanii*, a herb with a central stele. Lower Devonian, approx. 400 million years ago, Scotland.

100 µm

13.11 Cross section of *Paradoxylon leuthardii*, one of the Cordaitales without annual rings. Upper Triassic, approx. 220 million years ago, Vosges Mountains, France. Slide: R. Buxtorf.

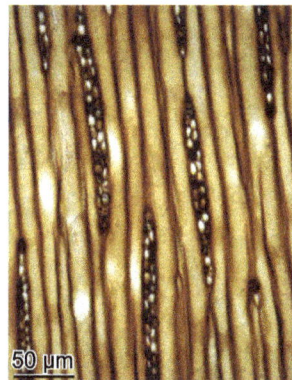

50 µm

13.12 Tangential section of *Paradoxylon leuthardii* with biseriate rays. Upper Triassic, approx. 220 million years ago, Vosges Mountains, France. Slide: R. Buxtorf.

50 µm

13.13 Radial section of *Paradoxylon leuthardii* with multiseriate bordered pits on tracheids. Upper Triassic, approx. 220 million years ago, Vosges Mountains, France. Slide: R. Buxtorf.

Taxonomic classification of petrified wood

13.14 *Sequoia* sp. with distinct and indistinct rings. Early Oligocene, approx. 34 million years ago, Florissant, Colorado, USA.

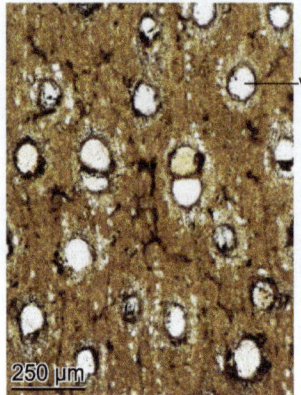

13.15 Mimosoideae, Fabaceae, without annual rings, with paratracheal parenchyma. Oligocene, approx. 28 million years ago, Vosges Mountains, France. Slide: R. Buxtorf.

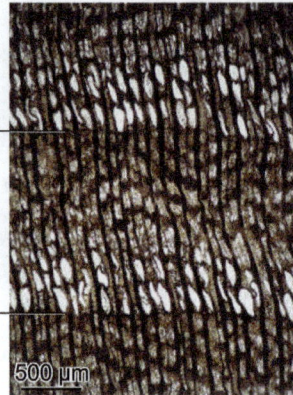

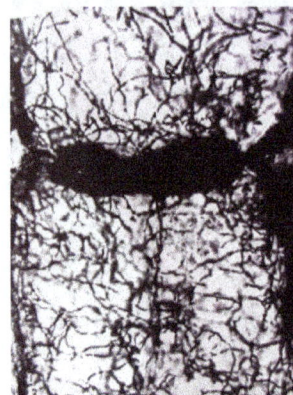

13.16 *Celtixylon* cf., Ulmaceae, with distinct annual rings. Oligocene, approx. 28 million years ago, Vosges Mountains, France. Slide: R. Buxtorf.

13.17 *Palmoxylon*, Arecaceae, with distinct lignified parts of vascular bundles. Oligocene, approx. 28 million years ago, Vosges Mountains, France. Slide: R. Buxtorf.

Growth rings in petrified wood

13.18 Cross section of *Paradoxylon leuthardii*, without growth rings, indicating tropical rain forest as habitat? Upper Triassic, approx. 220 million years ago, Vosges Mountains, France. Slide: R. Buxtorf.

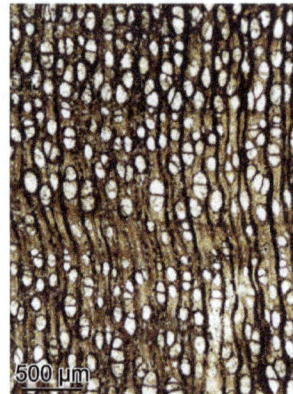

13.19 Cross section of *Paradoxylon leuthardii*, with growth rings, indicating seasonal climate. Upper Triassic, approx. 220 million years ago, Vosges Mountains, France. Slide: R. Buxtorf.

13.20 Lauraceae cf. with growth rings, seasonal climate. Oligocene, approx. 28 million years ago, Vosges Mountains, France. Slide: R. Buxtorf.

13.21 Growth rings in a petrified, ring-porous oak stem in Miocene deposits, approx. 15 million years ago, Oregon, USA.

Indicators of aerobic decay before petrification

13.22 Barrier zones in a petrified stem of a eucalypt tree. Australia, age unknown.

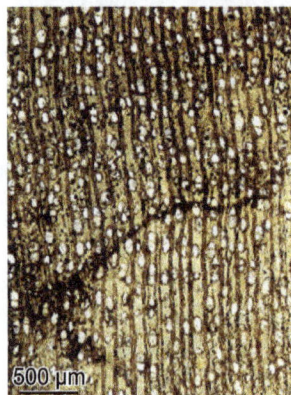

13.23 Barrier zones in a stem of a Lauraceae cf. tree. Oligocene, approx. 28 million years ago, Vosges Mountains, France. Slide: R. Buxtorf.

13.24 Hyphae in a vessel of a Meliaceae tree. Reprinted from Selmeier 2015.

13.25 Coprolites of a beetle in a Lauraceae tree. Reprinted from Selmeier 2015.

Indicators of anaerobic decay before petrification

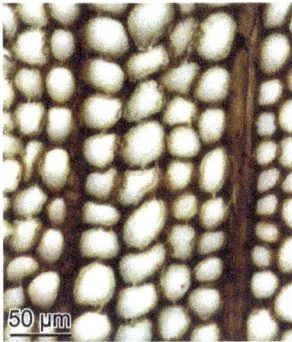

13.26 Decayed cell walls in *Paradoxylon leuthardii*. Preserved are the thick primary walls and thin tertiary walls. Slide: R. Buxtorf.

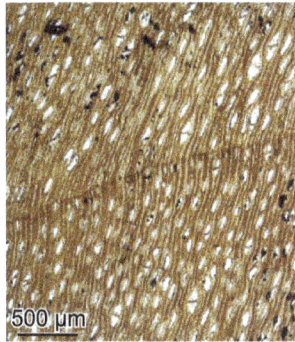

13.27 Radially compressed dicotyledonous wood. Slide: R. Buxtorf.

13.28 Heavily tangentially compressed wood of *Paradoxylon leuthardii*. Slide: R. Buxtorf.

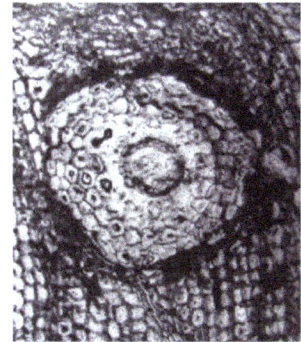

13.29 The root of a herb squeezed the soft, anaerobically decayed xylem tissue. Reprinted from Selmeier 2015.

Permineralization under polarized light

13.30 Single crystals in individual tracheids in *Paradoxylon leuthardii*. Slide: R. Buxtorf.

13.31 Many crystals in large earlywood vessels in Ulmaceae wood. Slide: R. Buxtorf.

13.32 Irregular crystallization of silicates without any wood anatomical context in Ulmaceae wood. Slide: R. Buxtorf.

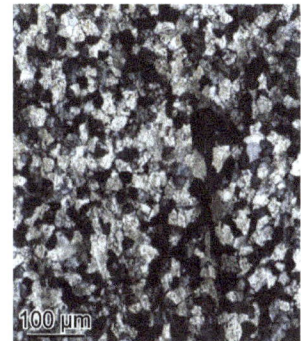

13.33 Crystallization of different minerals relating to groups of tracheids in *Paradoxylon leuthardii*. Slide: R. Buxtorf.

13.2 Permineralization of archaeological artifacts

Permineralization also takes place on metallic archaeological artifacts. The scabbards and handles of iron swords, bronze daggers and axes often contain small mineralized remains of wood. Transverse and longitudinal breaks of particles clearly show the anatomical structure. Species identification is therefore possible. If highly concentrated liquids of iron, copper or sulfur soak the wood, minerals precipitate on cell walls and preserve their structure.

Permineralization of archaeological artifacts

13.34 Wooden scabbard permineralized by dissolved iron. Photo: W. Tegel.

13.35 Wooden scabbard permineralized by dissolved copper. Photo: W. Tegel.

13.36 Microscopic structure of archaeological *Quercus* sp. wood, permineralized by iron. Photo: W. Tegel.

13.37 Microscopic structure of archaeological *Alnus* sp. wood, permineralized by sulfur. Photo: W. Tegel.

13.3 Coalification

Coalification is a very slow process in which wood is transformed into coal at geological time scales. The anaerobic decay of organic substances is the first stage of coalification (see Chapter 12.2). The ratio of carbon increases during anaerobic decay processes, and the wood loses its stability. The process starts with soft wood, which under high pressure and high temperatures turns into brown coal, bituminous coal and finally anthracite. An increase in carbon content defines these grades—wood contains 50%, brown coal 70% and anthracite 90% carbon. When softened logs come under mechanical pressure by sediments or ice, the thin-walled earlywood cells collapse, and the stems get deformed. Early stages of coalification are known from stems in medieval glacier deposits and late glacial clays (subfossil wood).

Remnants of wood (xylite, lignite) with well-preserved structures are frequently found in brown coal layers (Eocene to Oligocene). Genera or even species can be identified as long as the earlywood zones are not too much distorted (Dolezych 2005).

Tertiary wood is often so well preserved that even stages of cell wall decay can be recognized. However, in most cases the earlywood zones are compressed, and the corresponding ray features are difficult to observe (Dolezych 2005).

Fossil tree stump

Taxonomically and climatologically relevant features in cross sections of conifers

13.38 Stumps from a brown coal bed. Photo: Elbwestfale 2003, via Wikimedia Commons, CC BY-SA 3.0.

13.39 Conifer *Taxodium taxodii* cf. without growth zones. Oligocene, approx. 34–23 million years ago. Material: M. Dolezych.

13.40 Conifer *Sciadopityoxylon wettsteinii* without resin ducts and distinct earlywood and latewood zones. Miocene, approx. 25–5 million years ago. Material: M. Dolezych.

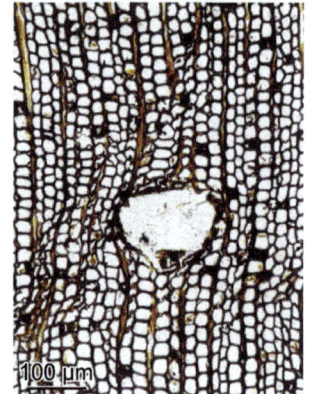

13.41 Conifer *Doliostroboxylon priscum* with resin ducts. Eocene, approx. 56–38 million years ago. Material: M. Dolezych.

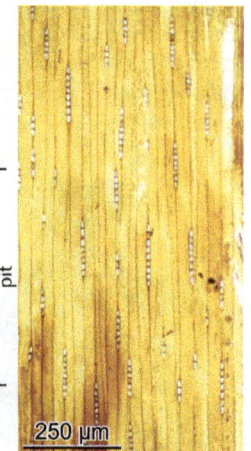

Taxonomically relevant features in radial and tangential sections of conifers

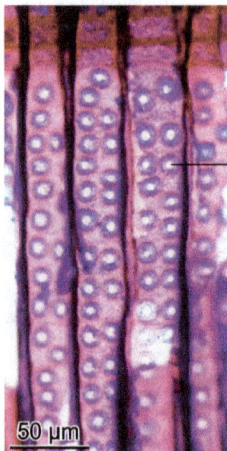

13.42 Conifer *Larix decidua* with biseriate tracheid pits. Interglacial period, Italian Alps, >50,000 years ago.

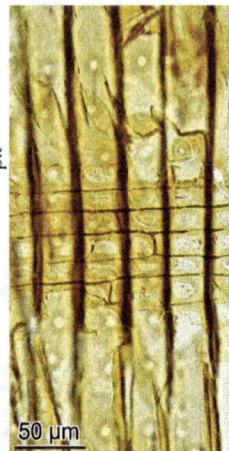

13.43 Fenestrate vessel-ray pits in the earlywood of *Sciadopityoxylon wettsteinii*. Miocene, approx. 25–5 million years ago. Material: M. Dolezych.

13.44 Small round pits in the latewood of *Taxodium gypsacum*. Miocene, approx. 25–5 million years ago. Material: M. Dolezych.

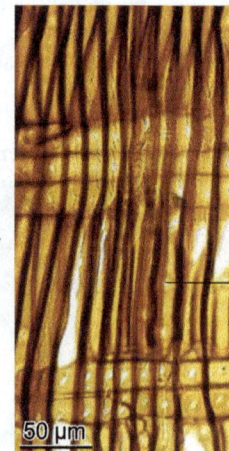

13.45 Large round pits in the early- and latewood of *Doliostroboxylon priscum*. Eocene, approx. 56–38 million years ago. Material: M. Dolezych.

13.46 Uniseriate rays with three to ten cells in *Taxodium gypsacum*. Miocene, approx. 25–5 million years ago. Material: M. Dolezych.

Cell wall decay in latewood

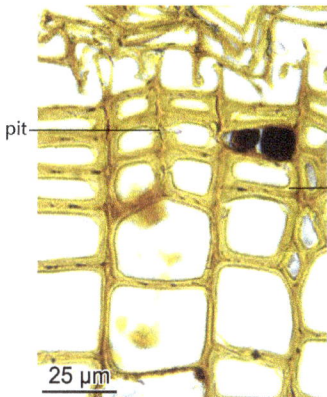

13.47 Perfectly preserved cell walls in latewood tracheids of *Glyptostoboxylon rudolfii*. Middle Miocene, approx. 15 million years ago. Material: M. Dolezych.

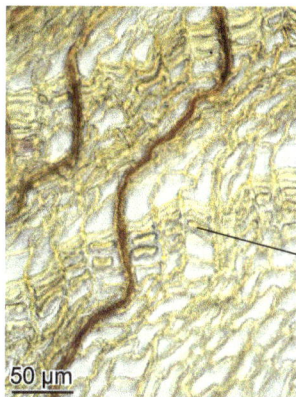

13.48 Anaerobically decayed cell walls in latewood tracheids of *Larix decidua* from the moraine of an alpine glacier. Middle Ages, approx. 1400 AD. Material: M. Dolezych.

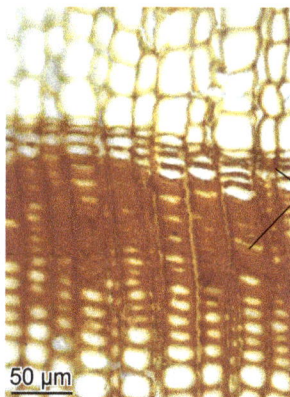

13.49 Various stages of anaerobic decay in the latewood zone of *Sciadopityoxylon wettsteinii*. Miocene, approx. 15 million years ago. Material: M. Dolezych.

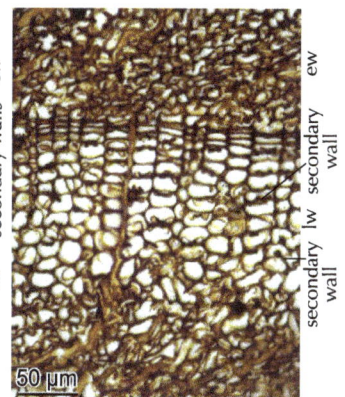

13.50 Primary and tertiary cell walls are preserved, secondary walls are completely decayed in latewood tracheids of *Taxodium taxodii* cf. Oligocene, approx. 34–23 million years ago. Material: M. Dolezych.

Mechanical deformation of cells

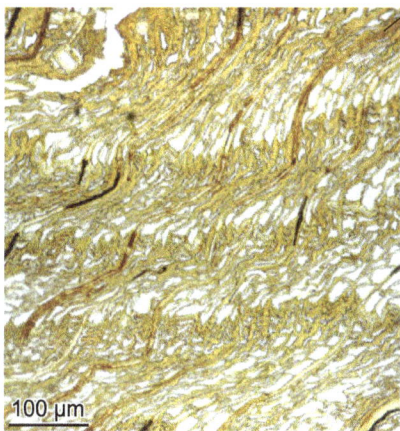

13.51 Wood of *Larix decidua*, heavily radially compressed by a glacier. Moraine of an alpine glacier, Middle Ages, approx. 1400 AD.

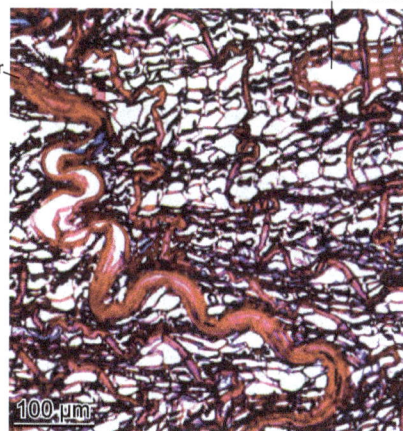

13.52 Radial compression is indicated by bent rays in wood of *Larix decidua*. Interglacial period, Italian Alps, >50,000 years ago. Slide stained with Astrablue/Safranin.

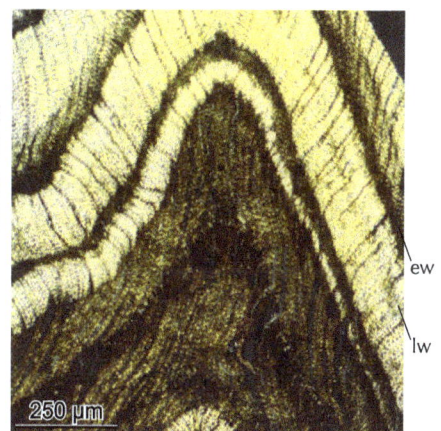

13.53 Compression in a branch of *Abies alba* cf., polarized light. Interglacial period, Switzerland, approx. 45,000 years ago.

13.54 Compression of two weak zones in ringless wood of the conifer *Doliostroboxylon priscum*. Eocene, approx. 56–37 million years ago. Material: M. Dolezych.

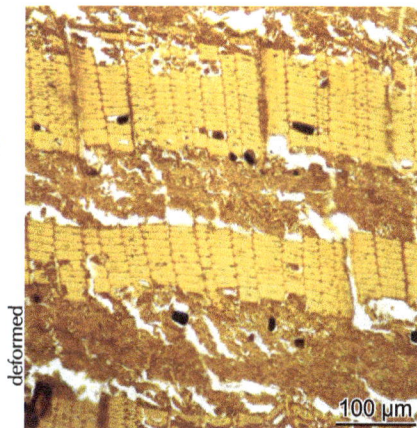

13.55 Intensively compressed earlywood zones between thick-walled latewood zones in *Taxodium gypsacum*. Miocene, approx. 15 million years ago. Material: M. Dolezych.

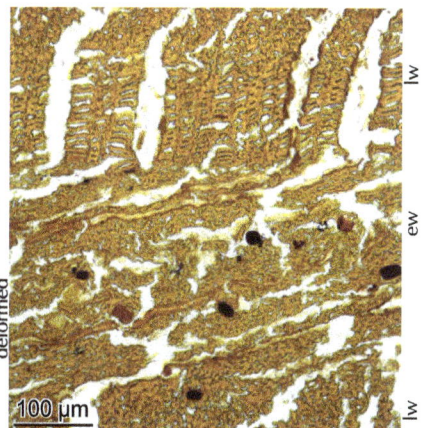

13.56 Intensively compressed earlywood and latewood zones in *Sequoiadendroxylon* sp. Czech Republic, middle Miocene, approx. 15 million years ago.

13.4 Carbonization

Carbonization is the process of transforming wood into charcoal by pyrolysis. Charcoal used to be one of the most important energy resources in pre-industrial times because it was easy to transport. Most human civilizations produced charcoal in kilns, with the wood losing 50–70% of its weight. Charcoal burning and grazing were the main reasons for the depletion of ancient forests all over the world. With the switch to fossil fuels like coal and coal oil, charcoal lost some of its importance. It also used to be important for the production of gun powder (black powder). Today, charcoal is still widely used, e.g. for local metallurgic processes, chemical filtering, barbecues and charcoal crayons.

The carbonization process does not destroy the species-specific anatomical structures of the material. Also, charcoal is very stable against biodegradation in both dry and wet environments. Charcoal is therefore of relevance to historical studies,

as ancient prehistoric fires—changing human behavior and vegetation patterns from the tropics to the arctic—can be documented with the identification of particles as small as a cubic millimeter.

During the carbonization process, cell walls shrink by approx. 50% and also lose their fibrillose structure. However, the structure of pits and perforations, and of artifacts caused during decay before the carbonization process, remain. Longitudinal cracks and tangential cell collapses are a sign of vapor pressure during the heating phase.

Binocular stereomicroscopes and episcopic microscopes facilitate the observations. Thin sections, embedded in two component epoxy resin, are the basis for photographic presentations (see Chapter 2).

Charcoal production

13.57 Establishment of a charcoal kiln in the Black Forest. Photo: T. Ludemann.

Carbonization artifacts

13.58 Radial cracks in this conifer are caused by vapor pressure during the heating phase.

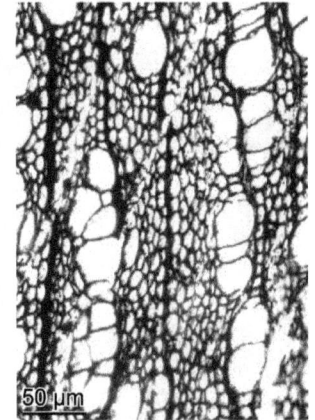

13.59 Thin carbonized cell walls in *Alnus* sp.

Species identification from anatomical structures preserved in charcoal

13.60 Conifer with an insect gallery, filled with coprolites.

13.61 Charcoal cross section of *Fraxinus excelsior*.

13.62 Charcoal cross section of *Alnus* sp.

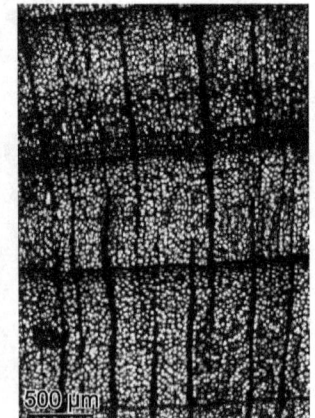

13.63 Charcoal cross section of *Fagus sylvatica*.

13.5 Wet wood conservation

Woods in anoxic sediments can keep their form over millennia. However, decomposition takes place on a cellular level. Alterations become obvious when logs or archaeological artifacts dry out. Radial shrinking and weight loss are macroscopic signs of intensive decomposition, cell collapses are visible on a microscopic level. Despite many deformation artifacts, parts such as roots from plants along lake shores can be identified.

Natural preservation occurs in the heartwood of logs of pines soaked in resin, and of oaks with high levels of phenols. The central parts of stems resist decomposition at the air, but the sapwood decays. The situation is much worse in wood of lake dwellings. Most wooden artifacts are of small dimensions,

and made mostly of wood of deciduous trees. Five-thousand-year-old Neolithic stems of deciduous trees with a diameter of 10 cm lose approximately 50% of their original weight, and shrink radially when dehydrated up to 80%. Therefore a large part of human cultural heritage would be lost without artificial preservation. Archaeologists preserve the form of large artifacts mainly with polyethylene glycol (PEG), e.g. the Swedish warship Vasa. Smaller wooden instruments are preserved by freeze drying or dimethyl ether treatment. The form and superficial traces of human treatment of ancient wet wood can therefore be kept in museums. The cell wall structures of wet wood do survive preservation techniques.

Macroscopic changes on wet wood after dehydration

13.64 Late glacial stems of *Pinus sylvestris* that have been exposed to air. The stump surfaces decay and scientific information about the life of those trees is lost.

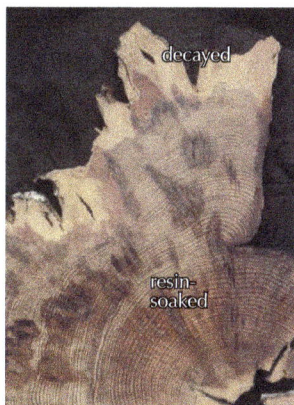

13.65 Disc of an air-exposed *Pinus sylvestris* stem. The large central part is preserved due to natural impregnation with resin. The peripheral zone without resin decays.

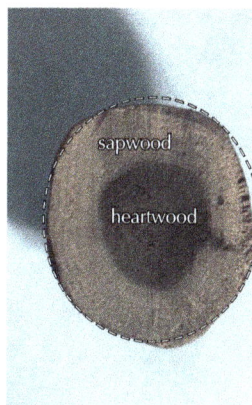

13.66 Disc of a water-saturated archaeological *Quercus* sp. Its form is perfectly preserved.

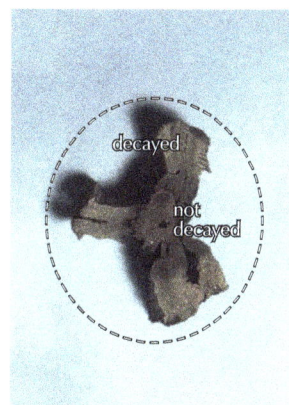

13.67 Disc of the same *Quercus* stem, but dehydrated. Intensively decayed parts shrink dramatically, as indicated by the original outline.

Microscopic changes on wet wood after dehydration

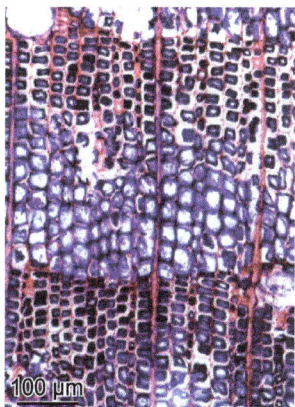

13.68 Not dehydrated, heavily degraded wood of late glacial *Pinus sylvestris*. The cell form and primary cell walls (red) remain. Secondary walls lost their original structure and are delignified (blue).

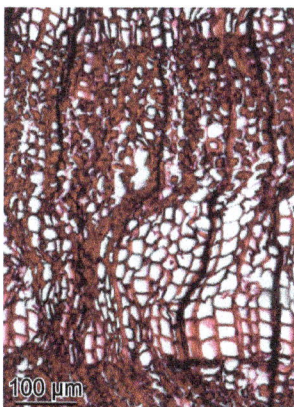

13.69 Dehydrated, heavily degraded wood of late glacial *Pinus sylvestris*. Cells lost their original form due to contraction, and degraded secondary cell walls appear as unstructured bodies.

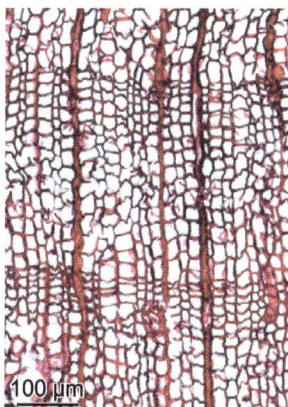

13.70 Dehydrated, heavily degraded wood of late glacial *Pinus sylvestris*. The cell forms remain, but the secondary walls are gone.

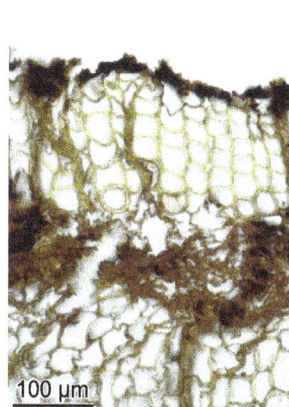

13.71 Dehydrated, heavily degraded wood of *Pinus sylvestris*. Latewood cells are collapsed, earlywood cells of the outer ring kept their form. The tree died during earlywood formation in early summer approx. 13,000 years ago.

Microscopic changes on wet wood after dehydration   Material: Swiss National Museum.

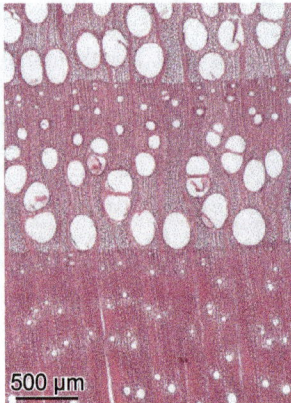

13.72 Modern wood of *Fraxinus excelsior*, stained with Safranin. All microscopic structures can be recognized.

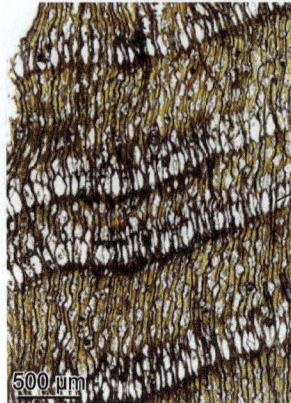

13.73 Dehydrated, heavily degraded wood of Neolithic *Fraxinus excelsior*. Lateral contraction deformed the original structure, but major anatomical characteristics for the species remain.

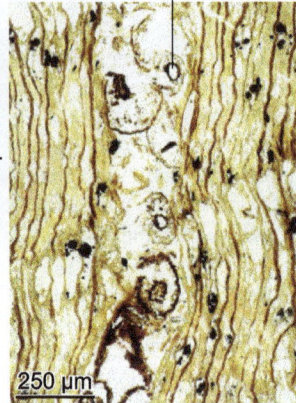

13.74 Dehydrated, heavily degraded wood of Neolithic *Fraxinus excelsior*. It maintained its anatomical structure despite some lateral deformation by roots from plants of the lake shore.

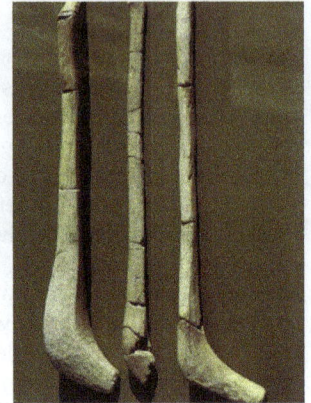

13.75 Dehydrated, heavily degraded wooden artifacts, made from *Fraxinus excelsior*, maintained their original form after conservation treatment by freeze drying.

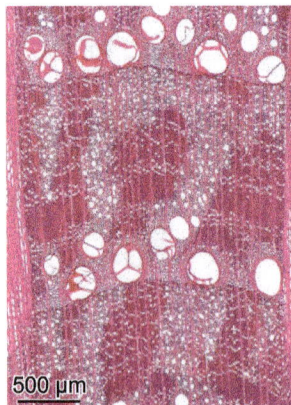

13.76 Modern wood of *Quercus petraea*, stained with Safranin. All microscopic structures can be recognized.

13.77 Dehydrated, heavily degraded wood of Neolithic *Quercus* sp. Lateral contraction deformed the original structure, but major anatomical characteristics for *Quercus* remain e.g. ring porosity, large rays, dark heartwood.

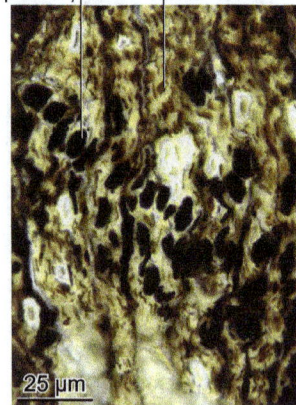

13.78 Dehydrated, heavily degraded wood of Neolithic *Quercus* sp. Alterations due to dehydration are specific to cell types. Fibers are heavily compressed, parenchyma cells with cell contents less so.

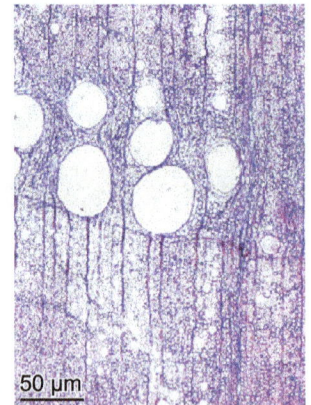

13.79 Well preserved anatomical structure of wood of Neolithic *Quercus* sp. after conservation treatment by freeze drying and polyethylene. Cell walls are decomposed.

Effect of conservation

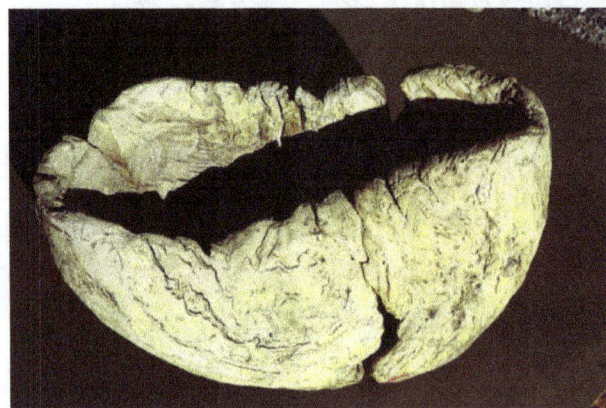

13.80 Neolithic wooden bowl conserved with diethyl ether resin.

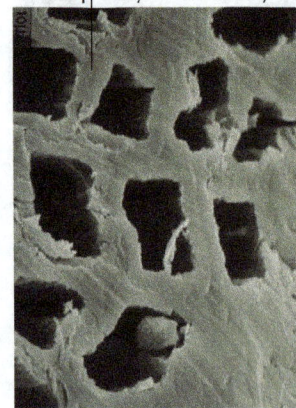

13.81 Cell walls without structure after freeze drying. Electron-optical photograph. Reprinted from Bräker *et al.* 1979.

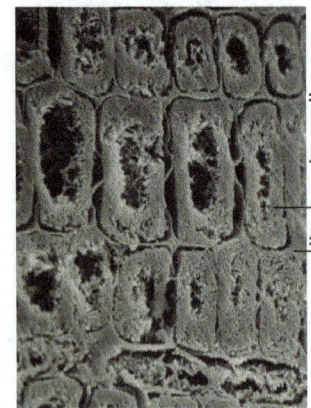

13.82 Partially preserved cell wall structure after diethyl ether treatment. Electron-optical photograph. Reprinted from Bräker *et al.* 1979.

14

Engineered Wood

This chapter gives an extremely brief overview over a few frequently used products made from technically altered wood—plywood, particleboards, granulated cork stoppers, hardboards, pencils and paper.

Naturally grown wood used to be one of the most valuable removable resources in pre-industrial times. Today, large industries modify wood from its natural form and structure, and change its physical and chemical properties to create products that suit modern human needs. In construction, shrinking and swelling of materials is an undesirable feature. This can be avoided with the production of plywood, particleboards and hardboards in which the fiber direction of the wood is artificially altered. Fibers in plywood run horizontally and vertically, particleboards and cork stoppers have fibers running in all directions, and hardboards feature compressed wooden structures. These processed products do not react in an anisotropic way anymore. For other products, the formal design has priority, e.g. in wooden pencils. Paper industries chemically disaggregate wood species with long fibers.

Plywood with two fiber directions

longitudinal section   cross section   longitudinal section   cross section

glued joint

500 µm

14.1 Macroscopic aspect of glued plywood.

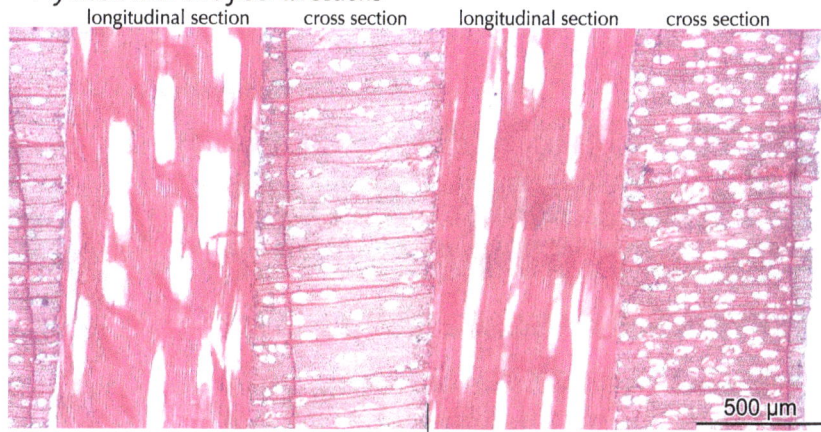

14.2 Microscopic aspect of a multi-layered plywood. Longitudinally and vertically oriented layers stabilize the board.

Chipboards and particleboards with various fiber directions

500 µm   500 µm

14.3 Macroscopic aspect of various chipboards, consisting of small wood particles and synthetic products.

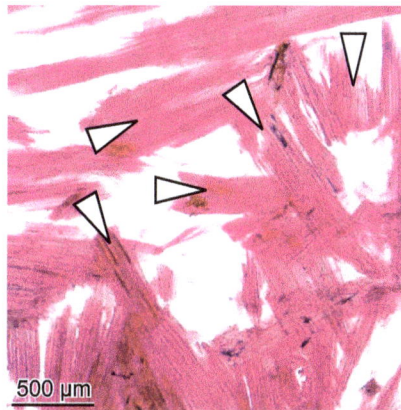

14.4 Chipboard with large particles and large spaces with synthetic fillers (white).

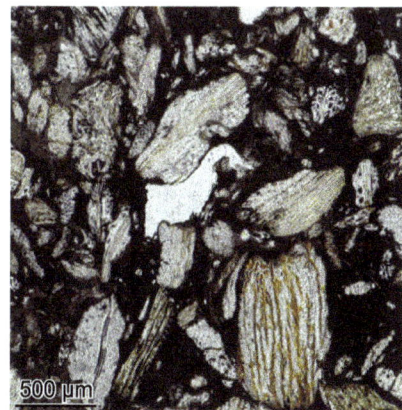

14.5 Particleboard with small particles in polarized light. The synthetic fillers are not birefringent.

Hardboards

compression

14.7 Compressed wooden structure of *Populus* sp. The intensity of the compression is reflected in the bending of the rays.

Granulated cork stoppers

14.8 Wine cork stoppers.

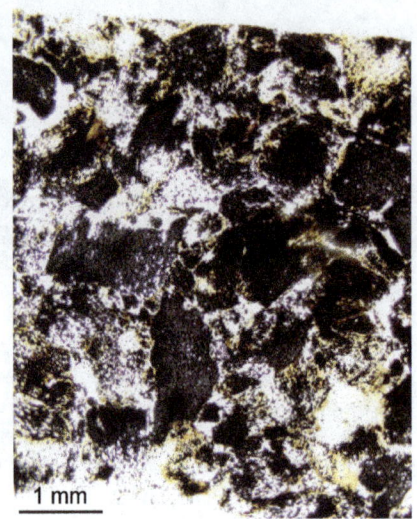

14.9 Chips in a cork stopper. Cells of small particles are filled with air (dark).

Wooden pencils

14.10 Wooden pencils with paint coating. The lead is embedded in a holder of *Thuja*, *Juniperus* and many other species.

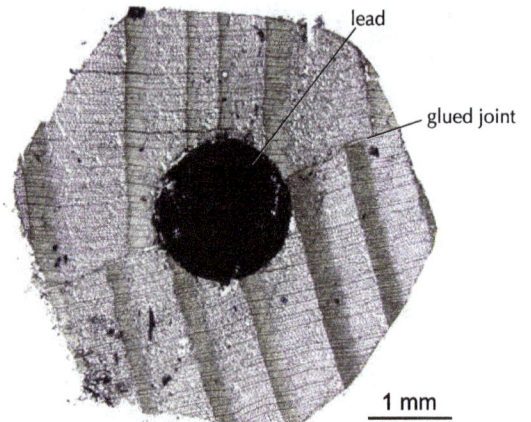

lead

glued joint

14.11 Cross section of a lead pencil with a holder of *Larix* wood.

Paper

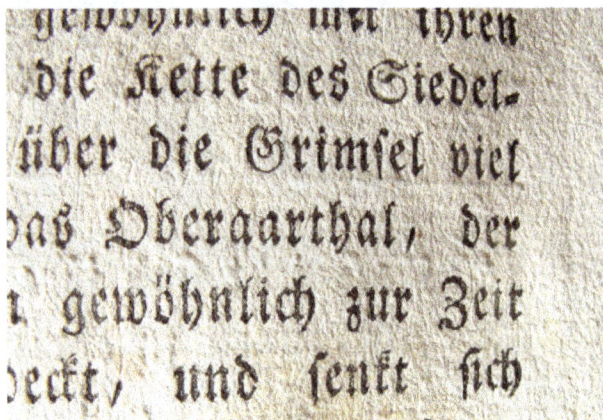

14.12 The ground-up wood pulp is visible in this old book page.

14.13 Macerated conifer tracheids with a length of up to 4 mm.

References and Recommended Reading

Aeschimann D., Lauber K., Moser D.M. and Theurillat J.-P. 2004. *Flora alpina*. Haupt Verlag Bern Stuttgart Wien.

Bailey I.W. 1923. The cambium and its derivative tissues. IV. The increase in girth of the cambium. *Am. J. Bot.* 10(9): 499–509.

Bailey I.W. 1944. The development of vessels in angiosperms and its significance in morphological research. *Am. J. Bot.* 31(7): 421–428.

Beck C.B. 1962. Reconstruction of *Archaeopteris* and further consideration of its phylogenetic position. *American J. of Botany* 49: 373–382.

Beck B.C. 2010. *An introduction to plant structure and development. Plant anatomy for the twenty-first century.* 2nd ed. Cambridge Univ. Press.

Bräker O.U., Schoch W. and Schweingruber F.H. 1979. Ergebnisse des Nassholzkonservierungsversuches; naturwissenschaftliche Wertung. *Z. Schweiz. Archäol. Kunstgesch.* 36: 102–120.

Bresinksy A., Strasburger E., Körner C., Kadereit J.W., Neuhaus G. and Sonnewald U. 2008. *Strasburger. Lehrbuch der Botanik.* Spektrum Akademischer Verlag Heidelberg.

Carlquist S. 1992. Wood, bark and pith anatomy of old world species of *Ephedra* and summary for the genus. *Aliso* 13: 255–295.

Carlquist S. 1996. Wood, bark and stem anatomy of Gnetales: A summary. *Int. J. Plant Sci.* 157 (6 Suppl.): S58–AS77.

Carlquist S. 2001. *Comparative Wood Anatomy. Systematic, ecological and evolutionary aspects of Dicotyledon wood.* Springer Verlag. Series in Wood Science. Berlin Heidelberg.

Carlquist S. 2011. *Equisetum* xylem: SEM studies and their implications. *Am. Fern J.* 101: 133–141.

Carlquist S. 2012. How wood evolves: a new synthesis. *Botany* 90: 901–940.

Carlquist S. and Gowans D.A. 1995. Secondary growth and wood histology of *Welwitschia*. Bot. J. Linnean Soc. 118: 197–121.

Christenhusz M.J.M. and Byng J.W. 2016. The number of known plants species in the world and its annual increase. *Phytotaxa* 261: 201–217.

Crivellaro A. and Schweingruber F.H. 2015. *Stem Anatomical Features of the Dicotyledons. Xylem, Phloem, Cortex and Periderm Structures for Ecological and Taxonomical Analyses.* Verlag Kessel. *www.forstbuch.de*

Cutler D.F., Bothah T. and Stevenson D.W. 2008. *Plant Anatomy: An Applied Approach.* Blackwell Publishing.

Dernbach D. and U. 2006. *Petrified forests. The world's 31 most beautiful petrified forests.* D'Oro-Verlag, Heppenheim.

Dolezych M. 2005. *Koniferenhölzer im 2. Lausitzer Flöz und ihre ökologische Position.* LPP Contribution Series No. 19, LPP Foundation Utrecht.

Eschrich W. 1995. *Funktionelle Pflanzenanatomie.* Springer Berlin, Heidelberg, New York.

Ellenberg H. and Mueller-Dombois D. 1967. A key to Raunkiaer plant life forms with revised subdivisions. *Ber. Geobot. Inst. ETH Zürich* 37: 56–73.

Evans P.D., Thay P.D. and Schmalzl K.J. 1996. Degradation of wood surfaces during natural weathering. Effects on lignin and cellulose on the adhesion of acrylic latex primers. Wood Sci. Technol. 30: 411–422.

Evert R.F. 2006. Esau's plant anatomy. Meristems, cells, tissues of the plant body. Their structure, function n and development. 3rd ed. John Wiley & Sons Hoboken New Jersey.

Fahn A. 1990. *Plant anatomy.* 4th ed. Pergamon press Oxford, New York, Toronto, Xydney, Paris, Frankfurt.

Fink S. 1999. *Pathological and regenerative plant anatomy.* Gebrüder Bornträger Berlin Stuttgart.

Fromm J. (ed.) 2013. Cellular aspects of wood formation. *Plant cell monographs.* Springer Verlag Berlin Heidelberg.

Gärtner H. 2003. Holzanatomische Analyse diagnostischer Merkmale einer Freilegungsreaktion von Koniferenwurzeln zu Rekonstruktion geomorphologischer Prozesse. *Dissertationes Botanicae* 378. J. Cramer in der Gebrüder Bornträger Verlagsbuchhandlung Berlin Stuttgart.

Gärtner H. and Schweingruber F.H. 2013. *Microscopic preparation techniques for stem analysis.* Verlag Kessel. *www.forstbuch.de*

Gärtner H., Lucchinetti S. and Schweingruber F.H. 2015. A new sledge microtome to combine wood anatomy and tree-ring ecology. IAWA Journal 36(4): 452–459.

Gardiner B., Barnett J., Saranpää P. and Gril J. 2014. *The biology of reaction wood.* Springer Verlag Heidelberg New York Dordrecht London.

George B., Suttie E., Merling A. and Deglise X. 2005. Photodegradation and photostabilisation of wood – the state of the art. *Polymer Degradation and Stability* 88: 268–274.

Ghislain B. and Clair B. 2017. Diversity in the organisation and lignification of tension wood fibre walls – a review. *IAWA J.* 38: 245–265.

Godet J.-D. 2011. *Baumrinden vergleichen und bestimmen*. Arboris Verlag Hinterkappelen.

Gorissen I. 2004. *Die Zwergstrauchheiden Europas vom Atlantik bis zum Kaukasus und Ural*. Verlag Gorissen Siegburg.

Greguss P. 1945. *The identification of Central-European dicotyledonous trees and shrubs based on xylotomy*. Hungarian Museum of Natural History, Budapest.

Greguss P. 1955. *Xylotomische Bestimmung der heute lebenden Gymnospermen*. Akademiai Kiado Budapest.

Greguss P. 1968. *Xylotomy of the living Cycads*. Akademiai Kiado Budapest.

Grosser D. and Liese W. 1971. On the anatomy of Asian Bamboos, with special reference to their vascular bundles. *Wood Sc. and Tech.* 5: 290–312.

Henes E. 1959. *Fossile Wandstrukturen untersucht am Beispiel der Tracheidenwände paläozoischer Gefässpflanzen*. Gebrüder Bornträger Berlin-Nikolassee. *www.schweizerbart. de/9783443390044*

Herendeen P.S., Wheeler E.A. and Baas P. 1999. Angiosperm wood evolution and the potential contribution of paleontological data. *Bot. Rev.* 65: 278–300.

Hirmer M. 1927. *Handbuch der Paläobotanik*. Verlag von R. Oldenbourg München Berlin.

Holdheide W. 1951. Anatomie mitteleuropäischer Gehölzrinden. Mit mikrofotographischem Atlas. In: Freund H. *Handbuch der Mikroskopie in der Technik*. Umschau Verlag Vol. V, part 1.

Hough R.B. 2002. *The Woodbook*. Reprint of *The American woods* (1883–1913, 1928). Taschen Verlag Köln London New York Paris Tokyo.

Huber B. 1961. Grundzüge der Pflanzenanatomie. Versuch einer zeitgemässen Neudarstellung. Springer Verlag Berlin Göttingen Heidelberg.

Junikka L. 1994. Survey of English macroscopic bark terminology. *IAWA J.* 15(1): 3–45.

Kaennel M. and Schweingruber F.H. 1995. *Multilingual glossary of dendrochronology*. Haupt Verlag Bern Stuttgart Wien.

Keller W., Wohlgemuth T., Kuhn N., Schütz M. and Wildi O. 1998. Waldgesellschaften der Schweiz als floristische Grundlage. *Mitt. Eig. Forschungsanstalt WSL* 73: 93–357.

Klötzli F., Dietl W., Marti K., Schubiger-Bossard C., Walther G.-R. 2010. *Vegetation Europas. Das Offenland im vegetationskundlich-ökologischen Überblick*. Ott Sachbuchverlag Bern.

Larson P.R. 1994. *The vascular cambium. Development and structure*. Springer Verlag Berlin Göttingen Heidelberg.

Liese W. and Köhl M. (eds) 2015. *Bamboo. The Plant and its Uses*. Springer International Publishing Switzerland.

Mauseth J.D. 1988. *Plant anatomy*. The Benjamin/Cummings Publishing Company Inc., Menlo Park CA.

Metcalfe C.R. 1971. *Anatomy of Monocotyledons. Vol. V Cyperaceae*. Oxford at the Clarendon Press.

Nabors M.W. 2004. *Introduction to botany*. Pearson Benjamin Cummings San Francisco.

Nogler P. 1981. Auskeilende und fehlende Jahrringe in absterbenden Tannen (*Abies alba* Mill.). *Allg. Forst Z.* 128: 709–711.

Onaka F. 1949. Studies on compression and tension wood. *Wood Research* 1: 1–88. (With English summary.)

Pfandenhauer J.S. and Klötzli F. 2014. *Vegetation der Erde. Grundlagen, Ökologie, Verbreitung*. Springer Spektrum Berlin Heidelberg.

Piermattei A., Crivellaro A., Carrer M. and Urbinati C. 2015. The "blue ring": anatomy and formation hypothesis of a new tree-ring anomaly in conifers. *Trees* 29(2): 613–620.

Schweingruber F.H. 2007. *Wood structure and environment*. Springer Verlag Berlin Heidelberg.

Schweingruber F.H. and Berger H. 2017. Anatomy of grass culms. Atlas of Central European Poaceae. Verlag Kessel. *www.forstbuch.de*

Schweingruber F.H., Börner A. and Schulze E.-D. 2008. Atlas of woody plant stems. Springer Verlag Berlin Heidelberg.

Schweingruber F.H., Börner A. and Schulze E.-D. 2011 and 2013. *Atlas of stem anatomy in herbs, shrubs and trees*. Vol. 1 and 2. Springer Verlag Berlin Heidelberg.

Schweingruber F.H. and Poschlod P. 2005. Growth rings in herbs and shrubs: life span, age determination and stem anatomy. *Forest, Snow and Landscape Res.* 79: 195–415.

Selmeier A. 2015. *Anatomy of tertiary silicified woods from the North Alpine Foreland Basin*. Holzforschung München TUM.

Shigo A.L. 1989. *A new tree biology. Facts, photos and philosophies on trees and their problems and proper care*. Shigo and Trees Ass. Durham, New Hampshire.

Spicer R. and Groover A. 2010. Evolution of development of vascular cambia and secondary growth. *New Phytologist* 186: 577–592.

Stern W.L. 2014. *Anatomy of Monocotyledons. Vol. X Orchidaceae*. Oxford Univ. Press New York.

Taylor N.T., Taylor E.L. and Krings M. 2009. *Paleobotany. The biology and evolution of fossil plants*. Academic Press Elsevier Amsterdam.

Timell T.E. 1986. *Compression wood in gymnosperms*. 3 Vols Springer Verlag Berlin Heidelberg New York Tokyo.

Tomlinson P.B., Horn J.W. and Fisher J.B. 2011. *The anatomy of Palms*. Oxford Univ. Press New York.

Wermelinger B. 2017. *Insekten im Wald. Vielfalt, Funktionen und Bedeutung*. Haupt Verlag Bern Stuttgart Wien.

Wheeler E.A., Baas P. and Rodgers S. 2007. Variations in dicot wood anatomy: A global analysis based on the INSIDE WOOD DATABASE. *IAWA J.* 28: 229–258.

White R.A. 1963. Tracheary elements of the ferns II Morphology of tracheary elements; conclusions. *Am. J. of Bot.* 50(6) 1: 514–522.

Zimmermann W. 1959. Die Phylogenie der Pflanzen. Gustav Fischer Verlag Stuttgart.

Permissions

All chapters in this book were first published in TPSAMA, by Springer by Fritz H. Schweingruber and Annett Börner; hereby published with permission under the Creative Commons Attribution License or equivalent. Every chapter published in this book has been scrutinized by our experts. Their significance has been extensively debated. The topics covered herein carry significant information for a comprehensive understanding. They may even be implemented as practical applications or may be referred to as a beginning point for further studies.

We would like to thank the editorial team for lending their expertise to make the book truly unique. They have played a crucial role in the development of this book. Without their invaluable contributions this book wouldn't have been possible. They have made vital efforts to compile up to date information on the varied aspects of this subject to make this book a valuable addition to the collection of many professionals and students.

This book was conceptualized with the vision of imparting up-to-date and integrated information in this field. To ensure the same, a matchless editorial board was set up. Every individual on the board went through rigorous rounds of assessment to prove their worth. After which they invested a large part of their time researching and compiling the most relevant data for our readers.

The editorial board has been involved in producing this book since its inception. They have spent rigorous hours researching and exploring the diverse topics which have resulted in the successful publishing of this book. They have passed on their knowledge of decades through this book. To expedite this challenging task, the publisher supported the team at every step. A small team of assistant editors was also appointed to further simplify the editing procedure and attain best results for the readers.

Apart from the editorial board, the designing team has also invested a significant amount of their time in understanding the subject and creating the most relevant covers. They scrutinized every image to scout for the most suitable representation of the subject and create an appropriate cover for the book.

The publishing team has been an ardent support to the editorial, designing and production team. Their endless efforts to recruit the best for this project, has resulted in the accomplishment of this book. They are a veteran in the field of academics and their pool of knowledge is as vast as their experience in printing. Their expertise and guidance has proved useful at every step. Their uncompromising quality standards have made this book an exceptional effort. Their encouragement from time to time has been an inspiration for everyone.

The publisher and the editorial board hope that this book will prove to be a valuable piece of knowledge for students, practitioners and scholars across the globe.

Index

www.ingramcontent.com/pod-product-compliance
Lightning Source LLC
Chambersburg PA
CBHW050458200326
41458CB00014B/5228

* 9 7 8 1 6 4 1 1 6 7 5 4 3 *